Exchange Bias

Monograph Series in Physical Sciences

Exchange Bias
From Thin Film to Nanogranular and Bulk Systems

Edited by
S. K. Sharma

CRC Press
Taylor & Francis Group
Boca Raton London New York

CRC Press is an imprint of the
Taylor & Francis Group, an **informa** business

CRC Press
Taylor & Francis Group
6000 Broken Sound Parkway NW, Suite 300
Boca Raton, FL 33487-2742

First issued in paperback 2020

ISBN-13: 978-1-4987-9723-8 (hbk)
ISBN-13: 978-0-367-78187-3 (pbk)

Library of Congress Cataloging-in-Publication Data

Names: Sharma, Surender K. (Professor of physics), editor.
Title: Exchange bias : from thin film to nanogranular & bulk systems / edited by, Surender K. Sharma.
Description: Boca Raton, FL : CRC Press, Taylor & Francis Group, [2018] | Series: Monograph series in physical sciences
Identifiers: LCCN 2017015450| ISBN 9781498797238 | ISBN 1498797237
Subjects: LCSH: Nanostructured materials--Magnetic properties. | Thin films--Magnetic properties. | Magnetic materials. | Nanoparticles. | Ferromagnetic materials. | Antiferromagnetism.
Classification: LCC TA418.9.N35 E93 2018 | DDC 620.1/1597--dc23
LC record available at https://lccn.loc.gov/2017015450

Visit the Taylor & Francis Web site at
http://www.taylorandfrancis.com

and the CRC Press Web site at
http://www.crcpress.com

Contents

Monograph Series in Physical Sciences

This monograph series brings together focused books for researchers and professionals in physical sciences. They are designed to offer expert summaries of cutting-edge topics at a level accessible to nonspecialists. As such, the authors are encouraged to include sufficient background information and an overview of fundamental concepts, together with presentation of the state-of-the-art theory, methods, and applications. Theory and experiment are both covered. This approach makes these titles suitable for some specialty courses at the graduate level as well. Subject matter addressed by this series includes condensed matter physics; quantum sciences; atomic, molecular, and plasma physics; energy science; nanoscience; spectroscopy; mathematical physics; geophysics; environmental physics; and other areas.

Proposals for new volumes in the series may be directed to Lu Han, senior publishing editor at CRC Press, Taylor & Francis Group (lu.han@taylorandfrancis.com).

Preface

This book has been developed as an overview of exchange-coupled magnetic systems for an audience of graduate students and scientists in both academic and industrial sectors. It provides a broad overview of several existing studies concerned with the understanding of exchange bias in magnetic nanostructures, presenting key examples from thin films to nanogranular and bulk systems.

This contributed volume shares the up-to-date progress in the area of exchange bias with an in-depth investigation of the cutting-edge developments on this particular subject. The first chapter serves as an introductory material, thereby enabling it to serve as a first text for researchers working in the area of magnetism and exchange bias.

This book has been intended as a forum for the critical evaluation of many aspects of exchange bias phenomenon that are at the forefront of research in the magnetism industry. Chapter authors have also been encouraged to present the highlights from the extensive literature on the topic, including the latest research in this field.

Editor

S. K. Sharma, PhD, is an assistant professor in the department of physics and group leader and coordinator of the Functional Nanomaterials Laboratory at the Universidade Federal do Maranhão (UFMA), São Luis, Brazil. Dr. Sharma obtained his PhD degree in July 2007 from H. P. University, Shimla, India. He has previously worked several years in research/academic positions in Brazil, France, Czech Republic, India, and Mexico, focusing on the area of nanomagnetism and functional nanomaterials. At present, he is an active member of postgraduate research program at UFMA and actively involved in research, teaching, and supervising research students at undergraduate/postgraduate levels. He has been awarded FAPEMA Senior Researcher grants and to date has published more than 60 peer-reviewed articles, 2 books as a single author, 3 book chapters and 4 articles in national level proceedings and has attended numerous international conferences.

Contributors

Chun-Gang Duan
Key Laboratory of Polar Materials and
  Devices
Ministry of Education
East China Normal University
Shanghai, People's Republic of China

and

Collaborative Innovation Center of
  Extreme Optics
Shanxi University
Shanxi, People's Republic of China

E. Eftaxias
Institute of Nanoscience and
  Nanotechnology
National Centre of Scientific Research
  "Demokritos"
Attiki, Greece

I. Fita
Institute of Physics
Polish Academy of Sciences
Warsaw, Poland

Matthias Hudl
Department of Chemical Physics
Stockholm University
Stockholm, Sweden

A. Kovalev
B. Verkin Institute for Low Temperature
  Physics and Engineering
National Academy of Sciences of Ukraine
Kharkov, Ukraine

and

V. N. Karazin Kharkiv National University
Svobody, Ukraine

Gabriel C. Lavorato
Div. Resonancias Magnéticas
Centro Atómico Bariloche, CONICET
Bariloche, Argentina

Enio Lima Jr.
Div. Resonancias Magnéticas
Centro Atómico Bariloche, CONICET
Bariloche, Argentina

Xiao-Min Lin
Center for Nanoscale Materials
Argonne National Laboratory
Argonne, Illinois

G. Margaris
Institute of Nanoscience and
  Nanotechnology
National Centre of Scientific Research
  "Demokritos"
Attiki, Greece

V. Markovich
Department of Physics
Ben-Gurion University of the Negev
Beer-Sheva, Israel

Roland Mathieu
Department of Engineering Sciences
Uppsala University
Uppsala, Sweden

Per Nordblad
Department of Engineering Sciences
Uppsala University
Uppsala, Sweden

Quy Khac Ong
Institute of Materials
École Politechnique Fédérale de Lausanne
Lausanne, Switzerland

M. Pankratova
Institute of Physics
Faculty of Sciences
Pavol Jozef Šafárik University
Košice, Slovakia

Mariana P. Proenca
IFIMUP and IN—Institute of
    Nanoscience and Nanotechnology
Departamento de Física e Astronomia
Faculdade de Ciências
Universidade do Porto
Porto, Portugal

R. Puzniak
Institute of Physics
Polish Academy of Sciences
Warsaw, Poland

Sarveena
Department of Physics
Himachal Pradesh University
Shimla, India

Jyoti Sharma
Department of Physics
IIT Bombay
Mumbai, India

S. K. Sharma
Department of Physics
Federal University of Maranhão
São Luis, Brazil

and

Department of Physics
Himachal Pradesh University
Shimla, India

Navadeep Shrivastava
Department of Physics
Federal University of Maranhão
São Luis, Brazil

and

Institute of Chemistry
University of São Paulo
São Paulo, Brazil

M. Singh
Department of Physics
Himachal Pradesh University
Shimla, India

K. G. Suresh
Department of Physics
IIT Bombay
Mumbai, India

Wen-Yi Tong
Key Laboratory of Polar Materials and
    Devices
Ministry of Education
East China Normal University
Shanghai, People's Republic of China

K. N. Trohidou
Institute of Nanoscience and
    Nanotechnology
National Centre of Scientific Research
    "Demokritos"
Attiki, Greece

M. Vasilakaki
Institute of Nanoscience and
    Nanotechnology
National Centre of Scientific Research
    "Demokritos"
Attiki, Greece

João Ventura
IFIMUP and IN—Institute of
    Nanoscience and Nanotechnology
Departamento de Física e Astronomia
Faculdade de Ciências
Universidade do Porto
Porto, Portugal

Alexander Wei
Department of Chemistry
Purdue University
West Lafayette, Indiana

Elin L. Winkler
Div. Resonancias Magnéticas
Centro Atómico Bariloche, CONICET
Bariloche, Argentina

A. Wisniewski
Institute of Physics
Polish Academy of Sciences
Warsaw, Poland

M. Žukovič
Institute of Physics
Faculty of Sciences
Pavol Jozef Šafárik University
Košice, Slovakia

Roberto D. Zysler
Div. Resonancias Magnéticas
Centro Atómico Bariloche, CONICET
Bariloche, Argentina

1

The Basis of Nanomagnetism: An Overview of Exchange Bias and Spring Magnets

Navadeep Shrivastava, Sarveena, M. Singh, and S. K. Sharma

1.1 Introduction

In the last two decades, new terms with the prefix *nano* have been added to the scientific vocabulary and are reflected in the plethora of research published, for example, nanoparticle (NP), nanostructure, nanomaterial, nanocluster, nanochemistry, nanocolloids, nanoreactor, and nanotechnology. Nanomaterials already have a significant commercial impact, which will certainly increase in the future. The enhanced interest of researchers in nano-objects is due to their unique properties that arise due to finite-size and surface effects. Finite-size effects are related to manifestation of so-called quantum-size effects, which arise in the case where the size of the system is commensurable with the de Broglie wavelengths of the electrons or phonons and excitons propagating in them. Surface effects can be related, in simplest case, to the symmetry breaking of the crystal structure at the boundary of each particle but can also be due to different chemical and magnetic structures of internal ("core") and surface ("shell") parts of a NPs (Gubin et al. 2005). Among the fundamental scientific disciplines that have contributed to extraordinary recent advances in nanoscience and nanotechnology, magnetism holds a prominent place. Finite-size and surface effects in the magnetic properties of matter have unraveled interesting new magnetic phenomena in nanomaterials, not manifested in the bulk (Papaefthymiou 2009). The properties of magnetic nanomaterials are as a result of both the intrinsic properties of the particles and the interactions between particles. Nanomagnetic system displays behaviors different from those in the bulk, pertaining to magnetic ordering, magnetic domains, magnetization reversal, and so on. Magnetic nanomaterials can now be regarded as being indispensable in modern technology. Magnetism has come a long way and has led to new physics and new technological applications in a range of multidisciplinary fields in present and future nanotechnologies such as ultrahigh-density magnetic recording, non-volatile magnetic memory, improved nanocomposite permanent magnet materials, giant-magnetoresistance phenomena, spintronics, quantum tunneling of magnetization, biomedicine and health science, and emerging technologies such as spin logic, spin torque nano-oscillators, and magnonic crystals (Bedanta et al. 2013; Papaefthymiou 2009).

The magnetic anisotropy (MA) is one of the most important properties of magnetic materials from the technological viewpoint. This is the quantity that determines the easy magnetization direction of a magnet, and it is also decisive of the magnetization reversal in external fields. Depending on the type of application, material with high, medium, or low MA will be required. The magnetic properties of NPs are determined by many factors; the key of these factors include the chemical composition, the type, and the degree of defectiveness of the crystal lattice; the particle size and shape; the morphology (for structurally inhomogeneous particles); the interaction of the particle with the surrounding matrix; and the neighboring particles (Gubin 2009). By changing the size, shape, composition, and structure of NPs, one can control the magnetic characteristics of the material based on them to an extent.

The evolution of this chapter starts from fundamental terminology and discussion over the magnetic ordering, along with the description of magnetic properties, for example, hysteresis and anisotropy. Section 1.3 deals with the various nanoscale phenomena, as characteristic scale length, superparamagnetism, exchange bias (EB), interaction phenomenon, and maximum energy product $(BH)_{max}$. The MA and its role have been discussed in detail in Section 1.4, providing enough background for other chapters. Next generations of advanced permanent magnets, exchange-spring magnets (ESMs), and exchange-biased magnets (EBMs) have been taken as examples in the next section. Section 1.6 insights the short description of research and development (R&D) in nanomagnetism.

1.2 Magnetism Description and Prerequisite Terminology

1.2.1 Fundamental of Magnetism

Magnetism is a result of moving charges. The macroscopic magnetic properties of materials are a consequence of magnetic moments associated with individual electrons. On the basis of elementary knowledge of atomic structure, two possible origins may be imagined for the atomic magnetic moment. From the atomic point of view of matter, there are two electronic motions: orbital motion of electron and spin motion of electron. In addition to these two, there is negligible contribution due to the nuclear magnetic moment, and this will not be discussed here. The two electronic motions are a source of the macroscopic magnetic phenomena in materials.

1.2.1.1 Origin of Magnetism

The orbital motion of an electron around the nucleus is analogous to the current in a loop of wire (see Figure 1.1). The magnetic moment of a current-carrying conductor is given by:

$$\mu_{orb} = IA \tag{1.1}$$

where:
    I is the current in ampere
    A is the area of the closed loop

The current I can be expressed in terms of the charge $-e$ and period of rotation T, as:

$$I = \frac{-e}{T} = \frac{-e}{2\pi r/v} = \frac{-ev}{2\pi r} \tag{1.2}$$

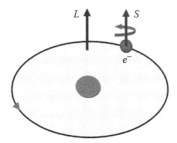

FIGURE 1.1   Origin of magnetism.

where:
  r is the radius of orbit
  $-e$ is the electronic charge
  v is the velocity of the electron.

Using Equation 1.2 in 1.1:
  The magnetic moment of an electron in orbit is given by:

$$\mu_{orb} = \frac{-evr}{2} \tag{1.3}$$

Multiplying and dividing the right-hand side (RHS) by m, the mass of electron, we get:

$$\mu_{orb} = \frac{-e}{2m} mvr \tag{1.4}$$

where mvr is the orbital angular momentum of the electron. Hence, the magnetic moment and angular momentum are related by:

$$\mu_{orb} = \frac{-e}{2m} \times \text{orbital angular momentum} \tag{1.5}$$

$$\mu_{orb} = \frac{-e}{2m} L \tag{1.6}$$

where L is the orbital angular momentum and is the smallest possible orbital magnetic moment of the electron. Similarly, the smallest possible magnetic moment due to spin motion of electron is $\mu_{spin} = (e/m)S$, where S is the spin angular moment $(\pm 1/2).(\hbar/2\pi)$.

The constant $e/2m$ in Equation 1.6 is called the gyromagnetic ratio and is represented by the symbol γ. For the atoms being the quantum system, the above equation can be generalized to include the quantum nature of the angular momentum of the atom. Angular momentum L is quantized, and it can be taken as the sum of the individual angular momentum of all the electrons in the atom. The general expression for total magnetic moment of the atom, including both orbital and spin angular momentum contributions, is written as:

$$\mu = -g\frac{e}{2m} J \tag{1.7}$$

where J is the total angular momentum $(L + S)$ of the system and g is a constant called the spectroscopic g factor or Lande's g factor, given as:

$$g = 1 + \frac{J(J+1) + S(S+1) - L(L+1)}{2J(J+1)} \tag{1.8}$$

J, L, and S are total, orbital, and spin quantum numbers, respectively.

The g factor arises due to precession of electrons similar to the precession of a top in a gravitational force, and the value of g tells us whether the origin of magnetic moment is spin or orbital motion of electrons.

The component of magnetic momentum in the direction of the applied field (taken as Z-direction) is of interest. Hence the Z-component of the total angular momentum is given by:

$$\mu_z = -g\frac{e\hbar}{2m} m_J \tag{1.9}$$

where $\hbar = h/2\pi$, where h is the Planck's constant and m_J is the magnetic number and takes values $J, J{-}1, J{-}2, \ldots {-}J$.

For a single isolated electron, the total magnetic moment is due to spin only. The Z-component of the magnetic moment will be obtained by using $L = 0$ and $S = 1/2$ in Equations 1.7 through 1.9 to get $g = 2$ and by taking $J = 1/2$, so that $m_J = \pm 1/2$. With the application of an external magnetic field, the Z-component of the magnetic moment will be:

$$\mu_z(\text{spin}) = -2\left(\frac{e\hbar}{2m}\right)(\pm 1/2)$$

$$\mu_z = \pm\frac{e\hbar}{2m} \tag{1.10}$$

μ_z is the most fundamental magnetic moment; it is called Bohr Magneton, μ_B.

$$\mu_B = \left|\frac{e\hbar}{2m}\right| = 9.27 \times 10^{-24}\,\text{Am}^2 \tag{1.11}$$

All other atomic moments are expressed in terms of this fundamental unit.

1.2.1.2 Magnetic Parameters

The response of a material when subjected to an external magnetic field is the root of magnetism. There are common terms for ratios between some of these different quantities, because the magnetic properties of a material are often defined by how they vary with an applied magnetic field. Important magnetic parameters that will be frequently used to characterize the magnetic properties of materials are as follows:

- Magnetic flux, φ_m
- Magnetization vector, $\bar{M}$
- Magnetic permeability, μ
- Magnetic susceptibility, χ

1.2.1.3 Magnetic Flux, φ_m

The magnetic flux, defined as magnetic field times an area, is given by:

$$\varphi_m = \oint_s \bar{B}.\overline{ds} \tag{1.12}$$

where B is the magnetic field and ds is an infinitesimal area element, where we may consider the field to be constant. Since the isolated magnetic pole never exists in nature, if we consider a closed surface in a magnetic field, the net magnetic flux across the closed surface is always zero. Hence, Equation 1.12 becomes:

$$\oint_s \bar{B}.\overline{ds} = 0 \tag{1.13}$$

Equation 1.13 is called Gauss's law for magnetism.

1.2.1.4 Magnetization Vector, $\bar{M}$

When any material is subjected to an external magnetizing field, the dipoles, which are in a random manner, tend to align partially, giving rise to a net magnetic moment in the direction of the magnetizing field in a small volume of the material. This magnetic moment per unit volume of the material is defined as magnetization (Givord 2007). The tendency of the dipoles to align also depends on the external applied magnetic field. It is called intensity of magnetization. The unit of magnetic moment is Am^2. Hence, the unit of magnetization is A/m.

1.2.1.5 Magnetic Permeability, μ

Magnetic permeability of the material is the measure of the number of magnetic flux lines that can penetrate the material on subjecting it to an external magnetic field (H). It is the degree of the magnetization capability of a material. It is defined as the ratio of the magnetic flux density to the external magnetic field and is given by:

$$\mu = \frac{\text{Magnetic flux density}}{\text{Magnetic field}} = \frac{B}{H} \tag{1.14}$$

The unit of magnetic permeability is Henry/m.

Another important term in magnetism is known as the relative permeability, μ_r. It is the ratio of the permeability of a specific medium, μ, to the permeability of free space, μ_0.

$$\mu_r = \frac{\mu}{\mu_0} \tag{1.15}$$

where, $\mu_0 = 4\pi \times 10^{-7}$ Henry/m. μ_r is a dimensionless quantity and is the ratio of two similar quantities.

1.2.1.6 Magnetic Susceptibility, χ

Magnetic susceptibility is the quantitative measure of the ease with which a material is magnetized in the presence of an external magnetic field. It is the ratio of magnetization vector (M) to external magnetic field (H).

$$\chi = \frac{M}{H} \tag{1.16}$$

Since M and H have the same unit, χ is a dimensionless quantity.

1.2.1.7 Relation between Magnetic Parameters

When a material is subjected to external magnetic field, H, the magnetic induction in the material B is given by:

$$B = \mu_0\mu_r H$$

B, M, and H are related as $B = \mu_0(H + M)$ and $\mu_0\mu_r = \mu_0(H + M)$; $\mu_r = \left(1 + \dfrac{M}{H}\right)$

Therefore,

$$\mu_r = 1 + \chi \tag{1.17}$$

It is a relation connecting the permeability and susceptibility of the medium.

1.2.2 Magnetic Order

The magnetic orders in magnetic materials according to their response to the external magnetic field are divided into three major categories: diamagnetism, paramagnetism, and ferromagnetism. All the magnetic orders are as follows (Cullity and Graham 2009):

- Diamagnetism
- Paramagnetism
- Ferromagnetism
- Antiferromagnetism
- Ferrimagnetism

1.2.2.1 Diamagnetism

Diamagnetism is a universal phenomenon exhibited by all substances. It is a weak magnetism in which magnetization is exhibited opposite to the direction of the applied field. Diamagnetism characterizes substances that have only non-magnetic atoms. Diamagnetism arises due to reluctant and the non-cooperative behavior of orbiting electrons when exposed to an applied external magnetic field. The susceptibility of a diamagnetic material is negative, and the order of the magnitude is usually about 10^{-5}. Owing to this, the material gets weakly repelled in the magnetic field. Magnetization becomes zero once the magnetic field is removed. The atoms do not possess magnetic moment when field is zero, and when field is applied, the atoms acquire induced magnetic moment opposite to the applied field, resulting in negative susceptibility (Figure 1.2a). The susceptibility is independent of temperature (Figure 1.2a). This magnetism is so weak that it is observed only if the paramagnetic or ferromagnetic effect does not mask the weak diamagnetic effect (Buschow and Boer 2003). Some examples of diamagnetic materials are copper, gold, mercury, silver, lead, and zinc.

1.2.2.2 Paramagnetism

Paramagnetism occurs in materials with permanent magnetic moment. In the absence of the external magnetic field, the magnetic moments are randomly oriented; this makes the net magnetic moment and hence the magnetization zero. On application of the external magnetic field, the individual magnetic moments tend to align, so as to produce weak magnetization parallel to applied magnetic field (Figure 1.2b). The expected behaviors of paramagnetic materials with an applied external magnetic field and with temperature are shown in Figure 1.2b. At finite temperatures, the magnetic moments are thermally agitated and disturb the alignment of the magnetic moments. Hence, a large field is required to attain the same magnetization. The susceptibility in this case is inversely proportional to the absolute temperature. In general, the response of a material to a magnetic field, that is, the magnetic susceptibility (χ), is of the order

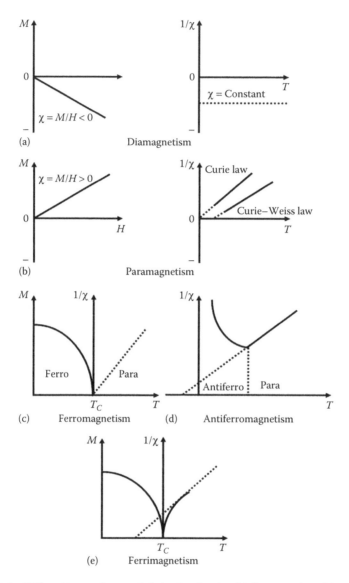

FIGURE 1.2 Different types of magnetic behavior showing (a) diamagnetism, (b) paramagnetism, (c) ferromagnetism, (d) antiferromagnetism, and (e) ferrimagnetism.

of 10^{-3}–10^{-5}. The temperature dependence of paramagnetic susceptibility is described by the Curie's law, given as:

$$\chi = \frac{C}{T} \tag{1.18}$$

where C is called the Curie constant. This type of temperature dependence (called the Curie's law) is found at any temperature in paramagnetic materials or above a certain temperature of magnetic order in ferromagnetic and antiferromagnetic materials (in these cases, called the Curie–Weiss law, as described in Section 1.2.2.3). The examples of paramagnetic materials are sodium, titanium, aluminum, and chromium.

1.2.2.3 Ferromagnetism

Ferromagnetism is the existence of spontaneous magnetization; that is, the atomic magnetic moments are aligned even in the absence of an external magnetic field. This is the indication of the strong internal field within the material that enables the magnetic moments to align with each other. The atomic dipole moments in these materials are characterized by very strong positive interactions produced by electronic exchange forces, which result in a parallel alignment of atomic moments. Two distinct characteristics of ferromagnetic materials are their (1) spontaneous magnetization and (2) the existence of magnetic ordering temperature. They have relatively large susceptibilities. Pure iron, cobalt, nickel, and many transition metal alloys exhibit this property. Figure 1.2c shows a graph between magnetization and the applied magnetic field.

The Curie temperature and hysteresis are two important characteristics of ferromagnets. As the temperature increases, the thermal agitation disturbs the arrangement of the magnetic moments, thus resulting in temperature dependence of spontaneous magnetization. Above the Curie temperature, susceptibility obeys the Curie–Weiss law, given as:

$$\chi = \frac{C}{T - T_C}$$

where:
C is a constant characteristic of the material
T is absolute temperature
T_C is Curie temperature

The Curie temperature is the temperature above which exchange forces cease to be present; that is, above that temperature, a ferromagnetic material randomizes owing to the thermal energy, as in paramagnetic systems (Figure 1.2c). Hence, T_C is the ferromagnetic transition temperature above which spontaneous magnetization vanishes and materials behave like paramagnetic materials, and below T_C, ferromagnetic materials have spontaneous magnetization.

1.2.2.4 Antiferromagnetism

Antiferromagnetism arises when the magnetic moments are aligned antiparallel to each other, resulting in a net zero magnetic dipole. Antiferromagnetism, macroscopically similar to paramagnetism, is a weak form of magnetism, with a weak and positive susceptibility. The temperature dependence of the susceptibility of antiferromagnetic material is characterized by Néel temperature (T_N). The susceptibility of an antiferromagnetic substance is not infinite at $T = T_N$, but it has a weak cusp (Figure 1.2d). Antiferromagnetic property is observed only below a certain temperature known as Néel temperature due to ordered antiparallel alignment of spin magnetic moments. Antiferromagnetic materials are very similar to ferromagnetic materials, but the exchange interaction between neighboring atoms leads to antiparallel alignment of the atomic magnetic moments. Therefore, the magnetic field cancels out, and the material appears to behave in the same way as a paramagnetic material.

Above Néel temperature, antiferromagnetic materials become paramagnetic. This behavior is exhibited only in a very few elements such as oxides of manganese, chromium, iron (ferrous), and nickel oxide. The antiparallel arrangement of the magnetic moments results from the interaction between the neighboring atoms called negative-exchange interaction. These interactions work against the effect of the applied field, which would tend to align all moments in parallel. The exchange interaction is such as to cause antiparallel alignment of the moments. At higher temperature, thermal agitation overcomes interaction effects, and thermal variation of susceptibility similar to paramagnets is observed.

1.2.2.5 Ferrimagetism

Ferrimagnetism describes the magnetism in ferrites. Ferrimagnetism characterizes a material that is antiferromagnetic-like microscopically but the moments are unequal and spontaneous magnetization remains. The net moment arises due to the difference in moments on the two types of sites. As the temperature increases, the arrangement of the spins is disturbed by thermal agitation, which in turn is accompanied by a decrease of spontaneous magnetization. Ferrimagnetic materials, like ferromagnetic materials, have spontaneous magnetization below a critical temperature called the Curie temperature (T_C). At T_C, spontaneous magnetization vanishes, and the material exhibits paramagnetism above T_C (Figure 1.2e). Ferrimagnetism can be treated as a special case of ferromagnetism, as the interactions of the net moments of the lattices are continuous throughout the crystal. Ferrimagnetic materials exhibit all properties of ferromagnetic materials such as high permeability, saturation magnetization, and hysteresis. Difference arises in the way the moments are aligned in the crystal. Within these materials, the exchange interactions lead to parallel alignment of atoms in some of the crystal sites and anti-parallel alignment of others.

The magnitude of magnetic susceptibility (χ) of ferro-/ferrimagnetic materials is identical, but the alignment of magnetic dipole moments is drastically different. The material breaks down into magnetic domains, just like a ferromagnetic material, and the magnetic behavior is also very similar; however, ferrimagnetic materials usually have lower saturation magnetizations. The most important ferrimagnetic substances are certain double oxides of iron and other metals, called ferrites (although not all oxide ferrites are ferrimagnetic). These are based on the spinel structure; the prototypical example is magnetite, Fe_3O_4.

1.2.3 Properties of Magnetic Materials

1.2.3.1 Magnetic Domains

Magnetic order observed in a ferromagnetic material arises essentially from the action of the exchange interaction. The presence of other interactions such as anisotropy, dipolar interactions, and magnetoelastic interactions leads to the formation of magnetic domains. Under an applied magnetic field, the boundaries between these domains—the domain walls—are displaced, which in turn change the magnetization of the ferromagnet. To understand the magnetic properties and process of magnetization, one requires to study the domain itself, particularly with respect to the structure and orientation of the walls that surround these domains.

The sample of every magnetic material is not necessarily a magnet. In ferromagnetic materials, despite the existence of a spontaneous magnetization below T_C, a piece of ferromagnetic materials is not necessarily magnetized; its magnetic moment can be zero. The material is said to be demagnetized. Weiss explained this by making an assumption: a ferromagnet in the demagnetized state is divided into a number of small regions called domains; these are the regions where the atomic magnetic moments point approximately in the same direction. Each domain is spontaneously magnetized to the saturation value M_S, but the directions of magnetization of the various domains are such that the resulting magnetic moment (and the average magnetization) remains nearly zero.

The process of magnetization involves the process of converting the specimen from a multi-domain state into one in which it is a single domain magnetized in the direction of the applied field (Cullity and Graham 2009). This process is illustrated schematically in Figure 1.3. The dashed line in Figure 1.3a encloses a portion of a crystal, in which there are parts of two domains; the boundary separating them is called a domain wall. The net magnetization of this part of the crystal is zero, because the two domains are spontaneously magnetized in opposite directions. In Figure 1.3b, when field H is applied, the upper domain in the direction of the applied field tends to grow at the expense of the lower one by downward motion of the domain wall, until in Figure 1.3c, the wall moves right out of the region considered. Finally, at still higher applied fields, the magnetization rotates the domains to become parallel with the applied field, and the material is saturated, as in Figure 1.3d. During this entire process, there was no change in the magnitude of magnetization of any region, only in the direction of magnetization. The Weiss theory therefore contains two essential postulates: (1) spontaneous magnetization and (2) division into domains.

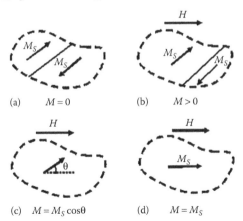

FIGURE 1.3 The magnetization process in a ferromagnet depicting domain walls under magnetic fields (a) $M = 0$, (b) $M > 0$, (c) $M = M_S \cos\theta$, (d) $M = M_S$. (Cullity, B.D. and Graham, C.D.: *Introduction to Magnetic Materials.* 2nd ed. 2009. Copyright Wiley-VCH Verlag GmbH & Co. KGaA. Reproduced with permission.)

Domains are created because their existence reduces the magnetostatic energy. As the number of domains is increased, the magnetostatic energy is reduced. This cannot continue indefinitely, since the presence of the transition region between the domains, the domain walls, also brings about an increase in exchange and anisotropy energy. Let us consider a 180° domain wall of thickness of N atoms, each one with spin S; the average angle between neighbor spins is π/N, and the energy per pair of neighbors is given by (Guimaraes 1998) as:

$$E_{ex}^{pair} = JS^2 \left(\frac{\pi}{N} \right)^2 \tag{1.19}$$

where J is the exchange parameter.

A line of atoms with $N+1$ neighbors perpendicular to the domain wall has the energy:

$$E_{ex} = N E_{ex}^{pair} = \frac{JS^2 \pi^2}{N} \tag{1.20}$$

The condition for the energy E_{ex} to be minimum is that N grows indefinitely. If the separation between the atoms is a, a unit length of the domain wall crosses $1/a$ lines of atoms; a unit area of wall is crossed by $1/a^2$ lines. Then, the exchange energy per unit area is:

$$E_{ex} = \pi^2 \frac{JS^2}{Na^2} \tag{1.21}$$

The anisotropy energy per unit volume of a uniaxial crystal is $E_K = K \sin^2\theta$. Since a wall of unit area has a volume Na, the anisotropy energy per unit area is:

$$e_K = \overline{K \sin^2 \theta} Na \approx KNA \tag{1.22}$$

The total energy per unit area is $e = e_{ex} + e_K$ (exchange plus anisotropy), and the condition that minimizes e is:

$$\frac{de}{dN} = \frac{-\pi^2 JS^2}{N^2 a^2} + Ka = 0 \tag{1.23}$$

and the number N of the atoms that satisfies this condition is:

$$N = \frac{\pi S}{a^{3/2}} \sqrt{\frac{J}{K}} \tag{1.24}$$

Therefore, the domain wall's thickness is:

$$\delta = Na = \frac{\pi S}{a^{1/2}} \sqrt{\frac{J}{K}} \tag{1.25}$$

In conclusion, the domain wall's thickness is directly proportional to $\sqrt{J}$ and inversely proportional to $\sqrt{K}$.

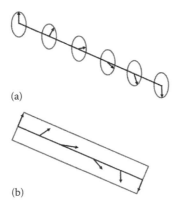

(a)

(b)

FIGURE 1.4 Two main types of magnetic domain walls, exemplified for a 180° wall: (a) Bloch wall and (b) Néel wall.

The formation of interfaces (walls) between the domains leads to an increase in energy due to MA and to the exchange interaction. The width of the domain wall is defined by the competition between the anisotropy energy and the exchange energy; the former is reduced for narrow walls, and the latter is reduced for thick walls. Domain walls can have many different forms; however, there are two main types, known as Bloch walls and Néel walls. They are distinguished from one another based on the way the atomic magnetic moments in the wall turn. In Bloch domain walls, magnetization turns outside the plane of the magnetization of the neighbor domains, and in Néel walls, the moments turn in the same plane as that of the domain moments. Bloch domain wall and Néel wall arrangements are illustrated in the case of a 180° magnetic domain wall in Figure 1.4.

1.2.3.2 Hysteresis

Hysteresis is a feature of ferromagnets and ferrimagnets. At a temperature well below the T_C, despite the existence of the spontaneous magnetization on microscopic scale, a specimen as a whole is not necessarily spontaneously magnetized. The magnetic moment may be very much less than the saturation moment, and the external magnetic field may be required to saturate the specimen. This results from the fact that actual specimens are composed of small regions called domains, within each of which the local magnetization is saturated. From one domain to another, the moment direction varies in such a way that the resulting magnetic moment of the whole specimen is zero. However, the application of external magnetic field results in an increase in the gross magnetic moment of a ferromagnetic specimen by two independent processes (Kittel 1996):

    a. In weak applied magnetic field, the domains that are oriented in directions favorable to (or nearly aligned with) the applied field grow at the expense of those that are unfavorably oriented (domain wall's motion).

    b. In the strong applied field, the domain's magnetization rotates toward the field direction.

The technical magnetization curve is shown in Figure 1.5. It shows variation of magnetization as a function of applied field. Initially, for small fields, the variation is linear, and

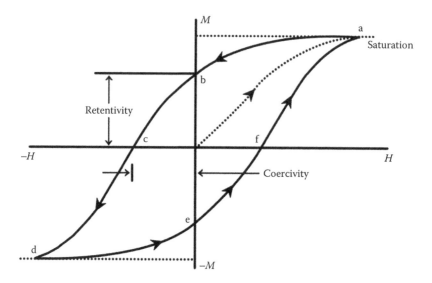

FIGURE 1.5 The magnetization curve.

then, the curve becomes non-linear, as shown by the dashed line. At a certain magnetic field, the magnetization reaches a maximum and does not increase further with increasing field; the curve becomes flat (point a). This is due to the formation of the single-domain formation, which results due to the growth of favorably oriented domains in the field direction. At this stage, the material has reached the point of magnetic saturation. Further increase in the field does not increase the magnetization any more, as the moments are already in the direction of the applied field. When the field is gradually reduced, the magnetization decreases, but it does not retrace the dashed curve; it takes the path ab. This means that the magnetization M lags behind the applied field H. At this point, it can be seen that some magnetization remains in the material, even though the applied field is zero. This is called the remanent magnetization, and it indicates the remanence or level of residual magnetism in the material in the absence of the applied field. As the field is reversed, the curve moves to point c, where the magnetization has been reduced to zero. This is called the coercivity. It is the field that has to be applied in the reverse direction to completely demagnetize the material.

As the field is now increased in the negative direction, the material will again become magnetically saturated but in the opposite direction (point d). Reducing field to zero brings the curve to point e. It will have a level of residual magnetism equal to that achieved in the other direction. The curve did not return to the origin of the graph, because some force is required to remove the residual magnetism. The curve will take a different path from point f back to the saturation point, completing the loop. This closed curve is called the hysteresis loop. The area under the hysteresis loop is a measure of the energy loss. The energy loss will be due to the dissipated energy required to push the domain walls back and forth. The most technological applications are based on the existence of this loop.

1.2.3.3 Magnetic Anisotropy

One factor that may strongly affect the shape of the hysteresis loop is MA. This term simply means that the magnetic properties depend on the direction in which they are measured. A magnetic material is said to possess MA if its internal energy depends on the direction of its spontaneous magnetization with respect to the crystallographic axes. Depending on the orientation of the field with respect to the crystal lattice, one would need a lower or higher magnetic field to reach the saturation magnetization (M_s). The energy that directs the magnetization along certain crystallographic axes, called the direction of easy magnetization, is called the magnetocrystalline or anisotropy energy (Baberschke 2001).

For example, in cubic crystal, let M_s make angles a, b, and c with the crystal axes, and let α_1, α_2, and α_3 be the cosines of these angles, which are called direction cosines. Then (Frey 2008),

$$E = K_0 + K_1(\alpha_1^2\alpha_2^2 + \alpha_2^2\alpha_3^2 + \alpha_3^2\alpha_1^2) + K_2(\alpha_1^2\alpha_2^2\alpha_3^2) + ... \tag{1.26}$$

where, K_0, K_1, K_3 ... are constants for a particular material and are expressed in erg/cm³ (cgs) or J/m³ (SI unit). It has been experimentally found that the crystal anisotropy can be described by the first two or three terms of an infinite power series in the direction cosines of the magnetization vector with respect to the crystal axes. The first term, K_0, is independent of angle and is usually ignored. Thus, for cubic crystal, it becomes:

$$E = K_1(\alpha_1^2\alpha_2^2 + \alpha_2^2\alpha_3^2 + \alpha_3^2\alpha_1^2) + K_2(\alpha_1^2\alpha_2^2\alpha_3^2) + ... \tag{1.27}$$

Higher powers are generally neglected; sometime, K_2 is so small that the terms involving it can be neglected. The physical origin of the MA energy is the interaction of the mean exchange field and the orbital angular momenta of the atoms (ions) in the lattice by means of spin–orbit coupling. Other types of anisotropy also exist, such as shape anisotropy, stress anisotropy, exchange anisotropy, and induced anisotropy. The detailed description of bulk contribution and surface contribution of different MA has been discussed in Section 1.4.

1.2.3.4 Magnetostriction

When a material is subjected to a magnetic field, its dimensions change; this effect is called magnetostriction. Generally, change in dimensions occurs when the material is subjected to a change in the applied magnetic field. A change in magnetization accompanies a small change in size and volume of the medium. Alternatively, it undergoes a change in its magnetic state under the influence of an externally applied mechanical stress. The most common type of magnetostriction is the Joule magnetostriction, in which the dimensional change is associated with a distribution of distorted magnetic domains present in the magnetically ordered material. Joule showed that an iron rod increased in length when it was magnetized lengthwise by a weak field. The fractional change in length $\Delta l/l$ is simply a strain, and magnetically induced strain is given a special symbol λ to distinguish it from the strain ε caused by an applied stress (Cullity and Graham 2009):

$$\lambda = \frac{\Delta l}{l} \tag{1.28}$$

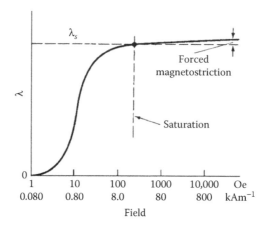

FIGURE 1.6 Dependence of magnetostriction on magnetic field (scale is logarithmic). (Cullity, B.D. and Graham, C.D.: *Introduction to Magnetic Materials.* 2nd ed. 2009. Copyright Wiley-VCH Verlag GmbH & Co. KGaA. Reproduced with permission.)

The value of λ measured at magnetic saturation is called the saturation magnetostriction λ_s.

The value of the saturation longitudinal magnetostriction λ_s can be positive, negative, or, in some alloys, at some temperature, zero. The value of λ depends on the applied field, and Figure 1.6 shows how *l* varies with *H* for a substance with positive magnetostriction. It is well known that ferromagnetic and ferrimagnetic materials are internally divided into domains. The process of magnetization occurs by two mechanisms: domain-wall motion and domain rotation. The rotation of small magnetic domains is the cause of magnetostriction. The rotation and re-orientation of the domain cause internal strains in the material structure. The strains in the structure lead to a change in the dimensions of the materials.

Considering any one of these domains, for materials with positive (negative) magneto-striction, the dimension along the magnetization direction is increased (decreased), while simultaneously, the dimension in the direction perpendicular to the magnetization direction is decreased (increased), keeping the volume constant (Buschow and Boer 2003). Between the demagnetized state and saturation, the volume of a specimen remains nearly constant. This means that there will be a transverse magnetostriction λ_t very nearly equal to one-half the longitudinal magnetostriction and opposite in sign (Cullity and Graham 2009):

$$\lambda_t = -\frac{1}{2}\lambda \qquad (1.29)$$

When technical saturation is reached at any given temperature, in the sense that the specimen has been converted into a single domain magnetized in the direction of the field, any further increase in the field further causes a small strain. This causes a slow change in λ with *H,* called forced magnetostriction, and the logarithmic scale of *H* in Figure 1.6 roughly indicates the fields required for this effect to become appreciable. It is caused by an increase in the degree of the spin order that very high fields can produce (the para process).

1.3 Nanoscale Phenomena: The Nano Perspective

The objective of nanomagnetism is to investigate new physical phenomena with improved effects compared with mesoscopic behavior. Obviously, main difference is scale length. Simply, any magnetic system of nanoscopic or mesoscopic scales is fundamentally affected by two major factors, dimensions comparable to characteristic lengths, for example, the limiting size of magnetic domains, and broken translation symmetry, which results in sites with reduced coordination number, with broken exchange bonds and frustration (Guimaraes 2009). Further, the magnetic characteristics in any kind of system are strongly affected by surface or interfacial atomic distribution, and a very close contact with other physical system (e.g., substrate for thin film or NPs compactly immersed in solid matrix or container) can raise a strong interaction with its close chemical/physical environment. Other more important facts that become more relevant in nanoscopic range are imperfections and defects in nanostructures, making the explanation of origin of magnetism more complicated. Some of the important formulations can be given as follows:

1. Dispersion relation for spin waves, $\hbar\omega = Ak^2$; wavelength for spin waves > lattice spacing.
2. If spin wave is comparable to the thermal energy, then $k = (k_B T/A)^{1/2}$.

Here, A is the stiffness constant and k is the magnitude of the spin wave vector. These two formulations show that spin wave spectra of nanoscopic structures are effectively modified in comparison with their bulk counterparts. In addition, the dynamical behavior at nanoscale also differs from the bulk nature due to enhanced thermal fluctuations under the usual experimental conditions. As an example, if the thermal fluctuation energy $k_B T$ is close, as comparable with the magnitude of the anisotropy energy of the particles, it gives rise to superparamagnetism (explained in Section 1.3.3), yielding effectively zero magnetic moment.

1.3.1 Characteristic Length Scales

Usually, the characteristic length scale of magnetism is given by the exchange length of order of a few nanometers, responsible for difference in the characteristic of nanoscale and meso- or microscale systems. Further, the local atomic symmetry is affected by the characteristic length and dimensionality, given by the crystal or the molecule structure, and the finite-size effects due to the increased influence of surfaces. The interfaces dominate the behavior and give rise to novel phenomena only present at the nanoscale. Combined, they create novel subjects as superparamagnetism; novel spin structures; special short-range coupling mechanisms; interaction between electrical current and magnetization, for example, magnetotransport; and spin-transfer torque, along with magnetization tunneling. The exchange length and the magnetic domain wall width are some of the characteristic lengths that are more relevant to the magnetic properties.

The effect of the characteristic lengths on the magnetic properties can be seen in the case of magnetic particles with dimensions smaller than or comparable to the critical magnetic single-domain diameter (D_{cr}). It is defined as the largest size that a ferromagnetic particle, over which it is more favorable to divide itself into two or more domains. Hence, these particles have a single domain (their lowest-energy configuration). Some of these characteristic lengths, which include the exchange interaction length, the domain wall width, and the spin diffusion length, are given as follows, together with their typical values, and can be surveyed in the reference by Guimaraes (2009):

Exchange interaction length (d_{ex}) ~ 10^{-1}–1; domain wall width (δ_0) ~ 1–10^2; exchange length (l_{ex}) ~ 1–10^2; and spin diffusion length (l_{sd}) ~ 1–10^2.

From this, it is shown that nano-objects have dimensions in the range of many of these characteristic lengths. The critical size for magnetic domains (D_{cr}) is given as:

$$D_{cr} = \frac{72\sqrt{AK}}{\mu_0 M_s^2} \qquad (1.30)$$

where A is the exchange stiffness parameter and K (>0) is the uniaxial anisotropy constant. It varies from 10 nm to a few microns.

The exchange length is formulated as:

$$l_{ex} = \sqrt{\frac{2A}{\mu_0 M_s^2}} \qquad (1.31)$$

The transition region between two magnetic domains is described by domain wall width parameter Δ, which is given by:

$$\Delta = \sqrt{\frac{A}{K}} \qquad (1.32)$$

Further, the domain wall width δ_0 is given by $\delta_0 = \pi \, \Delta$. Interestingly, the energy of the domain wall is also a function of the parameters A and K, for example, 180° wall of a cubic crystal (mentioned earlier in Section 1.2); the energy per unit area of the wall is:

$$\gamma = 4\sqrt{(AK)} \qquad (1.33)$$

It is evident that by varying different parameters of nanosystems, it is possible to probe and utilize different magnetic properties of the materials, with direct dependency on characteristic length.

Further, in every finite crystal, the translational symmetry can be lost or broken. In solids of nanometric size, which are also *called up* by characteristic length, enough proportion of the atoms are close to these symmetry sites. The broken/disturbed translation symmetry carries several consequences to the physical properties in these nanosystems: (1) the relation of the physical properties and dimensionality of nanomaterials (samples with quasi-0D, 1D, 2D, or 3D); (2) the change in coordination of the atoms at the interface; and (3) the effect of the increase in the proportion of surface (or interface) atoms in nanoscopic systems. Detailed discussion is beyond the scope of this chapter (Getzlaff 2008; Guimaraes 2009).

1.3.2 Magnetic Interactions at Nanoscale

In the single-domain limit, the inter-particle and intra-particle interactions play a crucial role in determining the magnetic nature and response for an ensemble of nanoparticles. The interaction strength between nanoparticles modifies the energy behavior, or these interaction strengths change according to their occupied volume concentration (Figure 1.7). Hence, it is essential to identify the different energy contributions, namely, anisotropy energy, Zeeman energy, dipolar interaction energy, and exchange interaction (Chandra 2013). The different types of magnetic interactions, which can be important in allowing the magnetic moments in a solid to interact with each other and may lead to long-range order, are explained in the following section.

1.3.2.1 Dipole–Dipole Interaction

Classically, two magnetic dipoles with moments m_1 and m_2 separated by a vector r have the potential energy w:

$$E = \left(\frac{\mu_0}{4\pi r^3}\right)\left[m_1{}^*m_2 - \left(\frac{3}{r^2}\right)(m_1{}^*r)(m_2{}^*r)\right] \tag{1.34}$$

This equation represents long-range interaction and anisotropic nature of the magnetic system. Further, this interaction energy E depends on separation r and alignment of magnetic moment. Dipolar interaction is too weak to account for the ordering of most magnetic materials, since most of the magnetic materials order at much higher temperature. The order of magnitude of the dipolar potential energy of two moments with $m_1 = m_2 = 1\ \mu_B$ separated by $r = 0.1$ nm can be easily estimated. It turns out to be $m^2/4\pi r^3 = 10^{-23}$ J, equivalent to about $1\ K$. However, in magnetic NP systems, moment, $m = 10^3 - 10^5\ \mu_B$ and the energy may correspond to an ordering temperature of a few tens or even hundreds of Kelvins. One great example can be found by (Djurberg et al. 1997) that shows the relaxation time of suspensions of nearly monodisperse 4.7 nm $Fe_{100-x}C_x$ particles ($x = 22$) in decaling as a function of temperature.

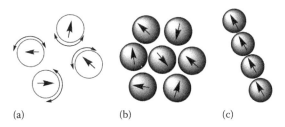

(a)   (b)   (c)

FIGURE 1.7 Example of ensemble of magnetic nanoparticles showing (a) isolated NPs dominated by SPM relaxation. (b) Interacting NPs forming a dipolar glass, and (c) NPs forming a chain with aligned dipole moments.

1.3.2.2 Exchange Interaction

The term *exchange interaction* arises from quantum electromagnetism, lying for long-range magnetic order (Figure 1.7). When the electrons on neighboring magnetic atoms undergo exchange interaction, this is known as direct exchange. Hence, direct-exchange interaction becomes the most important in a close assembly of NP. Monte Carlo simulations reveal that exchange along with dipolar interactions always suppresses the coercivity. The effect of exchange interaction is opposite that of the dipolar interaction over remanence and T_B of a 3D random assembly of ferromagnetic (FM) NPs. Again, it becomes important if both interactions are comparable. The blocking temperature is always enhanced due to interactions, except for the case that particles coalesce and the sample is above the exposed threshold. The crucial role of interparticle interactions in determining the response of an assembly of magnetic NPs to an externally applied field as well as the temperature dependence of the magnetic properties has been recognized long ago (Gaier 2009).

1.3.3 Superparamagnetism

The bulk ferromagnetic material (Fe, Co, Ni, and so on) has the tendency to split into magnetic domains, and the formation of magnetic domain has been discussed in Section 1.2. This splitting results in minimizing its internal energy (Figure 1.8). Basically, these domains are regions containing magnetic moments coupled in the same direction. In a single-domain particle, all spins can be considered in the same direction, except the surface ones; therefore, the particles a are considered uniformly magnetized (Figure 1.8). There are no domain walls to move; consequently, the magnetization can be reversed through spin rotation rather than through the motion of domain walls. Every single domain can be shown by using single magnetization vector representing all its magnetic moments per unit volume. Each domain is separated by domain wall, which usually originates due to defects, dislocations, or transformation in symmetry in the nanosamples. A small external magnetic field can make translation of domain wall. The change in the magnetization direction from one domain to another is in a gradual way due to *wall thickness*, depending on several energetic, crystallographic, and geometric factors. As the sample size becomes bigger, it increases the probability of magnetic domain formation, since the appearance of defects as nucleation sites for domain wall also increases.

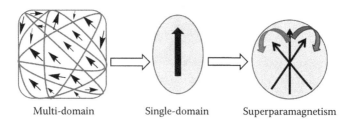

Multi-domain   Single-domain   Superparamagnetism

FIGURE 1.8   Formation of superparamagnetic structures.

As the dimensions of the materials are lowered, as explained in Section 1.3.1, like a system of nanoparticles, the energy stability due to formation of domains decreases enough. At very small size (approximately 10 nm) and under zero external fields, there are no domains and the nanosystem achieves a single-domain configuration, in which the sample is considered uniformly magnetized throughout its volume. Hence, it is represented by a single *superspin*. The existence of single-domain particles indicates a new phenomenon in magnetism (Ortega 2012).

A great deal of experimental evidence regarding the magnetic structure of single-domain small particles has been contributed thereafter, mainly using magnetic force microscopy (MFM; MFM images of 300 nm cobalt) and cross-checking the outcomes against micromagnetic simulations (Fernandez et al. 1998). It seems that the multi-domain to single-domain transition is not as well defined as it might be expected. As with other typical magnetic length scales, the single-domain size (R_{SD}) or the size for which a multidomain configuration is no longer stable, can be expressed as a function of the exchange length, l_{ex}, which gives an idea of the competition between dipolar and exchange interactions (Getzlaff 2008; Ortega 2012):

$$R_{SD} = 36 l_{ex}, \; l_{ex} = \sqrt{\frac{2A}{\mu_0 M_s^2}}, \; \kappa = \sqrt{\frac{K}{\mu_0 M_s^2}} \qquad (1.35)$$

where:

K is the first anisotropy constant, which will be introduced later in this chapter
κ is the dimensionless hardness parameter
A is the exchange stiffness constant
μ_0 is the permeability of the free space
M_s is the saturation magnetization

κ accounts for the balance between dipolar interactions and anisotropy energy. It would be worth comparing R_{SD} with other parameters of interest in nanoparticles, such as the superparamagnetic radius (R_{SPM}). The latter represents the size from which a single-domain particle will begin to undergo thermal fluctuations:

$$R_{SPM} = \sqrt[3]{\frac{6k_B T}{K}} \qquad (1.36)$$

The critical diameter typically lies in the range of a few tens of nanometers and depends on the material nature, and also, it is influenced by contribution from various anisotropy energy terms. The critical diameter of a spherical particle (D_C) below which it exists in a single-domain state is reached when $\Delta E_{MS} = E_{dw}$, which implies:

$$D_C \approx 18 \frac{\sqrt{A K_{eff}}}{\mu_0 M^2} \qquad (1.37)$$

where A is the stiffness constant, K_{eff} is the anisotropy constant, μ_0 is the vacuum permeability, and M is the saturation magnetization (Coey 2010).

There are at least two factors, in addition to freezing of the surface moments, which result in high coercivity of small nanoparticles: (1) spin rotation instead of domain wall motion and (2) shape anisotropy. The coercivity is smaller when the particles are spherical; shape anisotropy can also affect the estimation of the critical volume (below which the particle is single-domain). Spherical particles have small critical diameters as compared with those of large shape anisotropy. The second important phenomenon that takes place in nanoscale magnetic particles is the superparamagnetic limit.

Attempts to produce tiny dispersions of magnetic materials in metals have been done since 1930s. The first modelization of a nanometer-size particle was done by Kittel in 1946 (Kittel 1996). Since the very beginning, the models have been considered in which the magnetic moment would follow an Arrhenius law, with a characteristic relaxation time τ. Nevertheless, the determination of τ was only solved by Néel in 1946 (Kittel 1996). He supposed that each nanoparticle has been formed by rigidly aligned spins that rotate coherently during magnetization reversal, considering the case of uniaxial anisotropy when the energy barrier is much larger than the thermal energy of the system. The equivalence with a gyroscopic system allowed him to derive an expression for τ, as a function of gyromagnetic ratio, longitudinal magnetostriction constant, Young modulus, energy barrier, and thermal energy. He estimated characteristic time relaxation to be of the order of 10^{-10} s, in agreement with the available experimental data. When the size of a particle containing magnetic atoms is small enough, the energy necessary to divide itself into magnetic domains is higher than the energy needed to remain as a single magnetic domain or monodomain.

The magnetic properties of an assembly of monodomain particles are studied within the framework, the so-called superparamagnetic theory, a term coined by Bean and Livingston in analogy to paramagnetic systems. The first assumption of the superparamagnetic theory is to consider that all the magnetic moments within the particle rotate coherently; that is, the net magnetic moment can be represented by a single classical vector, with magnitude $\mu = \mu_{at}N$, where μ_{at} is the atomic magnetic moment and N is the number of magnetic atoms that constitute the corresponding particle (Gubin et al. 2005; Laurent et al. 2008).

The simplest assumption is to consider an effective uniaxial anisotropy, K, which leads to an energy barrier to the magnetization vector, which is proportional to KV, where V is the volume of the nanoparticle. However, it is important to mention that in ultrafine particle systems, surface effects can be very important. In such case, the magnetic relaxation does not proceed by coherent rotation of the spins within the particle. The spins in an isolated particle are held in a particular direction (not necessarily parallel to the applied field) by the MA energy (which is caused by spin–orbital interactions of the electrons and is responsible for holding the magnetic moments along a certain direction). If the particles are not isolated, other interactions will be involved. The anisotropy energy per particle is given by the equation:

$$E_a(\theta) = K_{eff} \, V \sin^2\theta \qquad (1.38)$$

which is the leading term of the series expansion, where V is the volume of the particle, K_{eff} is the effective anisotropy constant, and θ is the angle between the magnetization and the easy magnetization axis of the particle (Gubin et al. 2005; Laurent et al. 2008). The maximum energy barrier is $K_{eff} V$. This is the energy that separates the two energetically equivalent easy magnetization directions, that is, the energy barrier to magnetic moment reversal (the size of this energy depends on many factors, including magnetocrystalline and shape anisotropies). As the particle size decreases the values of V and hence E_a also decreases, as a result the $E_{th} = K_B T$, thermal energy, might exceed the E_a, which causes the particle magnetization to rotate freely resulting in the loss of magnetism in the absence of an applied magnetic field.

The temperature at which this spin flipping occurs is called the blocking temperature (T_B). The blocking temperature depends on the particle size and other factors. At $T > T_B$, the isolated (non-interacting) single-domain particle becomes superparamagnetic (Sarveena et al. 2016). In this state, the magnetic moment of the particle behaves as that of a single atom (like a paramagnet) but with much larger magnitude. The relaxation time of the magnetic moment of a particle, τ_N, is given by the Néel expression, where the factor $\tau_0 \approx 10^{-12} - 10^{-9} s$ is weakly temperature-dependent:

$$\tau_N = \tau_0 \exp\left(\frac{K_{eff} V}{K_B T}\right) \tag{1.39}$$

If the time window of the measurement (τ_m) is longer than the time needed for the particle's magnetic moment to flip, the particle is said to be in a superparamagnetic state. On the other hand, if the experimental time scale is shorter than the moment flipping time, the particle is said to be in the blocked state. The blocking temperature (defined as the mid point between these two states, where $\tau_N = \tau_m$) depends on several factors: (1) the size of the particles, (2) the effective anisotropy constant, K_{eff}, (3) the applied magnetic field, (4) the experimental measurement time, and (5) dipolar interactions.

At blocking temperature regime, the thermal energy and anisotropy energy have key roles in magnetic instability of single-domain magnetic nanoparticles. In the superparamagnetic state and due to thermal energy, the magnetic moments of the nanoparticles fluctuate around the easy axes of magnetization; thus, each one of the magnetic nanoparticles will possess a large magnetic moment that continuously changes orientation (Cahen and Kahn 2003; Coey 2010). When a magnetic field is applied, the magnetic nanoparticles in the superparamagnetic state display a fast response to the changes of the magnetic field without remanent (residual) magnetization and without coercivity (the magnetic field required to bring the magnetization back to zero). In addition, below the blocking temperature, no sufficient thermal energy is available to overcome the anisotropy energy; thus, the fluctuations in the orientations of the magnetic moments of the nanoparticles vanish, and as a result, all the moments are frozen in random orientations, and hysteresis is present on the magnetization cycles.

1.3.4 Single-Particle Phenomenon

The novel features of nanomagnets, as compared with bulk systems, can be categorized as being due to one, or a combination, of the following effects: (1) confinement or finite-size effects, (2) surface effects, and (3) interactions between nano-objects in assemblies of nanoparticles and their interactions with the hosting medium (Schmool and Kachkachi 2015). With regard to magnetic nanosystems, confinement effects are strongly related to surface effects, since the surface acts as a discontinuity of the magnetic object, and this defines the boundary conditions of the system (e.g., magnons). The boundary conditions and size confinement act in a similar way to define the allowed standing wave (magnon) states of the system. Such considerations are extremely important, for example, in ferromagnetic resonance (FMR) experiments or standing spin wave modes (easily not detected).

In nanomagnetism, there are two main characteristic length scales that are of importance: the domain wall width and the exchange length. In the simplest case of a ferromagnetic material with uniaxial anisotropy, the domain wall width has been expressed previously in Section 1.3.1. By definition, stiffness constant and anisotropy constant provide a measure of spin–orbit interaction and the strength, so that magnetization is maintained along its *easy axis*. An inspection of equations explained in Section 1.3.1 shows that these characteristic lengths are related to the (square roots of the) ratios of the exchange energy and the magnetocrystalline anisotropy (MCA) energy or the magnetostatic energy, respectively. These are obtained by minimization of the energies involved with respect to distance. Further, a ferromagnet forms multidomain exceeding over these dimensions; also, single domains are obtained below these characteristic lengths, suggesting the effect of size and shape of particles/samples on the magnetic properties (Schmool and Kachkachi 2015). Small particles with their enhanced surface effects show new features, because of the (intraparticle or intrinsic) surface effects and by the averaging procedure of the experimental probe.

In order to access the particle's intrinsic features, such as the local effective fields and spin configuration, some experimental groups are trying to develop some adequate measurement techniques for single particles, irrespective of the host material, besides theoretical calculation. Nevertheless, theoretical activity has been conducted for single isolated particles as well as nanoparticle assemblies on the basis of models mostly already utilized in thin films. The main reason for this is that the spatial arrangement of the surface atoms is not known, and thereby, no atomic quantum calculations have been performed on clusters of reasonable sizes. A more detailed explanation, including theoretical model of single-particle phenomenon, can be found in the reference (Schmool and Kachkachi 2015).

1.3.5 Exchange Bias

The EB effect usually occurs when a ferromagnet (FM) is in close contact with the antiferromagnet (AFM). In particular, the exchange coupling at an FM–AFM interface may induce unidirectional anisotropy in the FM when system is cooled (or grown) in a static magnetic field above the Néel temperature of the AFM (with the Curie temperature, T_C, of the FM lower than T_N), causing a shift in the hysteresis loop along the magnetic field axis, a phenomenon known as exchange bias (H_E). Exchange coupling between FM and AFM produces a system with a stable order and high anisotropy, owing to large exchange parameter of FM, which makes ferromagnetic order stable at high temperatures, particularly if the dimensions are in nanometers and there are large anisotropies and consequently very stable orientations of FM (Stamps 2001). In addition to FM–AFM

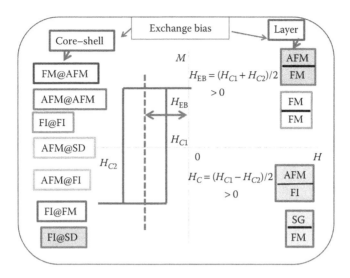

FIGURE 1.9 Schematic of the exchange bias (EB) phenomenon as a shift in the magnetic hysteresis loop at a low temperature when the sample is field-cooled (FC) from well above the Néel or blocking temperature. Different core–shell and layered (different kind of interfaces) magnetic systems show EB effects. SG and SD represent spin-glass and spin disorder, respectively. (Data from Phan, M.H. et al., Exchange bias effects in iron oxide-based nanoparticle systems, *Nanomaterials*, 6, 221, 2016.)

interfaces, EB and related effects have also been observed in other types of interfaces, for example, those involving ferrimagnets (FI): AFM–FI and FI–FM (Figure 1.9).

In such a system, the two coercive fields of the magnetic hysteresis loop are not symmetric and the center of the magnetic hysteresis loop shifts to the left or right. The majority of theoretical models dealing with EB attribute such effects to the formation of domains and pinning of domain walls, either in the FM layer or in the AFM layer. The sign of EB, positive or negative, refers to the shift of the loop with respect to the cooling field, H_{FC}. H_{FC} is the field applied to the sample as it is cooled through the Néel temperature T_N of the AF. If the loop is shifted in the direction of H_{FC}, then EB is positive, $+H_E$, or vice versa. Many FM–AF systems exhibit $-H_E$. However, some systems containing FeF_2, MnF_2, magnetic oxides (e.g., $La_{0.67}Sr_{0.33}MnO_3/SrRuO_3$), and ferromagnetic systems (e.g., FeSn/FeGd) exhibit positive EB, $+H_E$ (Phan et al. 2016). The loop shift associated with EB can be understood qualitatively by considering the intuitive picture of the AFM and the FM exchange couple at the interface (Zheng 2004).

The intuitive spin configuration, for an FM–AFM couple, is shown schematically in Figure 1.10 for different stages of a hysteresis loop. If a magnetic field is applied at a temperature T above the Néel temperature of the AFM, but below the Curie temperature of the FM, that is, $T_N < T < T_C$, all the FM spins will align parallel to H and the AFM spins are randomly oriented. When the FM–AFM couple is cooled through T_N, and $T < T_N$, the magnetic order in the AFM is set up. At the FM–AFM interface, the AFM spins couple with the FM spins and align them ferromagnetically, while the AFM spins orient antiparallel to each other, so as to give a zero net magnetization in the AFM. After the field-cooling process, the spins in both the FM and the AFM lie parallel to each

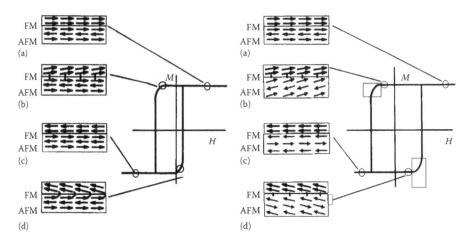

FIGURE 1.10 Schematic diagram of the spin configurations of an FM–AFM couple at the different stages of a shifted hysteresis loop for a system with large K_{AFM} (left) and with small K_{AFM} (right). (From Nogues, J. and Schüller, I.K., Exchange bias, *Journal of Magnetism and Magnetic Materials*, 192, 203–232, 1999. With permission.)

other at the interface (Figure 1.10a, left). When the magnetic field is reversed, the FM spins start to rotate. However, for sufficiently large AFM anisotropy, K_{AFM}, the AFM spins will remain unchanged. The exchange coupling between FM and AFM will exert a microscopic torque on FM spins to keep them in their original position (Figure 1.10b, left), resulting in a larger magnetic field required to switch the FM. As a result, the coercive field in the negative axis will increase (Figure 1.10c). Once the field is reversed to its original direction, the FM spins begin reversing at a smaller field because of the exchange couplings with the AFM spins, which now exert a microscopic torque in the same direction as the applied magnetic field (Figure 1.10d), resulting in a shifted hysteresis loop along the magnetic field axis, H_E. Hence, it signifies that the spins in the FM have only one stable configuration (i.e., unidirectional anisotropy).

For the low AFM anisotropy case, the situation is different from the above-described phenomenon (Figure 1.10, right). As in the previous case, after the field cooling, the spins in both FM and AFM layers are aligned in the same direction (Figure 1.10a, right). During the field reversal, when AFM anisotropy is low, spins in both the FM and the AFM rotate together (Figure 1.10b, right). This results in an enhanced coercivity. Similar behavior is observed after saturating in negative fields, when the hysteresis loop becomes broader (Figure 1.10c and d, right). When the temperature approaches T_N, an interesting phenomenon, such as enhancement of H_C, occurs close to T_N. Although this simple institutive description gives a good basic view of the EB phenomenology, it should be taken with caution, as this intuitive model neglects many parameters that have been shown to be important in EB, such as AFM or FM domains, interface roughness, and AFM spin structure. In other words, we can say that clear understanding of EB at the microscopic level is still lacking.

Although the main indication of the existence of EB effect is the observation of shifted hysteresis loops along the field axis after field cooling across the Néel temperature of the

AFM T_N, some other macroscopic effects usually accompany the observation of loop shifts. There are some systematic dependencies and related phenomena that occur in many EB systems:

1. Enhanced coercivity
2. Temperature dependence
3. Cooling-field dependence
4. FM layer thickness
5. AFM layer thickness
6. Particle size dependence
7. Influence of dilution
8. Vertical loop shifts
9. Training effects
10. Minor loops effect

All these effects and dependency on the EB phenomenon will be discussed in other chapters.

1.3.6 Maximum Energy Product $(BH)_{max}$

Advanced permanent magnets are materials used to convert mechanical energy to electrical energy (and vice versa) in many applications, due to their large maximum energy product $(BH)_{max}$ values, which describe the maximum amount of magnetic energy stored in a magnet. It is the single characteristic that best describes the suitability of a given material for use in permanent magnets. The *B-H* curve or the hysteresis loop, which characterizes each magnet material, is the basis of the magnet design. The second quadrant of the *B-H* curve is called the demagnetization curve. One quality measure of a permanent magnet is the size of this demagnetization curve. It concerns the product, is maximally attainable with a magnetic material, and is made out of flux density *B* and field strength *H*. The selection criterion used for the selection of permanent magnet is the energy product, *BH*. This is appropriately named since energy density is proportional to energy product. High energy product can be described as the maximum area swept by a rectangle in the second quadrant of a hysteresis loop (*B-H* curve), shown in Figure 1.11. It is given by $(BH)_{max} = \mu_0 M_s^2/4$ (Coey 2011). Hence, maximum energy product $(BH)_{max}$ is a measure of the maximum amount of useful work that a permanent magnet can do, outside the magnet.

A permanent magnet always operates as an open circuit. The presence of an air gap in the magnetic circuit creates free poles on surface and, consequently, a demagnetizing field H_d, which makes induction *B* lowers than the remanence B_r. When the magnet is manufactured, a strong magnetic field is applied to it, which makes the induction to follow the path as indicated in Figure 1.11. The intersection of the line OC, called the load line, with the second quadrant (also called the demagnetization curve) of the hysteresis defines the operating point *P* of the magnet. The best operating point *P* is chosen in the following way (Cullity and Graham 2009):

According to the Ampere's law:

$$\oint H dl = 0 \tag{1.40}$$

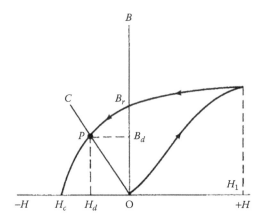

FIGURE 1.11 Initial magnetization and demagnetization curve of a permanent magnet. B_r is the residual induction; point P is the operating point. (Cullity, B.D. and Graham, C.D.: *Introduction to Magnetic Materials*. 2nd ed. 2009. Copyright Wiley-VCH Verlag GmbH & Co. KGaA. Reproduced with permission.)

Let H_g be the constant field strength provided by the magnet, B_m is the induction, and H_m is the field. Let l_m be the length of the magnet and l_g length of the gap between the poles. From Equation 1.1:

$$H_g l_g - H_m l_m = 0 \tag{1.41}$$

Since $B = H$ (cgs) or $\mu_0 H$ (SI) in the air gap, the magnetic flux (φ) is given as:

$$\varphi = B_g A_g = \mu_0 H_g A_g = B_m A_m \tag{1.42}$$

where A_g and A_m are the cross-sectional areas of the air gap and the magnet, respectively. Multiplying Equations 1.40 and 1.33 and using the fact that $B_g = \mu_0 H_g$:

$$(B_m H_m)V_m = \frac{B_g^2 V_g}{\mu_0} \tag{1.43}$$

where V_m and V_g represent the volumes of the magnet and the air gap, respectively.

The flux density in the gap is therefore maximized when the product $B_m H_m$ is maximum; hence, the energy product is a more specific criterion for the selection of the magnetic materials. For most efficient use of material, the magnet should be so shaped that the load line passes through the point at which BH has its maximum value $(BH)_{max}$. Although $(BH)_{max}$ is an index of material quality, it is not the only index or even the most suitable one for all applications. Magnets may operate under either static or dynamic conditions. In this respect, currently, the most actively pursued approach for improvement in the maximum energy product is the synthesis of exchange-coupled nanocomposites comprising ESMs and EBMs. This may provide an alternate for the $(BH)_{max}$.

1.4 Magnetic Anisotropy Energy

Anisotropy is most easily observed in single crystals of solid elements or compounds, in which atoms, ions, or molecules are arranged in regular lattices. The MA is the dependence of magnetic properties on a preferred direction. There are several different types of MAs, which are summarized here and discussed later in the subsections: (1) MCA: the atomic structure of a crystal introduces preferential directions for magnetization. (2) Shape anisotropy: when a particle is not perfectly spherical, the demagnetizing field will not be equal for all directions, creating one or more easy axes. (3) Magnetoelastic anisotropy: tension may alter magnetic behavior, leading to MA. (4) Exchange anisotropy: a relatively new type that occurs when antiferromagnetic and ferromagnetic materials interact.

Before going to details, we can revise a few key words: (1) An *easy axis* is an energetically favorable direction of spontaneous magnetization that is determined by the sources of MA. (2) *Hard axis* is the direction inside a crystal, along which large applied magnetic field is needed to reach the saturation magnetization. (3) The *magnetic moment* of a magnet is a quantity that determines the force that the magnet can exert on electric currents and the torque that a magnetic field will exert on it.

In the absence of an applied magnetic field, a magnetically isotropic material has no preferential direction for its magnetic moment, while a magnetically anisotropic material will align its magnetic moment with one of the easy axes. The two opposite directions along an easy axis are usually equivalent, and the actual magnetization direction can be along either of them. Most of the magnetic materials are magnetized to saturate along the same direction (magnetization direction). The anisotropic magnets can generate stronger magnetic field than the isotropic ones. The theory of ferro- and ferrimagnetism is based on electronic exchange forces. These forces are so strong that, naturally, these materials are spontaneously magnetized, even in the absence of an applied field (however, some field is required in laboratory). In some cases, the material in bulk form has a remanence of nearly zero. However, all ferro- or ferrimagnetic materials are not magnetized to saturation in the absence of a field. The ferromagnetic materials form domains that have been elaborated in previous sections. The net vector sum of all the domains therefore produces a total magnetization of near zero. As an example, ferromagnetic materials exhibit memory effects in their $M(H)$ dependence; this property is measured as a hysteresis loop (see Figure 1.5). By applying magnetic field in different cycles (Figure 1.5), we can understand saturation point, remanence (and remanence induction), coercive field (inverse relation with saturation magnetization), and so on. Energies (Hamiltonians) contributing to the magnetization process are:

$$\widehat{H_{\mathrm{ex}}} = -\sum_{i,j} J_{ij} \widehat{S_i} \widehat{S_j} \quad \text{and} \quad \widehat{H_{\mathrm{field}}} = \frac{\mu_B}{h} \sum_i \vec{B} \widehat{S_i} \sqrt{b^2 - 4ac} \tag{1.44}$$

These energy terms cannot explain the existence of the coercive field (H_c) and its relation with magnetic susceptibility ($\chi_m = M/H$). All these factors can arise due to *anisotropy*, the dependence of the magnetic properties on the direction of the applied field with respect to the crystal lattice. It results a lower or higher magnetic field to reach the saturation magnetization. In the absence of an external magnetic field, the magnetization M of a magnetic solid usually tends to lie along one or several axes. Energy is required to displace the magnetization from these preferential directions. The MA is defined as the energy that is necessary to turn M into any direction different from the preferred axes. Magnetic anisotropies might be caused by different mechanisms and are generally described as different contributions, E_{ani}, to the free energy density of a magnetic system

(Gaier 2009). For this, E_{ani} is advantageously expanded into a series of components α_i of the unit vector, pointing into the direction of magnetization:

$$E_{\text{ani}} = \sum_{i,j,k} K_{i,j,k} \alpha_1^i \alpha_2^j \alpha_3^k \tag{1.45}$$

The parameters $K_{i,j,k}$ in Equation 1.45 are the so-called anisotropy constants, which are experimental parameters sufficient to explain contributions from different anisotropies.

In this section, we will discuss, within the framework of the book, the bulk and surface contributions of anisotropy to the MA. Usually, MCA and shape anisotropy make a major contribution to the MA. In thin films and multilayers, strain effects can give rise to the so-called magnetoelastic anisotropy. Contributions of interface anisotropy become important, depending on the interface properties of the different layers. Magnetic nanoparticles, drastically different from those of their bulk counterparts, are advantageous for utilization in a variety of applications such as storage media and probes in the biomedical sciences. Their fundamental magnetic properties such as blocking temperature (T_B), spin life time (τ), coercivity (H_c), and susceptibility (χ) are strongly influenced by the nanoscaling laws. Hence, these scaling relationships can be used to control magnetism from ferromagnetic to superparamagnetic regimes. For example, life time of a magnetic spin is directly related to the MA energy and also to the size and volume of nanoparticles (Gaier 2009).

1.4.1 Bulk Contribution to Magnetic Anisotropy Energy

1.4.1.1 Magnetocrystalline Anisotropy

The most important type of anisotropy is the MCA, caused by the spin orbit interaction (basic principle) of the electrons, linked to the crystallographic structure. Owing to the interaction of orbits with the spins, spins prefer to align themselves along well-defined crystallographic axes, which are defined as easy magnetization axes. It is easier to use mathematical expressions (power series expansions, accounting for the crystal symmetry) and take the coefficients from the experiment. The MCA is usually small compared with the exchange energy. However, the direction of the magnetization is only determined by the anisotropy (MCA) as the exchange interaction just tries to align the magnetic moments parallel, no matter in which direction (Getzlaff 2008; Gaier 2009). The origin of the MCA lies in the spin–orbit interaction, in which the orbital motion represents a current loop that gives rise to a magnetic field in the center of the loop. This field interacts with the spin angular momentum, coupling the spin and orbital moments (Getzlaff 2008; Gaier 2009). In a quantum mechanical treatment, the spin–orbit interaction is described by the Hamiltonian equation:

$$H_{so} = -\frac{eh^2}{2m_e^2 c^2} \frac{1}{r} \frac{d\Phi(r)}{dr} \hat{s} * \hat{I} = \xi_n l(r) \tag{1.46}$$

The spin $\hat{s}$ and angular momentum $\hat{I}$ couple via the electrostatic potential of the nuclear charges, $\Phi(r)$, which has the largest gradient $d\Phi(r)/dr$ for small distance r from the nucleus. The expectation value of $\xi_{nl}(r)$ is called the spin–orbit coupling constant, or the spin–orbit parameter. Its value is of the order of 10–100 meV. The spin–orbit interaction is thus considerably weaker than the exchange interaction (≈ 1 eV).

In single crystalline materials, the bonding is anisotropic; that is, the overlap of the atomic wave functions depends on the crystallographic directions. This gives rise to anisotropy of the orbital magnetic moment, which results in different values of the spin–orbit energy in Equation 1.46 associated with different crystallographic directions. The symmetry of the MCA is apparently that of the crystal lattice (Getzlaff 2008). The MCA's contribution to cubic system has already been discussed in Equation 1.26. This equation can further be written in spherical polar coordinates as:

$$E = K_0 + K_1 \left(\frac{1}{4} \sin^2\theta \sin^2 2\varphi + \cos^2\theta \right) \sin^2\theta + \frac{K_2}{16} \sin^2 2\varphi \sin^2 2\theta \sin^2\theta + \ldots \quad (1.47)$$

The parameters K_1 and K_2 are the so-called first and second anisotropy constants for a cubic system. θ is the angle between magnetization and stacking direction of cubic close-packed plane (Cullity and Graham 2009). Usually, the interplay between these two anisotropy constants determines the direction of hard and easy axes.

1.4.1.2 Magnetoelastic Stress Anisotropy Energy

Magnetic strain (or *stress*) anisotropy refers to various magnetomechanical energy contributions, for example, an elastic deformation of a body due to mechanical pressure onto the system. The influence of mechanical stress onto the magnetic properties of embedded Fe NPs was investigated (Saranu et al. 2011). The stress is applied through expansion of tantalum (Ta) substrate by loading with hydrogen. A small modification of the slope of the hysteresis loop was observed. In another study, core–shell particles have been investigated. A ferroelectric $BaTiO_3$ shell exerts a strain on the ferrimagnetic Fe_3O_4 core, and by this, a magnetoelectric coupling is realized. In other words, the entire subject of strain in NPs is largely unexplored. The magnetoelastic effect also arises from the spin–orbit interaction. The spin moments are coupled with the lattice through the orbital electrons. If the lattice is changed by strain, the distances between the magnetic atoms are changed, resulting in change in interaction energies, thus creating magnetoelastic anisotropy (Chtchelkanova et al. 2003). Magnetostriction constants, λ_{hkl}, are defined for various crystal directions. For an elastically isotropic medium, with isotropic magnetostriction, the magnetoelastic energy per unit volume is given by:

$$E = -\frac{3}{2}\lambda\sigma\cos^2\theta \quad (1.48)$$

where σ is the stress and θ is the angle between magnetization and stress directions (Johnson et al. 1999). For positive λ, as in metallic iron, the easy magnetic direction will be along the direction of tensile stress or perpendicular to a compressive stress. Strain in thin films and multilayers can be produced by the growth conditions, for example, lattice mismatch between layers and thermal stress caused by differences in thermal expansion coefficients of proximity layers.

1.4.1.3 Dipolar Magnetostatic Anisotropy Energy

The magnetostatic (dipole) energy depends on the magnetization M and the magnetic-dipole moment per volume arising from the alignment of atomic magnetic dipoles. In a solid, the dipoles arise primarily from electron spins. Hence, the formation of single and multidomain specimens comes into existence as a consequence of energy minimization (Schmool and Kachkachi 2015).

Among the most important sources of the MA, in thin films, is the long-range magnetic dipolar interaction, which senses the outer boundaries (hence shape) of the sample. Neglecting the discrete nature of matter, the shape effect of the dipolar interaction in ellipsoidal ferromagnetic samples can be described, via an anisotropic demagnetizing field, H_d, given by $H_d = NM$. Here, M is the magnetization vector and N is the shape-dependent demagnetizing tensor. For a thin film, all tensor elements are zero, except for the direction perpendicular to the layer: $N^\perp = 1$; the magnetostatic energy can be expressed as:

$$E_d = -\frac{\mu_0}{2V}\int M^* H_d d\upsilon \tag{1.49}$$

Hence, an anisotropy energy contribution per unit volume V, of a film is:

$$E_d = \frac{1}{2}\mu_0 M_s^2 cos^2\theta \tag{1.50}$$

This expression assumes uniform magnetization magnitude M_s, subtending an angle θ with the film normal. According to this expression, the contribution favors an in-plane preferential orientation for the magnetization (Camarero et al. 2016). Because the thickness of the film does not interfere with the continuum approach used, it contributes only to some kV, responsible for the negative slope of the K_{eff} versus the t plot (Figure 1.12). This

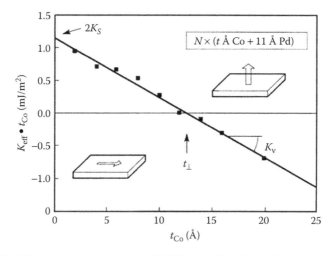

FIGURE 1.12 Magnetic anisotropic energy (MAE) times the individual Co layer thickness versus the individual Co layer thickness of Co/Pd multilayers. The vertical axis intercept equals twice the interface anisotropy, whereas the slope gives the volume contribution. (From den Broeder et al., Magnetic anisotropy of multilayers, *Journal of Magnetism and Magnetic Materials*, 93, 562–570, 1991. With permission.)

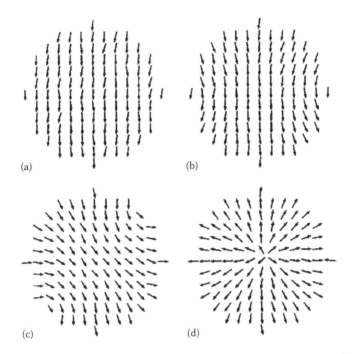

(a)

(b)

(c)

(d)

FIGURE 1.13 Spin structures from Monte-Carlo simulations shown at the central plane for various ratios of the surface-to-volume anisotropy: $K_s/K_V = 1$ (a), 10 (b), 40 (c), and 60 (d). (Reprinted with permission from Labaye et al., Surface anisotropy in ferromagnetic nanoparticles, *Journal of Applied Physics*, 91, 8715. Copyright 2002, American Institute of Physics.)

continuum approach is common in the analysis of the experimental data. However, when the thickness of the ferromagnetic layer is reduced to only a few monolayers (MLs), the film is not considered a magnetic continuum but considered a collection of discrete magnetic dipoles over a regular lattice. Calculations made on the basis of discretely summing the dipolar interactions for films in the range of 1–10 MLs lead to the following results (Draaisma 1988). Depending on the symmetry of the interface, the outer layers experience a dipolar anisotropy, which can be lower than that of the inner layers. For the inner layers, the dipolar anisotropy is rather close to the value based on the continuum approach. As a result, the average dipolar anisotropy can be phenomenological, expressed by a volume and an interface contribution (Figure 1.12). However, the magnitude of the dipolar interface contribution is of minor importance, and other sources of interface anisotropy, such as spin–orbit coupling, appear to be dominant (Figure 1.13).

1.4.2 Surface Contribution to the Magnetic Anisotropy Energy

1.4.2.1 Surface Anisotropy

Nanoparticles have remarkable magnetic properties different from their massive ferromagnets or ferrites (Frey 2008). The source of this difference is the surface effect. As the particles size decreases, the fraction of atoms that lie near or on the surface increases.

It is estimated that in a 3-nm-diameter particle, 80% are surface atoms. Surface effects can also be observed not only in small particles but also in larger-particle systems, depending on the chemical composition of the nanoparticles and their crystalline state. The MCA reflects the symmetry of the neighbors of each atom. The large perturbations to the crystal symmetry at surfaces should lead to MCAs of different magnitude and symmetry for surface sites (Bader 2006). Basically, surface anisotropy has a crystal-field nature, and it comes from the symmetry breaking at the boundaries of the particle announcing site-specific surface anisotropy of unidirectional character, from the broken exchange bonds. The low coordination number on the particle surface affects the magnetic behavior on the surface and could propagate within the particle. For instance, some experiments and Monte Carlo calculations show that the Curie temperature decreases with decreasing particle size. Therefore, the magnetization near and on the surface is lower than that in the core of the nanoparticle. This effect has been reported in magnetization measurements of Fe_2O_3 nanoparticles. High field magnetization measurements of Fe_2O_3 and Co show an increase in non-saturation tendency for small particle sizes (Jun et al. 2008). High-field open hysteresis loops have been observed and attributed to the high surface anisotropy. High-field magnetic relaxation (SG-like surface) state can be observed in nickel-ferrite particles. Many experiments have also been focused on the effect of surface oxidation on the magnetic properties of ferromagnetic particles. In materials with a larger core diameter, the coercivity is almost constant (~250 Oe for Fe case). Néel was the first to show that if uncompensated spins exist, antiferromagnetic particles can exhibit superparamagnetic behavior. Experimental work on Fe_2O_3 particles in rocks and on Cr_2O_3 particles has been followed by Néel's analysis. More recently, the experimental work has focused on iron oxides, which are used as high-density recording media. The magnetic properties of antiferromagnetic nanoparticles have been investigated by Monte Carlo simulations, showing that magnetization of a particle arises from the uncompensated spins attributed to the surface of the particle and the temperature variation of the magnetization, as well as the coercivity variations (size and uncompensated spins). Many of the systems that exhibit surface effects could be described by a core–shell model. This model considers each particle as a nanosystem composed of an internal magnetic (ferro- or antiferromagnetic) ordered core, described by the Stoner–Wohlfarth relaxation model, and a disordered shell of spins, interacting among them and the particle core. This model satisfactorily describes the observed features in magnetization measurements of ferromagnetic-like systems, such as FeNiB and CoNiB nanoparticles, and antiferromagnetic-like systems, such as NiO. Ferromagnetic resonance studies performed on these samples show a significant increment in the effective line width at low temperatures due to surface anisotropy. In addition, the resonance field falls with the increment in magnetization, owing to the polarization of the core by surface clusters. Thus, the field necessary to get the resonance condition becomes smaller than expected. For small nanoparticles, the surface effect must be considered to estimate the size dependence of the effective anisotropy.

Usually, it has been considered an effective anisotropy term, given approximately by the following phenomenological expression:

$$K_{eff} = K_V + \frac{6}{D} K_S \quad (1.51)$$

This equation has been extensively applied to take into account the effect of the surface on nanostructured systems, with rather good success in most of the cases. In fact, surface effects can be important even when dealing with larger particles, and a careful analysis is necessary to properly separate the effects of structural disorders, interparticle interactions, and surface contributions.

Effective anisotropy constant K_{eff} is divided into two parts, with K_V being the volume-dependent MCA constant and K_S being the surface-dependent MCA constant. The factor of the two is due to the creation of two surfaces. The second term exhibits an inverse dependence on the thickness $<D>$ of the system. Thus, it is only important for thin films. In order to illustrate the influence of the surface anisotropy, we will discuss the so-called *spin reorientation transition*. Rewriting results in:

$$d \cdot K_{eff} = d \cdot K_V + 2K_S \quad (1.52)$$

Plotting this dependence as a $d \cdot K_{eff}(d)$ diagram allows us to determine K_V as the slope of the resulting line and $2K_S$ as the zero-crossing, which is exemplarily shown for a thin Co layer with variable thickness d on a Pd substrate. Owing to the shape anisotropy, K_V is negative. This can directly be seen by the negative slope, which results in an in-plane magnetization. The zero-crossing occurs at a positive value K_S. This leads to a critical thickness d_c: $d_c = -2K_S K_V$, with $d < d_c$: perpendicular magnetization, and $d > d_c$: in-plane magnetization, due to the change of sign of K_{eff}. Thus, the volume contribution always dominates for thick films, with the magnetization being within the film plane. The relative amount of the surface contribution increases with the decreasing thickness, followed by a spin reorientation transition toward the surface normal below d_c.

1.4.2.2 Interface Anisotropy

There has been a significant interest in the properties of the interface between different-natured magnetic materials, observing interfacial magnetization and anisotropy. This interface anisotropy, at nanoscale, is controlled by several sources and is discussed in several chapters in the book, wherever required. The MA energy determines the direction of magnetization within a uniformly magnetized sample. It is one of the most noticeable and frequently measured features of ferromagnetism (Schmool and Kachkachi 2015). The theoretical calculations of the anisotropy energy for Fe, Co, and Ni materials pose several difficulties. The anisotropy energy, caused by the spin–orbit interaction, is much smaller than other electronic energies: for the elemental magnets, it is, at most, a few per atom (Djurberg et al. 1997). Non-elemental samples (super lattices) may have larger anisotropies. In any case, high numerical accuracy and convergences must be achieved despite the substantial loss of symmetry caused by the spin–orbit interaction (Camarero et al. 2016).

1.5 Phenomena in Nanostructured Magnetic Material

As mentioned before, the phenomenon of EB has been one of the most fascinating and complex effects that have been extensively studied both in theory and in experiments in the field of magnetism. The study of exchange coupling in nanostructured magnetic systems has drawn major attention due to remarkable EB interaction, especially interesting due to the fundamental role of this effect in the development of spin valves and tunneling devices. Research in permanent magnet development has historically been driven by the need to supply larger magnetic energy in ever-smaller volumes for incorporation in a variety of applications, including consumer products, transportation components, military hardware, and clean energy technologies such as wind-turbine generators and hybrid-vehicle regenerative motors (Lewis and Jimenez-Villacorta 2012).

Traditionally, the search for new permanent magnets has been focused mainly on the search for materials with large anisotropies, mainly based on rare earth elements. However, the ever-increasing demand for permanent magnets has triggered a shortage of rare earth raw materials, resulting in a significant increase in price. Exchange-biased magnetic nanocomposite systems are promising candidates for the next-generation permanent magnets and are a possible route for the development of rare-earth-free permanent magnets, overcoming the limitations of conventional permanent magnets (López-Ortega et al. 2014). In this context, the ESMs and EBMs are systems of major interest, with magnetic exchange coupling operation working between the two phases.

1.5.1 Exchange-Spring Magnets

The ESM is a material in which high-energy-density permanent magnets may be made as two-phase composite materials. The ESMs consists of nanoscale hard (high H_C) and soft (high M_S) magnetic phases coupled via interfacial exchange interaction, making possible the increase of the $(BH)_{max}$ product of the nanocomposite when compared with any one of the individual phases that form the nanocomposites (Leite et al. 2012). Hybrid ferrite-based exchange spring composites are promising for advanced permanent magnetic materials. The exchange-spring principle is believed to provide flexibility in material selection for the constituent hard and soft phases and represents the possible solution for creating high-performance permanent magnets by reducing or eliminating the expensive rare earth elements that are often necessary in the production of permanent magnetic materials. The soft phase can be a typical ferromagnetic transition metal such as Fe and Co, or their alloys, and the hard phase can also be a high-anisotropy material that does not contain rare earth elements (Jiang and Bader 2014). Typically, exchange-spring nanosystems are composed of phases with very different chemistries, such as $Nd_2Fe_{14}B$ combined with α-Fe or Fe_3B and Sm–Co combined with Co or Fe (Lewis et al. 2003). To attain a large value of $(BH)_{max}$, the ESM needs to have the volume fraction of the high-magnetization soft phase as large as possible. Still, some molecule-based materials composed either of hard magnets in direct contact with soft magnets or of an anisotropic molecule inside a soft magnetic lattice have also shown an ESM behavior (Prieto-Ruiz et al. 2015). Meanwhile, the short-ranged nature of magnetic exchange interaction sets the critical dimension for effective coupling of the soft phase, or exchange hardening, to the domain wall thickness, which is typically a few nanometers (Jiang and Bader 2014).

The ESMs are named so due to the largely reversible springy magnetic interaction that takes place between the magnetic components under the application and removal of an applied magnetic field. In the ideal case, the major demagnetization curve of an ESM is characteristic of a single magnetic phase, despite the existence of two phases with very different magnetic features (Figure 1.14). The largest possible maximum energy product of a given magnetic material with large coercivity ($H_C > M_S/2$) is determined by the intrinsic saturation magnetization (Jimenez-Villacorta and Lewis 2014):

$$(BH)_{max} \leq \frac{1}{4}\mu_0 M_S^2 \tag{1.53}$$

However, most of the high-coercivity permanent magnetic materials typically do not show very large saturation magnetization values (typically around $M_S \sim 1.0$–1.5 T). The nanoscale combination of high-magnetization soft ferromagnetic phase with a high-coercivity hard ferromagnetic phase to create a magnetic composite material could result in a much improved $(BH)_{max}$. To realize a strong interphase, exchange coupling in ESM on theoretical considerations describes a microstructure consisting of a large quantity of well-separated soft phase embedded in a hard matrix to provide high remanent

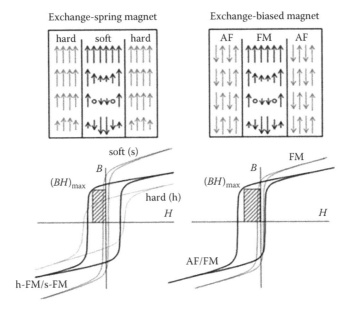

FIGURE 1.14 Schematic representation of the demagnetization mechanism (second quadrant) in exchange-coupled nanocomposite structures, for the case of hard–soft magnetic phases (exchange-spring magnets) and antiferromagnetic–ferromagnetic phases (exchange-biased magnets). (Reprinted from Lewis, L.H. and Jimenez-Villacorta, F., Perspectives on permanent magnetic materials for energy conversion and power generation, *Metallurgical and Materials Transactions A*, 44, 2–20, 2012. With permission.)

magnetization values. A refined expression of the maximum energy product that may be obtained in idealized exchange-coupled magnets is given as follows:

$$(BH)_{\text{max}} = \frac{1}{4}\mu_0 M_S^2 \left[1 - \frac{1}{2} \frac{\mu_0 (M_S - M_h) M_S}{K_h} \right] \tag{1.54}$$

where M_S and M_h are the magnetization values for the soft and hard phases, respectively, and K_h is the anisotropy constant of the hard phase. An estimated nucleation field H_N is the reversed applied field that causes the domain wall to enter the hard phase, thus initiating magnetic reversal. In this configuration, H_N is given by:

$$H_N = \frac{2(f_s K_s + f_h K_h)}{\mu_0 (f_s M_s + f_h M_h)} \tag{1.55}$$

where f_s and f_h are the fraction volumes of the soft and hard phases, respectively, and K_s is the anisotropy constant of the soft phase. Hence, with optimized proportions, arrangements, and dimensions of the hard and soft phases, the energy product of these exchange-coupled nanocomposites could surpass those of single-phase magnets (Lewis and Jimenez-Villacorta 2012).

Since the exchange coupling, being a two-phase phenomenon, is a combination of the individual intrinsic properties of each phase, there are some requirements that have to be fulfilled before any exchange coupling between the hard and soft magnetic materials can take place. First, both materials must be in intimate contact, for example, in the form of a core–shell structure, a layered structure, or as particulate composites. Second, the particle size of the soft phase should not exceed twice that of the domain wall width of the hard-phase material. Furthermore, in order to increase the $(BH)_{\text{max}}$ of the composite as much as possible, it is preferable that the particle size of the hard phase is at the limit of the single-domain particle size (to obtain the maximum coercivity) (Jenu et al. 2016).

The magnetic properties of ESMs and hard/soft magnetic heterostructures such as coupled bilayers and multilayers provide convenient model systems for better understanding of their magnetic properties. Combined with numerical modeling, these systems allow us to obtain greater insights into the coercivity mechanism and magnetization-reversal process in ESMs and hence help realistically estimate the ultimate gain in performance that can potentially be realized in permanent magnets based on the exchange hardening principle (Fullerton et al. 1999).

The exchange-spring behavior has been mostly investigated for metallic systems such as Nd-Fe-B/Fe and SmCo/Fe-Co because of their intrinsic high-energy product; these magnets suffer from low T_C, poor oxidation–corrosion resistance, and high price of rare earth elements. Recently, nanocrystalline rare-earth-free systems, that is, $CoFe_2O_4$-based spring magnets, have attracted considerable attention due to the fact that a partial reduction of cobalt ferrite leads to FeCo (soft)-$CoFe_2O_4$ (hard) composites, where crystallographic coherence, mandatory for effective coupling, is more likely than in systems fabricated by using two dissimilar materials (Quesada et al. 2016).

1.5.2 Exchange-Biased Magnetic Nanosystem

Analogous to ESMs, whereby a soft phase is exchange-coupled to a hard phase, EBMs are composed of an FM phase exchange-coupled to an AFM phase. Here, the FM component contributes high saturation magnetization (M_s), while the AFM component contributes high MA. The exchange interaction at the FM–AFM interface that pins the FM moments during the reversal process causes an increase of H_C and M_R and provides an enhanced $(BH)_{max}$ (Figure 1.14). Since the magnetic exchange interaction is active over very short (Ångstrom-level) dimensions, the physical proximity and condition of the AFM and FM phases are important (Jimenez-Villacorta and Lewis 2014). The EB phenomenon has been explained previously in Section 1.3. In addition to AFM–FM interfaces, EB is also observed in other types of interfaces involving an FI (e.g., FI–AFM and FI–FM) or a spin-glass (SG) phase (e.g., FM–SG, AFM–SG, and FI–SG) and an AFM/diluted FM semiconductor (AFM/DMS), reflecting its diverse origin (Nogues and Schüller 1999). The exchange field at the interface, H_E, may be expressed as (Berkowitz and Takano 1999):

$$H_E \approx \frac{\Delta\sigma_E^{FM-AFM}}{\mu_0 M_{FM} t_{FM}} \tag{1.56}$$

where $\Delta\sigma_E^{FM-AFM}$ is the interfacial exchange energy density. The EB can be observed when the following condition is satisfied:

$$K_{AFM} t_{AFM} \geq \Delta\sigma_E^{FM-AFM} \tag{1.57}$$

where K_{AFM} is the exchange anisotropy of the AFM phase. This provides a condition for the occurrence of EB and sets a minimum value for the anisotropy of the AFM phase:

$$K_{AFM} \geq H_E M_{FM} \left(\frac{t_{FM}}{t_{AFM}} \right) \tag{1.58}$$

Systems exhibiting EB have been studied extensively in recent years, particularly in thin-film forms, where the interface between the two phases is easier to control and characterize. Maximum interphase contact is created when the constituents are combined in nanocomposite form. However, to date, exchange anisotropy has not been widely investigated in bulk magnetic nanocomposite systems due to difficulty in obtaining suitable uniform phase separation and texturing during processing. Initial efforts to attain nanostructured exchange-biased permanent magnets have been limited to compacts of ferromagnetic transition-metal nanoparticles coated with their native antiferromagnetic oxide or analogous native sulfides and nitrides or to mixtures of mechanically alloyed components (Marion 2013). Other antiferromagnetic Mn-based alloys may be potential components of EBMs. In these Mn-based alloys, it is possible to tune the antiferromagnetic characteristics, such as the antiferro-MA constant K_{AF} or the Néel temperature T_N, through modification of the Mn concentration (Jiménez-Villacorta et al. 2014). In particular, the two-phase architecture inspired by the microstructure of AlNiCo has provided new insight into the character and phase relationships of the constituent FM (FeCo-rich) and the non-magnetic (NiAl-rich) phases. Optimization of the microstructure of these Fe-Co-Mn-based alloys is anticipated to be accompanied by improved magnetic performance that could contribute to the development of next-generation permanent magnets. Hence, AF phase will provide an additional source of anisotropy (exchange anisotropy) to enhance the coercivity and improve the properties of permanent magnetic.

1.6 Research and Development in Nanomagnetism

In recent decades, there has been a considerable interest in research and development in the field of nanoscale magnetism due the unique properties of NPs and their potential technological applications. This fact is reflected in the research in this area, whose focus has changed from microcrystalline to nanocrystalline size, and the optimum particle size varies according to the application. Nanoscale magnetism is one of the most prospective fields in the today's science and forms a ground for the new branches of high-tech industry. It includes study of properties and applications of magnetism of isolated nanoparticles, nanodots, nanowires, thin films, and multilayers, as well as macroscopic samples that contain nanoscopic particles. The intensive investigations in this field promoted considerable progress in technological and biomedical applications of magnetism in various areas such as information technology, magnetosensors, electronics, data storage, magnetic read heads of computer hard disks, magnetoelectronics, spintronics, microwave electronic devices, biomedicine, molecular biology, biochemistry, diagnosis, and catalysis.

The most outstanding breakthrough in the field of magnetism was the discovery of the EB effect. Since the discovery of exchange-biased giant magnetoresistance (GMR) spin valves, a variety of devices have been built and proposed, such as read heads, magnetic sensors, and magnetoresistive memories. The exchange-bias effect has a large impact on the development of magnetic storage and sensors. The EB effects in nanostructures are widely investigated for applications in nanometric spintronic sensors (Zhang and Krishnan 2016). Spintronics also opens the way for quantum mechanical information processing or quantum computing, which would offer unprecedented computational power, which will enable breaking of all current encryption schemes used for secure online transactions. Spin-based information processing will require much less power than its conventional electronics counterpart, which produces excessive heat. Voltage control of EB in multiferroics provides an energy-efficient way to achieve a rapidly 180° deterministic switching of magnetization, which has been considered a key challenge in realizing the next generation of fast, compact, and ultra-low-power magnetoelectric memories and sensors (Yang et al. 2017). Nanomagnetic Logic (NML) is an innovative technology, currently under investigation by the research community. It could be a possible alternative to traditional complementary metal-oxide-semiconductor (CMOS) devices. Among its different implementations, Perpendicular Nanomagnetic Logic (pNML) seems to be the most effective due to its low power consumption, non-volatility, and monolithic 3D integrability, which makes it possible to integrate memory and logic into the same device by exploiting the interaction of bi-stable nanomagnets with perpendicular MA (Riente et al. 2017).

The scientific community is seeking to exploit the intrinsic properties of magnetic NPs to obtain medical breakthroughs in targeting and delivering diagnostic contrast agents (CA) to enhance the contrast in magnetic resonance imaging (MRI), therapeutic agents, and pharmaceutical agents. Modulating motility of intracellular vesicles in cortical neurons with nanomagnetic forces on chip can have a huge impact on the development of new neurotherapeutic concepts. In a broader perspective, controlling vesicle motion locally and intracellularly by magnetic forces can bring many benefits to pharmacological treatments; nanomagnetic force applications will allow to wirelessly guide axons and dendrites by exogenously using permanent magnetic field gradients (Kunze et al. 2017). Smart hydrogels that respond to magnetic field have attracted increasing interest in recent years. This is due to their numerous potential applications such as tunable delivery vehicles, MRI contrast agents, and magnetic-heat therapy. Magnetic hydrogels are especially attractive because of their quick response properties by adjusting the external magnetic fields. The magnetism of magnetic hydrogels is related to the modification of magnetic nanoparticles, the arrangement of the nanoparticles, and other properties (Zhang et al. 2016).

Revolutionary developments have recently occurred in permanent magnetism. Permanent magnets are a vital part of many electromechanical machines and electronic devices, but they were usually hidden in sub-assemblies. Development of a rare-earth-free magnetic material with very high MCA would carry tremendous impact that ranges from the basic science realm all the way to advanced applications of great societal importance. Ferrite-based nanocomposites for permanent magnets have been exploiting nanostructure and artificial interfacing, with a larger energy product, in comparison with the currently used ferrites (Jenu et al. 2016). Magnetically hard materials are also being used to make permanent magnets, because in such magnets, high coercivity is a primary requirement, and a permanent magnet, once magnetized, must be able to resist the demagnetizing action of stray fields, including its own.

The understanding and control of magnetic nanostructures are required to achieve high-performance magnetic-based systems. Magnetic symmetry, dimensionality, and interfacial effects promote much of the properties observed in complex magnetic nanostructures. The competition between different anisotropy contributions can result in different magnetic configurations, reversal processes, and/or transport phenomena, which play a crucial role in device fabrication. The field of nanomagnetism is vast, and there are still a lot of unknown mysteries to explore.

1.7 Conclusion

The field of nanomagnetism, in layered structure or particles, is veritably bursting with day-by-day enhancement and cross-disciplinary innovation. Nanomagnetism is indispensable to fundamental studies and also to various technical devices; therefore, this chapter has covered some important concepts such as characteristic length scale, magnetic order, magnetic interactions, superparamagnetism, EB, domain formation, role of anisotropy, and spring magnetic concept, which help understand concept of nanomagnetism to a large extent. In this present chapter, the magnetic properties of some model nanostructured systems have been investigated. In the end, current research and development in the field of nanomagnetism is provided, summarizing its current status in the industrial research. The emphasis is on the use of nanomagnetism as a platform to illustrate some of the interest in the emerging field of nanoscience.

Acknowledgment

SKS and NS are grateful for financial support from CAPES, CNPq, and FAPEMA, Brazil. Sarveena and MS is thankful to DST, India.

References

Baberschke K (2001) Anisotropy in magnetism. In: *Band-Ferromagnetism* (Baberschke K, Nolting W, Donath M, Eds.), pp 27–45. Berlin, Germany: Springer.

Bader SD (2006) Colloquium: Opportunities in Nanomagnetism. *Reviews of Modern Physics* 78 (1): 1. doi:10.1103/RevModPhys.78.1.

Bedanta S, Barman A, Kleemann W, Petracic O, and Seki T (2013) Magnetic nanoparticles: A subject for both fundamental research and applications. *Journal of Nanomaterials* 2013: 1–22. doi:10.1155/2013/952540.

Berkowitz AE and Takano K (1999) Exchange anisotropy. *Journal of Magnetism and Magnetic Materials* 200: 552–570. doi:10.1016/S0304-8853(99)00453-9.

Buschow KHJ and Boer FR (2003) *Physics of Magnetism and Magnetic Materials*. New York: Springer.

Cahen D and Kahn A (2003) Electron energetics at surfaces and interfaces: Concepts and experiments. *Advanced Materials* 15: 271–277. doi:10.1002/adma.200390065.

Camarero J, Perna P, Bollero A, Teran FJ, and Miranda R (2016) Role of magnetic anisotropy in magnetic nanostructures: From spintronic to biomedical applications.

Chandra S (2013) Magnetization dynamics and related phenomena in nanostructures. PhD dissertation. Tampa, FL, University of South Florida.

Chtchelkanova et al. (2003) *Magnetic Interactions and Spin Transport*. New York: Springer.

Coey JMD (2010) *Magnetism and Magnetic Materials*. Cambridge, UK: Cambridge University Press.

Coey JMD (2011) Hard magnetic materials: A perspective. *IEEE Transactions on Magnetics* 47: 4671–4681. doi:10.1109/TMAG.2011.2166975.

Cullity BD and Graham CD (2009) *Introduction to Magnetic Materials*. 2nd ed. Somerset, UK: Wiley-IEEE Press.

den Broeder et al. (1991) Magnetic anisotropy of multilayers. *Journal of Magnetism and Magnetic Materials* 93: 562–570.

Djurberg C, Svedlindh P, Nordblad P, Hansen M, Bødker F, and Mørup S (1997) Dynamics of an interacting particle system: Evidence of critical slowing down. *Physical Review Letters* 79 (25): 5154–5157. doi:10.1103/PhysRevLett.79.5154.

Fernandez A, Gibbons MR, Wall MA, and Cerjan CJ (1998) Magnetic domain structure and magnetization reversal in submicron-scale co dots. *Journal of Magnetism and Magnetic Materials* 190 (1–2): 71–80. doi:10.1016/S0304-8853(98)00267-4.

Frey NA (2008) Surface and interface magnetism in nanostructures and thin films. PhD dissertation. Tampa, FL: University of South Florida.

Fullerton EE, Jiang JS, and Bader SD (1999) Hard/soft magnetic heterostructures: Model exchange-spring magnets. *Journal of Magnetism and Magnetic Material* 200: 392–404. doi:10.1016/S0304-8853(99)00376-5.

Gaier O (2009) A study of exchange interaction, magnetic anisotropies, and ion beam induced effects in thin films of Co_2-based Heusler compounds. PhD dissertation. Technical University of Kaiserslautern.

Getzlaff M (2008) *Fundamentals of Magnetism.* Berlin, Germany: Springer Science and Business Media.

Givord D (2007) Magnetism in nanomaterials. In: *Nanomaterials and Nanochemistry* (Brechignac C, Houdy P, Lahmani M, Eds.), pp. 101–134. Berlin, Germany: Springer. doi:10.1007/978-3-540-72993-8.

Gubin SP (2009) *Magnetic Nanoparticles.* Weinheim, Germany: Wiley-VCH Verlag GmbH & Co. KGaA.

Gubin SP, Koksharov YA, Khomutov GB, and Yurkov GY (2005) Magnetic nanoparticles: Preparation, structure and properties. *Russian Chemical Reviews* 74: 489–520. doi:10.1070/RC2005v074n06ABEH000897.

Guimaraes AP (1998) *Magnetism and Magnetic Resonance in Solids.* New York: John Wiley & Sons.

Guimaraes AP (2009) *NanoScience and Technology.* Berlin, Germany: Springer.

Jenu P, Topole M et al. (2016) Ferrite-based exchange-coupled hard-soft magnets fabricated by spark plasma sintering. *Journal of the American Ceramic Society* 99: 1927–1934. doi:10.1111/jace.14193.

Jiang JS and Bader SD (2014) Rational design of the exchange-spring permanent magnet. *Journal of Physics: Condensed Matter* 26: 64214. doi:10.1088/0953-8984/26/6/064214.

Jimenez-Villacorta F and Lewis LH (2014) Advanced permanent magnetic materials. In: *Nanomagnetism* (Gonzalez JM, Ed.), pp. 30. Manchester, UK: OCP Publishing Group.

Jiménez-Villacorta F, McDonald I, Heiman D, and Lewis LH (2014) Tailoring exchange coupling and phase separation in Fe-Co-Mn nanocomposites. *Journal of Applied Physics* 115: 17A729. doi:10.1063/1.4866704.

Johnson MT, Bloemen PJH, Den Broeder FJ, and Vries JJ (1999) Magnetic anisotropy in metallic multilayers. *Reports on Progress in Physics* 59 (11): 1409–1458. doi:10.1088/0034-4885/59/11/002.

Jun YW, Seo JW, and Cheon J (2008) Nanoscaling laws of magnetic nanoparticles and their applicabilities in biomedical sciences. *Accounts of Chemical Research* 41: 179–189. doi:10.1021/ar700121f.

Kittel C (1996) *Introduction to Solid State Physics.* 7th ed. New York: John Wiley & Sons.

Kunze A, Murray CT, Godzich C, Lin J, Owsley K, Tay A, and Carlo DD (2017) Modulating motility of intracellular vesicles in cortical neurons with nanomagnetic forces on-chip. *Lab Chip* 17 (5): 842–854. doi:10.1039/C6LC01349J.

Labaye et al. (2002) Surface anisotropy in ferromagnetic nanoparticles. *Journal of Applied Physics* 91: 8715. doi:10.1063/1.1456419.

Laurent S, Forge D, Port M, Roch A, Robic C, Elst LV, and Muller RN (2008) Magnetic iron oxide nanoparticles: Synthesis, stabilization, vectorization, physicochemical characterizations, and biological applications. *Chemical Reviews* 108: 2064–2110. doi:10.1021/Cr900197g.

Leite GCP, Chagas EF, Pereira R, Prado RJ, Terezo AJ, Alzamora M, and Baggio-Saitovich E (2012) Exchange coupling behavior in bimagnetic $CoFe_2O_4/CoFe_2$ nanocomposite. *Journal of Magnetism and Magnetic Materials* 324: 2711–2716. doi:10.1016/J.Jmmm.2012.03.034.

Lewis LH and Jimenez-Villacorta F (2012) Perspectives on permanent magnetic materials for energy conversion and power generation. *Metallurgical and Materials Transactions A* 44: 2–20. doi:10.1007/s11661-012-1278-2.

Lewis LH, Kim J, and Barmak K (2003) The CoPt system: A natural exchange spring. *Physica B: Condensed Matter* 327: 190–193. doi:10.1016/S0921-4526(02)01725-8.

López-Ortega A, Estrader M, Salazar-Alvarez G, Roca AG, and Nogués J (2014) Applications of exchange coupled bi-magnetic hard/soft and soft/hard magnetic core/shell nanoparticles. *Physics Reports* 553: 1–32. doi:10.1016/j.physrep.2014.09.007.1.

Marion J (2013) Towards rare-earth-free permanent magnets: Exchange bias in binary Mn-based alloys. MS Thesis. Boston, MA: Northeastern University.

Nogues J and Schüller IK (1999) Exchange bias. *Journal of Magnetism and Magnetic Materials* 192: 203–232. doi:10.1016/S0304-8853(98)00266-2.

Ortega D (2012) Structure and magnetism in magnetic nanoparticles. In: *Magnetic Nanoparticles: From Fabrication to Clinical Applications* (Tanh NTK, Ed.), pp. 3–44. Boca Raton, FL: CRC Press.

Papaefthymiou GC (2009) Nanoparticle magnetism. *Nano Today* 4: 438–447. doi:10.1016/j.nantod.2009.08.006.

Phan MH, Alonso J, Khurshid H, Lampen-Kelley P, Chandra S, Stojak Repa K, Nemati Z, Das R, Iglesias Ó, and Srikanth H (2016) Exchange bias effects in iron oxide-based nanoparticle systems. *Nanomaterials* 6: 221. doi:10.3390/nano6110221.

Prieto-Ruiz JP, Romero FM, Prima-García H, and Coronado E (2015) Exchange coupling in an electrodeposited magnetic bilayer of prussian blue analogues. *Journal of Material Chemistry C* 3 (42): 11122–11128. doi:10.1039/C5TC01926E.

Quesada Adrian, Granados-Miralles C, López-Ortega A, Erokhin S, Lottini E, Pedrosa J, Bollero A et al. (2016) Energy product enhancement in imperfectly exchange-coupled nanocomposite magnets. *Advanced Electronic Materials* 2: 1500365. doi:10.1002/aelm.201500365.

Riente F, Ziemys G, Mattersdorfer C, Boche S, Turvani G, Raberg W, Luber S, and Gamm SBV (2017) Controlled data storage for non-volatile memory cells embedded in nano magnetic logic. *AIP Advances* 7 (5): 055910. doi:10.1063/1.4973801.

Saranu et al. (2011) Effect of large mechanical stress on the magnetic properties of embedded Fe nanoparticles. *Beilstein Journal of Nanotechnology* 2: 268–275. doi:10.3762/bjnano.2.31.

Sarveena et al. (2016) Surface and interface interplay on the oxidizing temperature of iron oxide and Au–iron oxide. *RSC Advances* 70394–70404. doi:10.1039/C6RA15610J.

Schmool DS and Kachkachi H (2015) Single-particle phenomena in magnetic nanostructures. *Solid State Physics* 66: 301–423.

Stamps RL (2001) Mechanisms for exchange bias. *Journal of Physics D: Applied Physics* 34: 444. doi:10.1088/0022-3727/34/3/501.

Yang Q, Zhou Z, Sun NX, and Liu M (2017) Perspectives of voltage control for magnetic exchange bias in multiferroic heterostructures. *Physics Letters A* 1: 1–10. doi:10.1016/j.physleta.2017.01.065.

Zhang J, Huang Q, and Du J (2016) Recent advances in magnetic hydrogels. *Polymer International* 65 (12): 1365–1372. doi:10.1002/pi.5170.

Zhang W and Krishnan KM (2016) Epitaxial exchange-bias systems: From fundamentals to future spin-orbitronics. *Materials Science and Engineering R: Reports* 105: 1–20. doi:10.1016/j.mser.2016.04.001.

Zheng R (2004) Exchange bias in magnetic nanoparticles. PhD dissertation. Hong Kong University of Science and Technology.

2

Exchange-Coupled Bimagnetic Core–Shell Nanoparticles for Enhancing the Effective Magnetic Anisotropy

Gabriel C. Lavorato,
Elin L. Winkler,
Enio Lima Jr., and
Roberto D. Zysler

2.1 Introduction

Magnetic nanoparticles provide encouraging possibilities for adjusting the properties of magnetic materials, and in the last few years, they have opened new perspectives in materials for data and energy storage, biomedicine, and catalysis (Zeng et al. 2002;

Lu et al. 2007; Gao et al. 2009; Pankhurst et al. 2009). The main reason is associated with the dramatic increase in the surface (or interface)-to-volume ratio when the size of the nanostructure is reduced. For example, a nanoparticle with 3-nm diameter has around 70% of the atoms on its surface. At the same time, the interface between core and shell in a core–shell nanoparticle with an overall diameter of 5 nm and a shell thickness of 1 nm is occupied by nearby 50% of the total atoms. Given that surface and interface atoms, due to their broken symmetry, behave differently than bulk atoms, the examples listed above highlight the wide possibilities of tuning the magnetic properties in the nanoscale (Dormann et al. 1997; Kodama 1999).

Magnetic nanoparticles can be found naturally in some rocks or living organisms, and nowadays, many different types of synthetic nanoparticles can be fabricated by a number of chemical and physical methods. While the magnetic properties (hysteresis, magnetization reversal, domain structure, and so on) of relatively large structures are governed by the balance between the anisotropy, exchange, and magnetostatic interactions, when the size is reduced below a certain value (close to the expected domain wall width), single-domain structures are observed. One of the most interesting features of single-domain structures is superparamagnetism: as the volume is reduced, the energy barrier given by the anisotropy can be overcome by the thermal fluctuations, in which case the spins of the nanoparticle behave as paramagnetic superspins. Assemblies of interacting superspins have shown different interesting states, including superparamagnetism and other less-studied phenomena such as superferromagnetism and superspin glass (Knobel et al. 2008; Bedanta and Kleemann 2009; Mørup et al. 2010; Fiorani and Peddis 2014). The superparamagnetic fluctuation imposes a limit for the size reduction, because many applications require a thermally stable magnetic moment (Skumryev et al. 2003). A further important consequence of size reduction is the increased relevance of surface anisotropy, which may determine the spin configuration of a nanoparticle and govern the effective anisotropy and the magnetization reversal.

Furthermore, many remarkable topics of magnetic nanoparticles are related to interface effects. In this sense, when two distinct magnetic materials (e.g., an antiferromagnet [AFM] and a ferromagnet [FM] or ferrimagnet [FiM] are in contact, the coupling between them can rule the reversal and the thermal stability of the magnetic moment. Exchange-anisotropy phenomena are reflected by exchange-bias shifts, anisotropy enhancement, unidirectional anisotropy, or asymmetric hysteresis and were first observed by Meiklejohn and Bean (1956), who concluded that a new type of magnetic anisotropy was discovered.

In the last few years, numerous *bimagnetic* nanostructures were fabricated and studied, enriching the literature on exchange anisotropy-related phenomena and providing new tools for the design and development of novel magnetic materials, as these have been summarized in recent reviews (Mélinon et al. 2014; López-Ortega et al. 2015a). Many different systems were proposed, ranging from hard rare-earth-based nanoparticles to soft metallic or metal-oxide-based particles, including biocompatible and multifunctional materials. Researchers have shown that the coercivity (H_C) and the exchange bias (H_E) can be tuned by controlling their composition, size, morphology, and interactions. Therefore, bimagnetic nanoparticles are gaining industry attention due to the advantages resulting from the additional degree of freedom that enables a better control of their functionalities that may be interesting for diverse areas in materials engineering and nanobiomaterials.

This chapter focuses on the enhancement of the magnetic anisotropy due to the exchange coupling in bimagnetic nanoparticles synthesized by chemical methods. The second section is devoted to present various types of exchange-coupled bimagnetic nanoparticle systems, and the different reversal mechanisms are briefly discussed in view of the current literature on the subject. Then, the main aspects of the synthesis methods for obtaining core–shell nanoparticles are described, focusing on the *heat-up* thermal decomposition method. The fourth section is devoted to a case study of the anisotropy enhancement in AFM/FiM CoO-core/CoFe$_2$O$_4$-shell nanoparticles, where the general aspects of the CoO and CoFe$_2$O$_4$ are presented and the origin of the anisotropy increase as well as the size effects are analyzed.

2.2 Exchange Coupling in Bimagnetic Nanoparticle Systems

Multiple exchange-coupled bimagnetic nanoparticles were developed in the last few years, exhibiting outstanding properties. For example, Skumryev et al. demonstrated that the thermal stability of the magnetic moment of 4-nm Co nanoparticles can be enhanced up to room temperature by the additional anisotropy source given by the coupling with an AFM matrix (Skumryev et al. 2003). Core–shell particles formed by rigidly coupled soft-hard ferrites showed remarkable heat-generation properties that may be useful for future biomedical developments (Lee et al. 2011), and bimagnetic nanoparticles could also improve the performance of permanent magnets (Zeng et al. 2002).

Nano-heterostructures can combine AFM, FiM, and FM materials, and, from the viewpoint of the hysteretic characteristics, systems with distinct features were reported, including exchange-biased, enhanced coercivity or exchange-spring-like loops, as summarized in Figure 2.1. In many cases, these phenomena may be observed simultaneously, and careful studies are required in order to understand their origin and predict their properties.

Exchange-biased systems are usually identified by a field shift (occasionally also vertical shifts) in the hysteresis loop of the material and are typically found in FM(FiM)/AFM or AFM/FM(FiM) core–shell nanoparticles. Within the ideal Meiklejohn-Bean model (Meiklejohn 1962), single-domain AFM and FM-FiM with parallel easy axes and an ideal interface are considered. Then, exchange bias is observed when the anisotropy energy of the AFM phase is higher than the interface coupling energy ($K_{AFM}V_{AFM} > J_i$), where K_{AFM} and V_{AFM} are the anisotropy and the total volume of the AFM, and J_i is the exchange-coupling constant at the interface. If the material is cooled in a magnetic field from a temperature higher than the Néel temperature of the AFM (T_N), exchange-bias shifts will be observed due to the exchange coupling at the AFM/FM(FiM) interface. Although this effect was initially explored in Co/CoO fine particles (Meiklejohn and Bean 1956), the theory for exchange-coupled nanostructures was mostly developed for thin-film bilayers or multilayers, where only one dimension is being reduced and the interface characteristics are much more controlled (Nogués and Schuller 1999; Kiwi 2001; Nogués et al. 2005; Stamps 2000). A general approach to the interpretation of the exchange-bias-related phenomena in bimagnetic nanoparticles has not been established yet; however, valuable advances by Monte-Carlo simulations were reported (Eftaxias and Trohidou 2005;

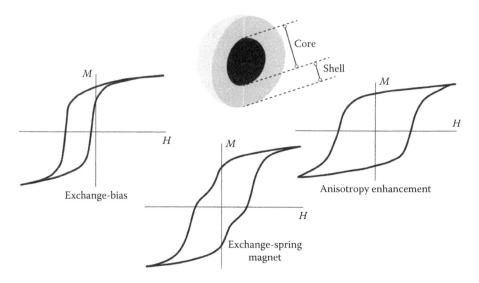

FIGURE 2.1 Summary of the multiple hysteretic characteristics observed in core–shell bimagnetic nanoparticles.

Iglesias et al. 2008a). Experimental research on exchange bias in bimagnetic nanoparticles was initially conducted for FM/AFM M/M_xO_y systems (M = Co, Fe, Ni), where the AFM is obtained by chemically modifying the surface of metallic nanoparticles (Nogués et al. 2005). Recently, as a consequence of the development of new synthesis techniques, more complex multicomponent systems such as various FM(FiM)/AFM nanoparticles (Baaziz et al. 2013; Binns et al. 2013), inverted AFM/FM(FiM) nanoparticles (Kavich et al. 2008; Sharma et al. 2011; Lima et al. 2012; Sun et al. 2012; Fontaiña-Troitiño et al. 2014), and doubly inverted AFM/FM(FiM) (where the Curie temperature of the FM(FiM) is lower than the Néel temperature of the AFM) were also explored (Salazar-Alvarez et al. 2007; Berkowitz et al. 2008; López-Ortega et al. 2010).

On the other hand, a different situation is expected in the limit case, in which the anisotropy energy of the AFM is lower than the interface-coupling energy (i.e., $K_{AFM} V_{AFM} < J_i$). Under such condition, a *rigid coupling* of both phases is promoted. Then, the exchange-bias shift will be no longer observed and, depending on the magnetic anisotropies of both phases, an increase in the effective anisotropy is typically expected. In thin films, this condition is easily fulfilled for bilayers with a very thin AFM (Jungblut et al. 1994) or for temperatures very close to the T_N, where a dramatic reduction of the AFM anisotropy occurs (Ali et al. 2003; Radu and Zabel 2008). In nanoparticles, increases in both exchange-bias shifts and anisotropy are usually found simultaneously, and the interface parameters responsible for these effects cannot be easily unveiled. However, some reports have shown a remarkable anisotropy enhancement and negligible exchange-bias shifts (Dobrynin et al. 2005; Feygenson et al. 2010; Lima et al. 2012). For example, Dobrynin et al. showed that Co/CoO nanoparticles obtained by oxidizing bare Co particles show a critical size for exchange bias. They found that 3-nm particles present an enhanced anisotropy but negligible exchange bias as a result of a high interfacial

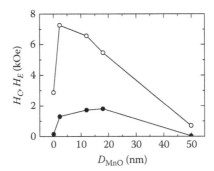

FIGURE 2.2 Dependence of the coercivity (H_C—empty symbols) and the exchange-bias shift (H_E—full symbols) with the core size for AFM/FiM MnO/Mn$_3$O$_4$ core–shell nanoparticles. (Reprinted with permission from Salazar-Alvarez G. et al., *J. Am. Chem. Soc.*, 129, 9102–9108. Copyright 2007 American Chemical Society.)

exchange coupling that dominates over other energies, as explained by a simple model (Dobrynin et al. 2005). Other researchers focused on the non-monotonic variation of exchange bias with the AFM size (Liu et al. 2003; Salazar-Alvarez et al. 2007). By studying AFM/FiM MnO/Mn$_3$O$_4$ nanoparticles, Salazar-Álvarez et al. (2007) found that very small (<2 nm) AFM sizes lead to a significant reduction of the exchange-bias shift due to the reduction of the total anisotropy energy of the AFM; at the same time, larger (>20 nm) AFM sizes also favor the reduction of the exchange bias due to lower domain sizes. Their results are summarized in Figure 2.2, which reflect the complex variation of the exchange-coupling interface phenomena in AFM/FiM nanoparticles.

In addition, the variation of exchange bias and coercivity were predicted through Monte Carlo simulations applied to bimagnetic nanoparticles with different sizes and variable-interface exchange coupling (Iglesias et al. 2008b). Detailed Monte Carlo simulations of the core diameter and shell thickness dependence of the magnetic properties were employed to demonstrate that the inverted AFM/FiM morphology can lead to an unexpected exchange-bias behavior that could be useful for the design of novel technologically suitable materials (Vasilakaki et al. 2015). The profuse experimental and theoretical studies on this topic reveal that multiple parameters (e.g. size, morphology and interface quality) need to be controlled in order to address a fine tuning of the anisotropy enhancement and the exchange bias fields.

Alternatively, if both phases are FM or FiM (and thus substantially contribute to the total magnetization), a magnetization-reversal mechanism called *exchange-spring* mechanism is usually identified (Soares et al. 2011; López-Ortega et al. 2015a). The conditions for the observation of an exchange-spring process are that (1) the anisotropy of the hard phase is significantly higher than that of the soft phase, (2) both phases are exchange-coupled (so that the soft-phase spins are pinned by the spins of the hard phase at the interface), and (3) the effective thickness of the soft phase is higher than the exchange length of the hard phase (Goto et al. 1965; Kneller and Hawig 1991; Lavorato et al. 2016). Under these assumptions, the magnetization reversal consists of the nucleation of a Bloch-type domain wall in the soft phase that will be compressed toward the soft/hard interface (Kneller and Hawig 1991; Fullerton et al. 1999). The rotation of the soft-phase magnetization will be reversible up to the irreversible switching field of the hard phase, and different nucleation fields for soft and hard will be observed. If the field is turned off at a lower value than the switching field of the hard phase, the magnetization of the soft phase will return to its original remanence value, resembling the behavior of a *spring*. If the effective size of the soft phase is, instead, lower than the exchange length of the hard phase, then a rigid-coupling reversal is expected—both phases will reverse at the same field and the magnetic properties of the material will be similar to the mean properties of its constituents. Hard/soft and soft/hard nanoparticles in which both components are FM or FiM have been reviewed in López-Ortega et al. (2015a), where applications are exhaustively discussed.

2.3 Synthesis Methods of Core–Shell Nanoparticles

The development of synthesis methods for nanostructured materials is one of the most important topics in current nanoscience and nanotechnology, since the control of the composition, size, and morphology of the nanostructures enables the fabrication of novel materials with basic and technological interest. In this section, we will give a brief overview of some synthesis methods of magnetic-oxides-based nanoparticles, focusing on surfactant-assisted colloidal routes. However, it is worth noting that similar procedures may be employed to produce other types of nanostructures such as nanowires or thin films.

2.3.1 Fundamentals of the Chemical Synthesis of Nanoparticles

Bottom-up techniques, in contrast to top-down methods, consist of the preparation of nanoparticles by the assembly of single atoms or small clusters. Different bottom-up techniques were proposed, depending, basically, on the synthesis medium in which the particles are obtained. In particular, several advances in the fabrication routes were promoted in the last years as a consequence of the reproducibility and the control of the microstructure, composition, shape and size attained by the colloidal chemistry. A good example is the synthesis of highly crystalline monodisperse magnetic nanoparticles (Hyeon et al. 2001; Sun and Zeng 2002), which has been particularly useful for the design of exchange-coupled nanoparticles.

The knowledge of the organic chemistry and the comprehension of the complexity of the chemical reactions involved allowed the production of nanocrystals with controlled characteristics in either aqueous or non-aqueous liquid media. One of the most widespread aqueous route consists of the co-precipitation of metal salts in an oxygen- and pH-controlled medium; this was extensively employed in the last decades to fabricate metal-oxides-based fine magnetic particles (Winkler et al. 2004; Gupta and Gupta 2005; Tobia et al. 2008; Winkler et al. 2008). Conversely, non-aqueous synthetic routes are based on the transformation of a precursor in an organic solvent. The oxygen required for the formation of the inorganic oxide can come from the solvent or from the precursor itself. While some processes are controlled by the solvent, others, called surfactant-assisted methods, involve the addition of surfactants. The former ones are generally simpler and less toxic, and the latter ones allow greater possibilities of controlling the nanocrystals' characteristics; however, they include more complex reactions (Pinna and Niederberger 2008; Niederberger and Pinna 2009).

Surfactant-assisted thermal decomposition methods are based on the pioneer works of LaMer and Dinegar (1950), where the basics of the fabrication of colloidal microparticles were given. The method essentially consists of separating the nucleation and growth stages. Without intending to give a full theoretical basis, in the following text, we will comment on some important aspects of surfactant-assisted colloidal methods in view of the rich literature on this topic.

The synthesis of a nanocrystal in a solution involves a solid material (nucleus) that acts as a seed in a crystallization process that takes place at the interface between the nucleus and solution (Klabunde and Richards 2009). If the nuclei are formed spontaneously, the process is called homogeneous nucleation (Mullin 2001), while in the heterogeneous nucleation, crystallization is induced by the introduction of an external material (seeds).

The duration of the nucleation process affects the size distribution, because the nuclei will have different growth times. Therefore, a fast nucleation, which can be achieved by employing a supersaturated solution, is desired (Hyeon et al. 2001). The growth is a diffusion-regulated process, and owing to the Ostwald ripening phenomenon, bigger particles grow at expenses of smaller ones (Kwon and Hyeon 2011). The use of appropriate precursors, surfactants, and solvent is proved to be crucial to adjust the growth process for a desired size or morphology.

2.3.2 The Heat-Up Method

The earliest surfactant-assisted methods suitable for the fabrication of monodisperse nanoparticles consist of a fast injection of the precursors to a mixture of solvent and surfactants, previously heated at temperatures of around 300°C, as was proposed by Murray et al. (1993) for the fabrication of semiconductor nanocrystals. Thanks to the high temperature and the fast injection, the decomposition of the precursors occurs in a short time, ensuring a rapid nucleation process. Murray's method was later extended to other oxide-based and metallic nanoparticles (Murray et al. 2001). Afterward, in a variation of the hot-injection method, the reactants were mixed at room temperature and the nucleation was promoted by heating the system to the final temperature. By this approach, simpler than the hot-injection method, FePt and iron oxides monodisperse nanoparticles were successfully obtained (Sun et al. 2000; Hyeon et al. 2001). The new synthesis route was called heat-up method and has been widely employed in the last years for the synthesis of iron oxides, metallic, oxide-based, and chalcogenide-based nanoparticles (Soon and Hyeon 2008; van Embden et al. 2015). For example, the mixture of $Fe(CO)_5$ at 100°C in ethyl ether and oleic acid leads to uniform γ-Fe_2O_3 nanoparticles when heated to 300°C (Hyeon et al. 2001). At the same time, Sun et al. fabricated monodisperse Fe_3O_4 nanoparticles with a similar procedure but by using iron acetylacetonate as a precursor (Sun and Zeng 2002). Subsequent modifications of the process, including the use of metal-oleate precursors, allowed the synthesis of large quantities (Park et al. 2004). Further theoretical and experimental studies were performed by analyzing, in detail, the precursor decomposition process and revealed the formation of intermediate species (Kwon et al. 2007; Niederberger and Pinna 2009). Regarding the employed reactants, most precursors consist of organometallic compounds, including metal acet-ylacetonates $M(acac)_n$ (M = Fe, Mn, Co, Ni, Cr; n = 2, 3; and acac = acetylacetonate), metal cupferronates or carbonyls, and surfactants are mostly oleic acid and long-chain amines, and the organic solvents are typically phenyl ether, benzyl ether, octyl ether, octadecene, and trioctylamine, among others (Lu et al. 2007; van Embden et al. 2015). Due to its simplicity and the possibility of obtaining many different oxides with low polydispersity and good dispersability in organic solvents, as well as due to the possibil-ity of tuning the composition, shape, and size of the nanocrystals, the heat-up method currently represents one of the most widespread nanoparticle synthesis methods.

Some essential aspects of the heat-up method have been accurately summarized in (van Emden et al. 2015), where the authors give a detailed description of the process by simulat-ing the evolution of the key parameters during the synthesis. The first step in Figure 2.3 is represented by an increase in the supersaturation (S) due to the decomposition of the pre-cursors, whose concentration (P) decreases as the temperature is increased. Then, when S reaches a critical value, the nucleation stage begins, and the number of nuclei (NCs) grows rapidly. By the end of the nucleation process, S drops and the size dispersion (SD) reaches a maximum. Later, during the growth stage, determined by a decrease in the nucleation rate, the number of crystals decreases, because some small clusters dissolve into the solution, allowing a reduction of SD and the stabilization of the final mean size <r> of the particles.

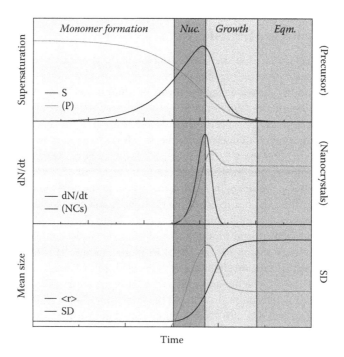

FIGURE 2.3 Simulations of the main parameters (supersaturation [S], precursors' concentration [P], number of nanocrystals [NCs], mean size <r>, and size dispersion [SD]) of the heat-up synthesis method. (Reprinted with permission from van Embden J. et al., *Chem. Mater.*, 27, 2246–2285, 2015.)

Once the new synthetic routes were established, one of the main challenges was to adjust the size and morphology of the nanocrystals in an easy and reproducible way. The first approaches were related to selective precipitation (Perales-Perez et al. 2002) and seed-mediated growth processes, where the obtained particles were employed as seeds in a subsequent nucleation-growth step (Sun and Zeng 2002). Simpler approaches were proposed later, including one-step methods, in which the size is controlled by the precursors' concentration (Hyeon et al. 2001), the heating rate (Guardia et al. 2010), the surfactant/precursors ratio (Vargas and Zysler 2005), or the solvent itself (Park et al. 2004). The control of the concentration of either precursors or surfactants can be effectively used to design nanocrystals with different shapes, including cubic, polyhedral, faceted, and elongated structures (Jana et al. 2004; Zeng et al. 2004b). The fabrication of size- and shape-controlled nanocrystals is a complex process that depends on many interacting variables, and it is still expected to be a fruitful topic for the design of novel nanostructures.

2.3.3 Fabrication of Multicomponent Nanoparticles

Recently, synthetic routes were adapted for the fabrication of multicomponent nanoparticles for two reasons: to get advantage of the interactions between different magnetic phases by improving or tuning specific magnetic properties (Zeng et al. 2004a), and for the development of multifunctional materials, particularly interesting in biomedical applications (Cho et al. 2011). The development of synthesis techniques to obtain multicomponent nanoparticles is thus essential for the fabrication of exchange-biased nanostructures, in which the interface interactions play a crucial role in determining the magnetic behavior of the system.

According to their morphology, multicomponent nanoparticles can be either core–shell or dumbbell-like (Schmid 2011). Their preparation is generally based on two or more steps that consist of the synthesis of monodisperse particles that are then used as seeds for a subsequent nucleation-growth process. If the grown material covers the whole surface of the seed, a core–shell structure is obtained, but if it grows on a certain facet of the nanocrystal seed, then a nanoparticle with dumbbell-like morphology will be obtained. The most critical step to be controlled is probably the nucleation of the precursors of the second step, in which a heterogeneous nucleation is required in order to avoid a material with two separated phases.

Since the first observation of exchange-bias shifts in Co/CoO particles, many core–shell nanoparticles were synthesized, aiming to study exchange-bias-related phenomena, as were briefly mentioned in the previous section. Currently, numerous types of multicomponent nanoparticles can be fabricated, and besides core–shell or dumbbell-type particles, other more complex structures were proposed: multicore–shell (Yang et al. 2005), elongated or cubic core–shell (Bodnarchuk et al. 2009; Mendoza-Reséndez et al. 2012), core-multishell (Salazar-Alvarez et al. 2011), and even branched nanostructures (Casavola et al. 2009). For example, Bodnarchuk et al. (2009) obtained core–shell $FeO/CoFe_2O_4$ nanoparticles with sizes between 11 and 25 nm by adjusting the decomposition temperature of Co^{2+} and Fe^{3+} oleates and the surfactants/precursors ratio. As it is shown in Figure 2.4, spherical particles were obtained by using oleic acid as a stabilizer, while cubic particles can be synthesized by employing Na-oleate.

Even if a broad variety of multicomponent nanoparticles have been reported, as illustrated in Figure 2.5, there is still a lack of systematic studies oriented to clarify the link

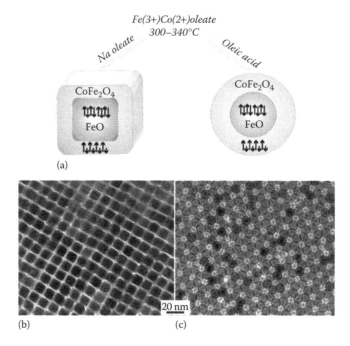

(a)

(b)   (c)

FIGURE 2.4   (a) Key conditions for the synthesis of cubic and spherical $FeO/CoFe_2O_4$ nanoparticles and (b,c) associated transmission electron microscope (TEM) images. (Bodnarchuk M.I. et al.: Exchange-coupled bimagnetic wüstite/metal ferrite core–shell nanocrystals: Size, shape, and compositional control. *Small.* 2009. 5. 2247–2252. Copyright Wiley-VCH Verlag GmbH & Co. KGaA. Reproduced with permission.)

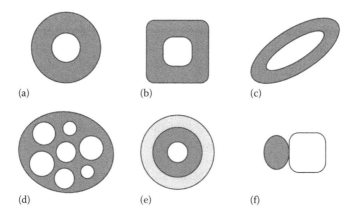

(a)   (b)   (c)

(d)   (e)   (f)

FIGURE 2.5   Schematic illustration of the different types of multicomponent nanoparticles: (a) core–shell, (b) cubic core–shell, (c) elongated core–shell, (d) multicore–shell, (e) core–multishell, and (f) dumbbell-type nanoparticles.

between the complex structure of this kind of nanoparticles and their physical properties. The following section intends to review some results obtained for $CoO/CoFe_2O_4$ AFM/FiM nanoparticles that focus on the role of the interface interaction and the size effects observed in such system.

2.4 Case Study: $CoO/CoFe_2O_4$ Bimagnetic Core–Shell Nanoparticles with Enhanced Effective Anisotropy

The properties of bimagnetic nanoparticles, being governed by interface effects, are expected to be strongly size-dependent and sensitive to the morphology and interactions. In the second section, we highlighted that if the interfacial exchange-coupling energy of an AFM/FM(FiM) multicomponent nanostructure is larger than the anisotropy energy of the AFM, we would expect an enhancement of the effective anisotropy and negligible or small exchange-bias fields. The study of such anisotropy increase will be discussed in the frame of our case study: AFM/FiM CoO-core/$CoFe_2O_4$-shell nanoparticles.

2.4.1 CoO-Core/$CoFe_2O_4$-Shell System

The magnetic hardness and the thermal stability of the magnetic moment can be increased by the combination of two nanostructured magnetic materials with different anisotropies. One of the most commonly employed AFM is cobalt monoxide (CoO) (Nogués et al. 2005; Fontaiña-Troitiño et al. 2014; Liu et al. 2015). It is an inorganic oxide that crystallizes in a cubic $Fm\overline{3}m$ structure (rock salt), with a lattice parameter of 4.26 Å and a density of 6.4 g/cm^3. O^{2-} ions form a face-centered cubic sublattice, where the Co^{2+} ions occupy interstitial octahedral sites. Owing to the superexchange interaction, CoO develops an antiferromagnetic order below a Néel temperature of 291 K (Coey 2010). Although its precise value is not known, the magnetocrystalline anisotropy of CoO is accepted to be higher than $3 \cdot 10^7$ erg/cm^3 (Tracy et al. 2005), making CoO an interesting material as a pinning phase in multicomponent nanostructures. However, its chemical stability is limited, and under certain conditions, it is oxidized to Co_3O_4 (Nam et al. 2010). Other AFM oxides typically employed in nano-heterostructures comprise FeO (Khurshid et al. 2013; Estrader et al. 2015), MnO (López-Ortega et al. 2010), NiO (Seto et al. 2005; Lee et al. 2006) (only the last one having a T_N higher than room temperature), α-Fe_2O_3 (Lyubutin et al. 2009), and Cr_2O_3 (Tobia et al. 2009).

Ferrites have been extensively investigated in their nanostructured forms because of the relatively easy fabrication routes and the common elements required. Given its low cost, biocompatibility, relatively high magnetic moment, and interesting electrical properties, Fe_3O_4 is probably the most studied material for magnetic nanoparticles, as it is reflected in the amount of publications devoted to the investigation of its magnetic, electrical, biological, or engineering functionalities. However, among ferrites, cobalt ferrite ($CoFe_2O_4$) is particularly interesting, since it presents a comparatively high magnetic anisotropy. It crystallizes in a cubic $Fd\overline{3}m$ structure (spinel), with a lattice parameter of 8.392 Å and a density of 5.3 g/cm^3. Each cell contains a total of 56 ions and 8 formula units. While O^{2-} ions form a face-centered cubic structure, the smaller metallic cations occupy either tetrahedral or octahedral sites, called A and B sites, respectively. Bulk $CoFe_2O_4$ presents an inverse structure, in which the divalent Co^{2+} cations occupy B sites, while Fe^{3+} cations are equally distributed between both sites. As the electronic orbitals of ferrites' magnetic ions are barely overlapped, the magnetic properties are determined by the hybridization of the *3d* orbitals with the *2p* oxygen orbitals, leading to the superexchange interaction. The sign and magnitude of the superexchange interaction are determined by the distance and the bond angle between the magnetic ions and the oxygen; the interaction is strongly antiferromagnetic when the magnetic ions and the oxygen are collinear. Thus, the main interaction in ferrites is the superexchange antiferromagnetic coupling between the AB ions, in which A-O-B angle is closer to 180° in comparison with the A-O-A and B-O-B angles, which are near 90°. The strong AFM AB interaction leads to a FiM structure, where moments of A ions are parallel to each other and antiparallel to the moments of B ions (O'Handley 2000). As a consequence, ferrites have relatively high Curie temperature and saturation magnetization: for example, the T_C of $CoFe_2O_4$ is approximately 793 K and the M_S is 80 emu/g at room temperature, reaching 90 emu/g at 10 K (Shenker 1957). In addition, the $CoFe_2O_4$ has high magnetocrystalline anisotropy ($K_{MC} = 2.7 \cdot 10^6$ erg/cm^3) and high magnetostriction originated by the non-quenched orbital momentum of Co^{2+}, which distinguish $CoFe_2O_4$ from other ferrites and make it an interesting material from a practical point of view (Bozorth et al. 1955; O'Handley 2000). In nanoparticles, variations of the inversion degree (typically analyzed by Mössbauer experiments) can significantly affect the observed magnetic properties (Concas et al. 2009), and the fabrication of nano-ferrites with dissimilar crystalline quality and size distribution are usually responsible for the wide dispersion in the reported properties.

Even if $CoFe_2O_4$ presents a high anisotropy, it is still an order of magnitude lower than the anisotropy of CoO, suggesting that the coupling with CoO can increase the effective anisotropy of the ferrite. Singly inverted AFM/FiM CoO-core/$CoFe_2O_4$-shell nanoparticles were first reported in Lima et al. (2012). The material was synthesized by the high-temperature decomposition of organometallic precursors, followed by a seed-mediated growth process, according to a modified Sun's method (Sun and Zeng 2002) and a subsequent thermal treatment (Lima et al. 2012). The authors demonstrated, through the analysis of dark-field and high-resolution TEM images, that particles with a core–shell morphology (~7-nm nanoparticles with a 2- to 3-nm-thick shell) can be obtained by a two-step heat-up method. Interestingly, the shell is not formed by a single crystal but consists of multiple partially aligned nanocrystals, as is inferred from TEM images shown in Figure 2.6. The most remarkable feature of this system is the increase in the coercivity and squareness ratio that are found to be, at 5 K, as high as 27.8 kOe and 0.79, respectively, much higher than the values obtained for single-phase $CoFe_2O_4$ nanoparticles (Chinnasamy et al. 2003; López-Ortega et al. 2015b; Pianciola et al. 2015). The thermal stability of the magnetic moment is also increased, and the blocking tem-perature is shifted up to temperatures close to the Néel temperature of CoO. However, no sign of exchange-bias shifts was found after field cooling up to 50 kOe. The coexistence of such lack of exchange bias and the remarkable anisotropy enhancement was ascribed to a very high interfacial coupling energy.

The crystalline matching between the cubic structures of core and shell and the ther-mal treatment could favor such high coupling energy at the interface. The structural

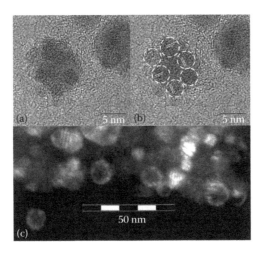

FIGURE 2.6 (a–b) High-resolution and (c) dark-field TEM images of CoO/CoFe$_2$O$_4$ core–shell nanoparticles, evidencing the core–shell morphology and the multigrain shell structure. (Reprinted with permission from Lima E. et al., *Chem. Mater.*, 24, 512–516. Copyright 2012 American Chemical Society.)

and magnetic properties of as-synthesized and annealed CoO/CoFe$_2$O$_4$ nanoparticles were compared, aiming to demonstrate that the inverted core–shell configuration (AFM core and FiM shell) confers chemical stability to the antiferromagnetic CoO. In fact, while bare CoO cores are transformed into Co$_3$O$_4$ when annealed in air, core–shell particles where the CoO is encapsulated within the CoFe$_2$O$_4$ ferrite do not show any sign of oxidation of the CoO phase, which exhibits, instead, high crystallinity (Lavorato et al. 2015a). The effects of a thermal annealing on the structural properties of CoO and CoO/CoFe$_2$O$_4$ nanoparticles can be exemplified by the X-ray diffraction (XRD) patterns shown in Figure 2.7. It can be observed that the CoO reflections in the bimagnetic system are better defined after the thermal treatment, and the CoFe$_2$O$_4$ reflections are broader and slightly shifted to higher angles for the annealed sample, suggesting that the ferrite may be subjected to structural distortions due to the interface mismatch.

Recently, the synthesis and magnetic properties of Co$_{0.3}$Fe$_{0.7}$O/Co$_{0.6}$Fe$_{2.4}$O$_4$ nanoparticles with a moderate coercivity enhancement ($\approx$20 kOe) and a very high exchange-bias shift ($\approx$8 kOe) were reported by Lottini et al. (2016). The authors showed that the exchange-bias shift decreases abruptly when reducing the core size, similarly to what has been shown in other core–shell systems (Salazar-Alvarez et al. 2007), and suggested that the quality of the interface (sharp interface) could be responsible for the observed properties. In fact, the synthesis method and the interface quality define the balance between the interface exchange-coupling energy and the AFM anisotropy energy and therefore determine the hysteretic behavior. Other CoO-based bimagnetic nanoparticles have also been explored in Fe$_3$O$_4$/CoO nanoparticles, where the AFM size was varied systematically. In such FiM/AFM system, Liu et al. (2015) showed that the exchange-bias vanishes when decreasing the shell thickness, while the coercivity reaches a maximum for the same values. Similar to the previous reports, the exchange bias was found to

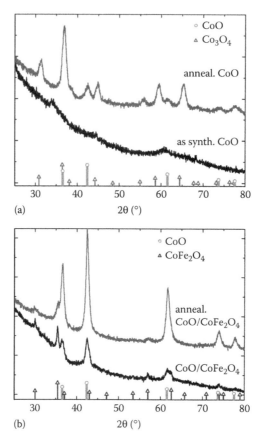

FIGURE 2.7 Powder X-ray diffraction patterns of as-synthesized and annealed (a) CoO and (b) $CoO/CoFe_2O_4$ core–shell nanoparticles. The symbols and vertical lines indicate the expected positions for the reflections of bulk CoO, Co_3O_4, and $CoFe_2O_4$. (Reprinted from *J Alloys Compd*, 633, Lavorato G.C. et al., Exchange-coupling in thermal annealed bimagnetic core–shell nanoparticles, 333–337, 2015a, with permission from Elsevier.)

vanish for very small Co/CoO nanoparticles (Dobrynin et al. 2005) as a result of the competition between the interface-coupling and the AFM energies. Furthermore, high exchange-bias fields were also found in inverted CoO/Fe_3O_4 nanoparticles with a total diameter as large as 40 and 98 nm (Fontaiña-Troitiño et al. 2014); however, in such case, the coercivity increase is moderate, probably because of the high anisotropy energy provided by the relatively large AFM core.

The multiple experimental works on CoO-based core–shell and inverted core–shell suggest that the critical size for the observation of either an exchange-bias shift or a coercivity increase is strongly dependent on the synthesis process, the composition, and the crystalline quality. Therefore, the core diameter, shell thickness, and interface quality should be simultaneously controlled to design exchange-coupled bimagnetic nanoparticles with the desired properties.

2.4.2 Origin of the Anisotropy Enhancement

The nature of the anisotropy enhancement in $CoO/CoFe_2O_4$ nanoparticles was analyzed in (Winkler et al. 2012) by employing a phenomenological model for the free energy of the multicomponent system. The authors fabricated $ZnO/CoFe_2O_4$ and $CoO/CoFe_2O_4$ nanoparticles with analogous size and morphology, in order to compare the effects of a diamagnetic (ZnO) or an AFM(CoO) core and to identify the different contributions to the anisotropy enhancement. By comparing the coercivities of both systems, as shown in Figure 2.8, it was confirmed that the remarkable increase in H_C cannot be attributed to the contribution of surface anisotropy (due to the shell morphology), but it is originated by the strong interaction between both phases.

Assuming that the magnetization rotates coherently with the field and that the easy axes of AFM and FiM are parallel, the free energy of an AFM/FiM coupled system (Winkler et al. 2012) can be written as:

$$E = -HM_{FiM}V_{FiM}\cos(\theta - \beta) - HM_{AFM}V_{AFM}\cos(\theta - \alpha)$$
$$+ K_{FiM}V_{FiM}\sin^2\beta + K_{AFM}V_{AFM}\sin^2\alpha - J_i\cos(\beta - \alpha)$$

(2.1)

where H is the applied field, M_{FiM} is the magnetization of the FiM, and M_{AFM} is the uncompensated magnetization of the AFM normalized by the total FiM (V_{FiM}) and AFM (V_{AFM}) volumes, respectively. K_{FiM} and K_{AFM} denote the magnetic anisotropy of each phase; J_i is the exchange-coupling constant at the interface; and α, β, and θ represent the angles between M_{AFM}, M_{FiM}, and H and the easy axis, respectively.

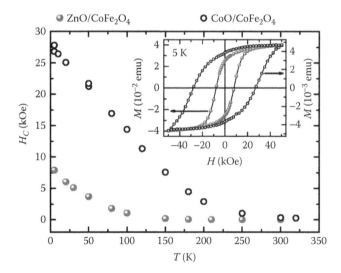

FIGURE 2.8 Temperature dependence of the coercivity (H_C) for $ZnO/CoFe_2O_4$ (full symbols) and $CoO/CoFe_2O_4$ (empty symbols) core–shell nanoparticles. The inset shows the hysteresis loops measured at 5 K. (Reprinted with permission from Winkler E.L. et al., *Appl. Phys. Lett.*, 101, 252405. Copyright 2012, American Institute of Physics.)

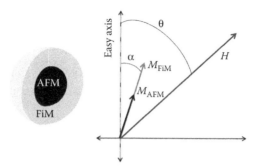

FIGURE 2.9　Schematic representation of the FiM magnetization and uncompensated AFM magnetization when applying a magnetic field according to a phenomenological model for the AFM/FiM core/shell nanoparticles.

The situation is schematically illustrated in the diagram of Figure 2.9. Given that no exchange bias is observed, $J_i > K_{AFM}V_{AFM}$ can be assumed. Then, $\beta - \alpha \cong 0$ is expected and an increase of the coercive field will be observed, according to:

$$H_C = \frac{2\left(K_{FiM}V_{FiM} + K_{AFM}V_{AFM}\right)}{M_{FiM}V_{FiM} + M_{AFM}V_{AFM}} \cong \frac{2K_{FiM}}{M_{FiM}} + \frac{2K_{AFM}V_{AFM}}{M_{FiM}V_{FiM}} \tag{2.2}$$

From Equation 2.2, two different contributions to the coercivity can be identified, where the second term represents the coercivity enhancement due to the coupling with the CoO core. Moreover, by comparing the coercivities of $ZnO/CoFe_2O_4$ and $CoO/CoFe_2O_4$, the authors estimated a K_{CoO} of 4.10^7 erg/cm^3, which is in agreement with the reported values (Winkler et al. 2012). A similar approach was employed to study the hysteresis of rigidly coupled $MnFe_2O_4/CoFe_2O_4$ core–shell nanoparticles, where the thickness of $CoFe_2O_4$ shell is kept below the exchange length. In such a case, the coercivity can be easily expressed as a function of the anisotropy of the hard phase and the volume fraction of the soft phase (Song and Zhang 2012), which seems a useful step to predict and carefully tune the magnetic properties of the system.

If we turn our attention to the temperature-dependent properties, the increase in anisotropy of $CoO/CoFe_2O_4$ nanoparticles is also reflected in the blocking process of the magnetic moment. While the average blocking temperature (T_B) of the bimagnetic sample is 271 K, $ZnO/CoFe_2O_4$ nanoparticles block at a mean value of 106 K (Lavorato et al. 2015b). In addition, the field dependence of the blocking temperature $T_B(H)$ evinces strong interactions for the bimagnetic sample. In fact, $T_B(H)$ of $CoO/CoFe_2O_4$ nanoparticles does not follow the usual $H^{2/3}$ relationship found for non-interacting nanoparticles (Dormann et al. 1997), which indeed describes $T_B(H)$ of $ZnO/CoFe_2O_4$ nanoparticles (Lavorato et al. 2015b). It is also worth noting that the coercivity of $ZnO/CoFe_2O_4$ (7.8 kOe) is lower than that observed for single-phase $CoFe_2O_4$ with analogous ferrite volume (around 15–20 kOe) (Pianciola et al. 2015). Such behavior can be explained by considering strong surface effects, inherent to the shell morphology of the ferrite, that are not present in the bimagnetic sample, probably because the AFM core stabilizes the surface spins of the ferrite (Lavorato et al. 2015b).

A less-analyzed feature also related to the stability of the magnetic moment is the temperature variation of the saturation magnetization, $M_S(T)$. In single-phase nanoparticles, the reduction of M_S with increasing temperature is associated with the presence of low-energy collective excitations (Kodama 1999) and can be analyzed by means of a modified Bloch model (Vázquez-Vázquez et al. 2010) according to $M_S(T) = M_0 \left(1 - BT^\alpha + A_0 e^{-T/T_f} \right)$, where M_0 is the saturation magnetization for $T = 0$, α and B depend on each magnetic system, and the last term is a correction resulting from the spin freezing of a magnetically disordered layer. A small B parameter corresponds to a high effective coordination number of the magnetic cluster, and therefore, being smaller than its bulk counterpart, the B parameter obtained for $CoO/CoFe_2O_4$ indicates an increased thermal stability associated with a reduction of the surface magnetic disorder due to the exchange coupling at the interface (Lavorato et al. 2015b).

The complexity of the core–shell structure hinders the interpretation of the link between structure and magnetic properties, and further magnetic characterization techniques are usually needed. Remanence and relaxation studies can provide complementary information regarding the magnetization-reversal process (Winkler et al. 2008; Binns 2014). According to Lavorato et al. (2015b), the activation (switching) volume (the minimum volume of the material that reverses its magnetization coherently) inferred from remanence and relaxation measurements is lower for $CoO/CoFe_2O_4$ than for $ZnO/CoFe_2O_4$ nanoparticles (~1/3 and ~1/5 of the total shell volume, respectively), suggesting that the interface exchange coupling also promotes a higher degree of incoherent rotation in the bimagnetic sample.

2.4.3 Size Effects

Given that the overall magnetic properties of bimagnetic core–shell nanoparticles are ruled by the interface coupling, the size should be playing a major role in the magnetic hardening. The $CoO/CoFe_2O_4$ nanoparticles with total diameter between 5 and 11 nm were synthesized by modifying the heating rate and surfactant/precursors ratio in a seed-mediated heat-up method, as reported in Lavorato et al. (2014a).

The authors found that when the nanoparticle size is reduced to 5 nm, the coercivity at 5 K is remarkably increased up to 30.8 kOe, which is more than 50% higher than the maximum coercivity found for single-phase $CoFe_2O_4$. Such anisotropy growth was attributed to the increased relevance of the interface for smaller sizes. However, despite their larger anisotropy, smaller nanoparticles also show a lower thermal stability, which is reflected in the blocking temperature values that are proportional to the volume of the particles. The opposite behavior was observed for 11-nm particles, which exhibit a much higher thermal stability, with a blocking temperature beyond room temperature and lower coercive field of 21.5 kOe. The main magnetic parameters as a function of the size of nanoparticles are summarized in Figure 2.10a. Interestingly, the general feature of the magnetic hardening in this kind of bimagnetic systems reflects major differences compared with single-phase nanoparticles. While coercivity and blocking temperature of single-domain $CoFe_2O_4$ increase with the size (Figure 2.10b), an opposite behavior is observed for bimagnetic nanoparticles, in which a compromise between coercivity and thermal stability is evinced in this particular size range.

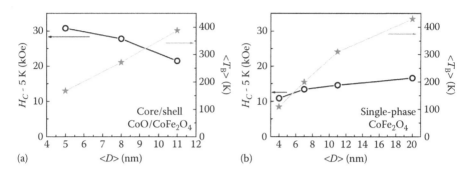

FIGURE 2.10 Effect of the nanoparticle diameter on the coercivity (H_C) and mean blocking temperature of (a) CoO/CoFe$_2$O$_4$ core–shell nanoparticles and (b) single-phase CoFe$_2$O$_4$ nanoparticles. (Panel (a): Adapted from Lavorato, G.C. et al., *Nanotechnology*, 2014b. http://iopscience.iop.org/journal/0957-4484/labtalk/article/58215. Panel (b): Reprinted with permission from López-Ortega, A. et al., *Chem. Mater.*, 27, 4048–4056. Copyright 2015b American Chemical Society.)

2.5 Concluding Remarks

New multicomponent nanoparticles were recently developed, and their increasing complexity requires comprehensive magnetic studies in order to elucidate the relationship between structure and physical properties at the nanoscale, which is a fundamental step for the engineering of materials with improved properties.

In this chapter, we reviewed some exchange-coupled bimagnetic nanoparticles and highlighted that different hysteretic behavior can be found, including exchange-biased loops, effective anisotropy increase, and exchange-spring magnets. The overview of the heat-up synthesis method given in Section 2.3 reflects that, by adjusting specific synthesis parameters, it is possible to effectively control the size and morphology of the nanoparticles, enabling the design of novel exchange-coupled materials. Although there are still few examples of non-conventional multicomponent nanostructures with a higher degree of complexity than conventional core–shell nanoparticles, it is likely that nanoparticles with, for example, a core-multishell structure may be important in the future. This should be particularly important for controlling the exchange coupling in multilayered nanoparticles. However, some problems that restrict the technological applications still persist, especially regarding the chemical methods involved and the temperature dependence of the physical properties. In this sense, it is desirable to achieve a better control of the interface quality or stoichiometry gradients usually observed in this kind of systems. In addition, the complexity of two-step methods limits the total amount of material that can be obtained, and simpler methods for fabricating bimagnetic nanoparticles are needed. In this sense, a full parameterization of multistep synthesis methods is required to scale up the fabrication process.

We showed some examples in which a non-monotonic behavior of the coercivity and exchange bias with the core and/or shell sizes are observed. Such studies highlight the complex dependence of the hysteretic behavior on the morphology and size of bimagnetic nanoparticles, and this still requires further experimental and theoretical studies. The AFM/FiM CoO-core/CoFe$_2$O$_4$-shell system was presented as a case study for evaluating the anisotropy increase due to the exchange coupling at the interface of bimagnetic nanoparticles. Since the AFM is located at the core, inverted core–shell structures enable a better control of the structural and chemical qualities of the CoO, while the thermal treatment promotes a strong interface coupling. The anisotropy increase is revealed by the remarkably large low-temperature coercivity, which approaches 30.8 kOe for 5-nm nanoparticles. However, the magnetic properties are deteriorated when approaching the Néel temperature of the CoO (close to room temperature), suggesting that other systems based on an AFM with higher ordering temperature might be explored. In fact, the possibility of achieving the same hardness enhancement (compared with single-phase nanoparticles) at room temperature could enable a further development of the energy product, interesting for the design of rare-earth-free permanent magnets.

It is worth noting that the general features (e.g., coercivity, exchange bias, and blocking temperature dependence with size or temperature) of bimagnetic nanoparticles differ from those usually observed in their single-phase counterparts. Even if occupying a small fraction of the total volume, it was shown that an AFM coupled to an FM(FiM) in inverted AFM/FM(FiM) nanostructures can lead to a strong increase in the effective anisotropy and the magnetic hardness. In addition, the possibility of tuning the size, crystallinity, and interface quality of those systems enables the control of the exchange-bias field and related properties, as it was also predicted by recent Monte Carlo simulations. Therefore, it can be concluded that the synergy between chemical methods and experimental or theoretical studies of the fundamental properties of interface-related phenomena in bimagnetic nanoparticles can reveal the key issues for designing novel materials suitable for applications.

References

Ali M et al. (2003) Antiferromagnetic layer thickness dependence of the IrMn/Co exchange-bias system. *Phys Rev B* 68:1–7.

Baaziz W et al. (2013) High exchange bias in Fe$_{3-\delta}$O$_4$@CoO core shell nanoparticles synthesized by a one-pot seed-mediated growth method. *J Phys Chem C* 117: 11436–11443.

Bedanta S, Kleemann W (2009) Supermagnetism. *J Phys D Appl Phys* 42:013001.

Berkowitz A et al. (2008) Antiferromagnetic MnO nanoparticles with ferrimagnetic Mn$_3$O$_4$ shells: Doubly inverted core-shell system. *Phys Rev B* 77:024403.

Binns C (2014) *Nanomagnetism: Fundamentals and Applications.* Jordan Hill, England: Elsevier Science.

Binns C et al. (2013) Exchange bias in Fe@Cr core-shell nanoparticles. *Nano Lett* 13: 3334–3339.

Bodnarchuk MI et al. (2009) Exchange-coupled bimagnetic wüstite/metal ferrite core/shell nanocrystals: Size, shape, and compositional control. *Small* 5:2247–2252.

Bozorth R, Tilden E, Williams A (1955) Anisotropy and magnetostriction of some ferrites. *Phys Rev* 99(6):1788.

Casavola M et al. (2009) Exchange-coupled bimagnetic cobalt/iron oxide branched nanocrystal heterostructures. *Nano Lett* 9:366–376.

Chinnasamy CN et al. (2003) Unusually high coercivity and critical single-domain size of nearly monodispersed $CoFe_2O_4$ nanoparticles. *Appl Phys Lett* 83:2862.

Cho N-H et al. (2011) A multifunctional core-shell nanoparticle for dendritic cell-based cancer immunotherapy. *Nat Nanotechnol* 6:675–682.

Coey JMD (2010) *Magnetism and Magnetic Materials*. Cambridge, UK: Cambridge University Press.

Concas G, Spano G, Cannas C, Musinu A, Peddis D, Piccaluga G (2009) Inversion degree and saturation magnetization of different nanocrystalline cobalt ferrites. *J Magn Mater* 321:1893–1897.

Dobrynin AN et al. (2005) Critical size for exchange bias in ferromagnetic-antiferro-magnetic particles. *Appl Phys Lett* 87:012501.

Dormann JL, Fiorani D, Tronc E (1997) Magnetic relaxation in fine-particle systems. *Advanced in Chemical Physics*, Vol. 98. New York: Wiley.

Eftaxias E, Trohidou K (2005) Numerical study of the exchange bias effects in magnetic nanoparticles with core/shell morphology. *Phys Rev B* 71:134406.

Estrader M et al. (2015) Origin of the large dispersion of magnetic properties in nano-structured oxides: Fe_xO/Fe_3O_4 nanoparticles as a case study. *Nanoscale* 7:3002.

Feygenson M, Yiu Y, Kou A, Kim K-S, Aronson MC (2010) Controlling the exchange bias field in Co core/CoO shell nanoparticles. *Phys Rev B* 81:195445.

Fiorani D, Peddis D (2014) Understanding dynamics of interacting magnetic nanoparticles: From the weak interaction regime to the collective superspin glass state. *J Phys Conf Ser* 521:012006.

Fontaiña-Troitiño N, Rivas-Murias B, Rodríguez-González B, Salgueirino V (2014) Exchange bias effect in $CoO@Fe_3O_4$ core-shell octahedron-shaped nanoparticles. *Chem Mater* 26(19):5566–5575.

Fullerton EE, Jiang JS, Bader SD (1999) Hard/soft magnetic heterostructures: Model exchange-spring magnets. *J Magn Mater* 200:392–404.

Gao J, Gu H, Xu B (2009) Multifunctional magnetic nanoparticles: Design, synthesis, and biomedical applications. *Acc Chem Res* 42:1097–1107.

Goto E, Hayashi N, Miyashita T, Nakagawa K (1965) Magnetization and switching char-acteristics of composite thin magnetic films. *J Appl Phys* 36:2951–2958.

Guardia P, Pérez-Juste J, Labarta A, Batlle X, Liz-Marzán LM (2010) Heating rate influ-ence on the synthesis of iron oxide nanoparticles: The case of decanoic acid. *Chem Commun* 46:6108–6110.

Gupta AK, Gupta M (2005) Synthesis and surface engineering of iron oxide nanopar-ticles for biomedical applications. *Biomaterials* 26:3995–4021.

Hyeon T, Lee SS, Park J, Chung Y, Na HB (2001) Synthesis of highly crystalline and monodisperse maghemite nanocrystallites without a size-selection process. *J Am Chem Soc* 123:12798–12801.

Iglesias Ò, Batlle X, Labarta A (2008a) Particle size and cooling field dependence of exchange bias in core/shell magnetic nanoparticles. *J Phys D Appl Phys* 41:134010.

Iglesias O, Labarta A, Batlle X (2008b) Exchange bias phenomenology and models of core/shell nanoparticles. *J Nanosci Nanotechnol* 8:2761–2780.

Jana NR, Chen Y, Peng X (2004) Size- and shape-controlled magnetic (Cr, Mn, Fe, Co, Ni) oxide nanocrystals via a simple and general approach. *Chem Mater* 16:3931–3935.

Jungblut R, Coehoorn R, Johnson MT, Aan de Stegge J, Reinders A (1994) Orientational dependence of the exchange biasing in molecular-beam-epitaxy-grown $Ni_{80}Fe_{20}/Fe_{50}Mn_{50}$ bilayers (invited). *J Appl Phys* 75:6659.

Kavich D, Dickerson J, Mahajan S, Hasan S, Park J-H (2008) Exchange bias of singly inverted FeO/Fe_3O_4 core-shell nanocrystals. *Phys Rev B* 78:174414.

Khurshid H et al. (2013) Synthesis and magnetic properties of core/shell FeO/Fe_3O_4 nano-octopods. *J Appl Phys* 113:17B508.

Kiwi M (2001) Exchange bias theory. *J Magn Magn Mater* 234(3):584–595.

Klabunde KJ, Richards RM (2009) Nanoscale Materials in Chemistry. Hoboken, NJ: John Wiley & Sons.

Kneller E, Hawig R (1991) The exchange-spring magnet: A new material principle for permanent magnets. *Magn IEEE Trans* 27:3588–3600.

Knobel M, Nunes WC, Socolovsky LM, De Biasi E, Vargas JM, Denardin JC (2008) Superparamagnetism and other magnetic features in granular materials: A review on ideal and real systems. *J Nanosci Nanotechnol* 8:2836–2857.

Kodama R (1999) Magnetic nanoparticles. *J Magn Mater* 200:359–372.

Kwon SG, Hyeon T (2011) Formation mechanisms of uniform nanocrystals via hot-injection and heat-up methods. *Small* 7:2685–2702.

Kwon SG et al. (2007) Kinetics of monodisperse iron oxide nanocrystal formation by "heating-up" process. *J Am Chem Soc* 129:12571–12584.

LaMer VK, Dinegar RH (1950) Theory, production and mechanism of formation of monodispersed hydrosols. *J Am Chem Soc* 72:4847–4854.

Lavorato GC et al. (2014a) Size effects in bimagnetic $CoO/CoFe_2O_4$ core/shell nanoparticles. *Nanotechnology* 25:355704.

Lavorato GC et al. (2014b) Tuning the magnetic properties of bimagnetic core/shell nanoparticles. *Nanotechnology.* http://iopscience.iop.org/journal/0957-4484/labtalk/article/58215.

Lavorato GC, Lima E, Troiani HE, Zysler RD, Winkler EL (2015a) Exchange-coupling in thermal annealed bimagnetic core/shell nanoparticles. *J Alloys Compd* 633:333–337.

Lavorato G, Winkler E, Rivas-Murias B, Rivadulla D (2016) Thickness dependence of exchange coupling in epitaxial $Fe_3O_4/CoFe_2O_4$ soft/hard magnetic bilayers. *Phys Rev B* 94:054405.

Lavorato GC et al. (2015b) Magnetic interactions and energy barrier enhancement in core/shell bimagnetic nanoparticles. *J Phys Chem C* 119:15755–15762.

Lee IS et al. (2006) Ni/NiO core/shell nanoparticles for selective binding and magnetic separation of histidine-tagged proteins. *J Am Chem Soc* 128:10658–10659.

Lee JH et al. (2011) Exchange-coupled magnetic nanoparticles for efficient heat induction. *Nat Nanotechnol* 6:418–422.

Lima E et al. (2012) Bimagnetic CoO core/$CoFe_2O_4$ shell nanoparticles: Synthesis and magnetic properties. *Chem Mater* 24:512–516.

Liu X et al. (2015) A systematic study of exchange coupling in core-shell $Fe_{3-\delta}O_4@CoO$ nanoparticles. *Chem Mater* 27(11):4073–4081.

Liu XS, Gu BX, Zhong W, Jiang HY, Du YW (2003) Ferromagnetic/antiferromagnetic exchange coupling in $SrFe_{12}O_{19}/CoO$ composites. *Appl Phys A Mater Sci Process* 77:673–676.

López-Ortega A, Estrader M, Salazar-Alvarez G, Roca AG, Nogués J (2015a) Applications of exchange coupled bi-magnetic hard/soft and soft/hard magnetic core/shell nanoparticles. *Phys Rep* 553:1–32.

López-Ortega A, Lottini E, de Julián Fernández C, Sangregorio C (2015b) Exploring the magnetic properties of cobalt-ferrite nanoparticles for the development of rare-earth-free permanent magnet. *Chem Mater* 27(11):4048–4056.

López-Ortega A et al. (2010) Size-dependent passivation shell and magnetic properties in antiferromagnetic/ferrimagnetic core/shell MnO nanoparticles. *J Am Chem Soc* 132:9398–9407.

Lottini E et al. (2016) Strongly exchange coupled core|shell nanoparticles with high magnetic anisotropy: A strategy toward rare-earth-free permanent magnets. *Chem Mater* 28:4214–4222.

Lu A-H, Salabas EL, Schüth F (2007) Magnetic nanoparticles: Synthesis, protection, functionalization, and application. *Angew Chem Int Ed Engl* 46:1222–1244.

Lyubutin IS, Lin CR, Korzhetskiy YV, Dmitrieva TV, Chiang RK (2009) Mössbauer spectroscopy and magnetic properties of hematite/magnetite nanocomposites. *J Appl Phys* 106(3):34311.

Meiklejohn WH (1962) Exchange anisotropy—A review. *J Appl Phys* 33:1328.

Meiklejohn WH, Bean CP (1956) New magnetic anisotropy. *Phys Rev* 102:1413.

Mélinon P et al. (2014) Engineered inorganic core/shell nanoparticles. *Phys Rep* 543: 163–197.

Mendoza-Reséndez R, Luna C, Barriga-Castro ED, Bonville P, Serna CJ (2012) Control of crystallite orientation and size in Fe and FeCo nanoneedles. *Nanotechnology* 23:225601.

Mørup S, Hansen MF, Frandsen C (2010) Magnetic interactions between nanoparticles. *Beilstein J Nanotechnol* 1:182–190.

Mullin JW (2001) *Crystallization*. Oxford, UK: Butterworth-Heinemann.

Murray CB, Norris DJ, Bawendi MG (1993) Synthesis and characterization of nearly monodisperse CdE (E= sulfur, selenium, tellurium) semiconductor nanocrystallites. *J Am Chem Soc* 115:8706–8715.

Murray CB, Sun S, Gaschler W, Doyle H, Betley TA, Kagan CR (2001) Colloidal synthesis of nanocrystals and nanocrystal superlattices. *IBM J Res Dev* 45:47–56.

Nam KM et al. (2010) Syntheses and characterization of wurtzite CoO, rocksalt CoO, and spinel Co_3O_4 nanocrystals: Their interconversion and tuning of phase and morphology. *Chem Mater* 22:4446–4454.

Niederberger M, Pinna N (2009) *Metal Oxide Nanoparticles in Organic Solvents*. London: Springer.

Nogués J, Schuller IK (1999) Exchange bias. *J Magn Mater* 192:203–232.

Nogués J et al. (2005) Exchange bias in nanostructures. *Phys Rep* 422:65–117.

O'Handley RC (2000) *Modern Magnetic Materials: Principles and Applications*. New York: Wiley.

Pankhurst QA, Thanh NTK, Jones SK, Dobson J (2009) Progress in applications of magnetic nanoparticles in biomedicine. *J Phys D Appl Phys* 42:224001.

Park J et al. (2004) Ultra-large-scale syntheses of monodisperse nanocrystals. *Nat Mater* 3:891–895.

Perales-Perez O et al. (2002) Production of monodispersed particles by using effective size selection. *J Appl Phys* 91:6958–6960.

Pianciola BN, Lima E, Troiani HE, Nagamine LCCM, Cohen R, Zysler RD (2015) Size and surface effects in the magnetic order of CoFe₂O₄ nanoparticles. *J Magn Mater* 377:44–51.

Pinna N, Niederberger M (2008) Surfactant-free nonaqueous synthesis of metal oxide nanostructures. *Angew Chemie Int Ed* 47:5292–5304.

Radu F, Zabel H (2008) *Magnetic Heterostructures*. Berlin, Germany: Springer.

Salazar-Alvarez G et al. (2011) Two-, three-, and four-component magnetic multilayer onion nanoparticles based on iron oxides and manganese oxides. *J Am Chem Soc* 133:16738–16741.

Salazar-Alvarez G, Sort J, Suriñach S, Baró MD, Nogués J (2007) Synthesis and size-dependent exchange bias in inverted core–shell MnO|Mn₃O₄ nanoparticles. *J Am Chem Soc* 129:9102–9108.

Schmid G (2011) *Nanoparticles: From Theory to Application*. Morlenbach, Germany: John Wiley & Sons.

Seto T, Akinaga H, Takano F, Koga K, Orii T, Hirasawa M (2005) Magnetic properties of monodispersed Ni/NiO core-shell nanoparticles. *J Phys Chem B* 109:13403–13405.

Sharma SK, Vargas JM, Pirota KR, Kumar S, Lee CG, Knobel M (2011) Synthesis and ageing effect in FeO nanoparticles: Transformation to core–shell FeO/Fe₃O₄ and their magnetic characterization. *J Alloys Compd* 509:6414–6417.

Shenker H (1957) Magnetic anisotropy of cobalt ferrite (Co₁.₀₁Fe₂.₀₀O₃.₆₂) and nickel cobalt ferrite (Ni₀.₇₂Fe₀.₂₀Co₀.₀₈Fe₂O₄). *Phys Rev* 107:1246–1249.

Skumryev V, Stoyanov S, Zhang, Hadjipanayis G, Givord D, Nogués J (2003) Beating the superparamagnetic limit with exchange bias. *Nature* 423:19–22.

Soares JM, Cabral FO, de Araújo JH, Machado FL (2011) Exchange-spring behavior in nanopowders of CoFe₂O₄–CoFe₂. *Appl Phys Lett* 98:072502.

Song Q, Zhang ZJ (2012) Controlled synthesis and magnetic properties of bimagnetic spinel ferrite CoFe₂O₄ and MnFe₂O₄ nanocrystals with core-shell architecture. *J Am Chem Soc* 134:10182–10190.

Soon GK, Hyeon T (2008) Colloidal chemical synthesis and formation kinetics of uniformly sized nanocrystals of metals, oxides, and chalcogenides. *Acc Chem Res* 41:1696–1709.

Stamps RL (2000) Mechanisms for exchange bias. *J Phys D Appl Phys* 33(23):R247.

Sun S, Murray C, Weller D, Folks L, Moser A (2000) Monodisperse FePt nanoparticles and ferromagnetic FePt nanocrystal superlattices. *Science* 287:1989–1992.

Sun S, Zeng H (2002) Size-controlled synthesis of magnetite nanoparticles. *J Am Chem Soc* 124:8204–8205.

Sun X, Huls NF, Sigdel A, Sun S (2012) Tuning exchange bias in core/shell FeO/Fe₃O₄ nanoparticles. *Nano Lett* 12:246–251.

Tobia D, Winkler E, Zysler R, Granada M, Troiani H (2008) Size dependence of the magnetic properties of antiferromagnetic Cr₂O₃ nanoparticles. *Phys Rev B* 78:104412.

Tobia D, Winkler E, Zysler RD, Granada M, Troiani HE, Fiorani D (2009) Exchange bias of Co nanoparticles embedded in Cr₂O₃ and Al₂O₃ matrices. *J Appl Phys* 106:3920.

Tracy J, Weiss D, Dinega D, Bawendi M (2005) Exchange biasing and magnetic properties of partially and fully oxidized colloidal cobalt nanoparticles. *Phys Rev B* 72:64404.

van Embden J, Chesman ASR, Jasieniak JJ (2015) The heat-up synthesis of colloidal nanocrystals. *Chem Mater* 27:2246–2285.

Vargas JM, Zysler RD (2005) Tailoring the size in colloidal iron oxide magnetic nanoparticles. *Nanotechnology* 16:1474–1476.

Vasilakaki M, Trohidou KN, Nogués J (2015) Enhanced magnetic properties in antiferromagnetic-core/ferrimagnetic-shell nanoparticles. *Sci Rep* 5:9609.

Vázquez-Vázquez C, López-Quintela M, Buján-Núñez MC, Rivas J (2010) Finite size and surface effects on the magnetic properties of cobalt ferrite nanoparticles. *J Nanoparticle Res* 13:1663–1676.

Winkler E, ZyslerRD, Fiorani D (2004) Surface and magnetic interaction effects in Mn_3O_4 nanoparticles. *Phys Rev B* 70:174406.

Winkler E et al. (2008) Surface spin-glass freezing in interacting core-shell NiO nanoparticles. *Nanotechnology* 19:185702.

Winkler EL et al. (2012) Origin of magnetic anisotropy in $ZnO/CoFe_2O_4$ and CoO/$CoFe_2O_4$ core/shell nanoparticle systems. *Appl Phys Lett* 101:252405.

Yang C, Wang G, Lu Z, Sun J, Zhuang J, Yang W (2005) Effect of ultrasonic treatment on dispersibility of Fe_3O_4 nanoparticles and synthesis of multi-core Fe_3O_4/SiO_2 core/shell nanoparticles. *J Mater Chem* 15:4252.

Zeng H, Li J, Liu JP, Wang ZL, Sun S (2002) Exchange-coupled nanocomposite magnets by nanoparticle self-assembly. *Nature* 420:395–398.

Zeng H, Li J, Wang ZL, Liu JP, Sun S (2004a) Bimagnetic core/shell FePt/Fe_3O_4 nanoparticles. *Nano Lett* 4:187–190.

Zeng H, Rice PM, Wang SX, Sun S (2004b) Shape-controlled synthesis and shape-induced texture of $MnFe_2O_4$ nanoparticles. *J Am Chem Soc* 126:11458–11459.

3

Exchange Bias in Dilute Magnetic Alloys

Per Nordblad,
Matthias Hudl, and
Roland Mathieu

3.1 Introduction

In 1956 and 1957, W.H. Meiklejohn and C.P. Bean published two articles with the same title "New magnetic anisotropy" (Meiklejohn and Bean, 1956, 1957); these articles reported the discovery of shifted hysteresis loops and unidirectional anisotropy after field-cooling a compact sample of oxidized Co particles. Shifted hysteresis loops after field-cooling the dilute mixed ferromagnetic–antiferromagnetic Ni(Mn) alloy were reported a couple of years later (Kouvel and Graham, 1959). The exchange bias of these systems was assigned to the existence of ferromagnetic-antiferromagnetic interfaces. Observations of shifted hysteresis loops after field-cooling the dilute magnetic alloy systems Cu(Mn) and Ag(Mn) were soon reported afterward (Kouvel, 1960). Numerous studies of these phenomena were published in the two to three decades that succeeded the 1950s. The theoretical and phenomenological understanding of the early observations of exchange bias in dilute magnetic alloys is comprehensively summarized in reviews by Kouvel, 1963 and Beck, 1980. An alternative approach to the existence of exchange bias in dilute magnetic alloys was based on the Dzyaloshinskii–Moriya (DM) interaction (Dzyaloshinsky, 1958, Moriya, 1960) and is concisely described by Levy et al., 1982. The simple sketches in Figure 3.1 summarize the essence of exchange-biased hysteresis loops in Mn-based dilute magnetic alloys, demonstrating the effects of a unidirectional and a uniaxial component of the anisotropy on a weak excess magnetization, ΔM.

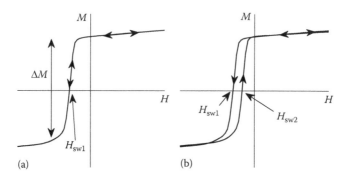

FIGURE 3.1 (a) An M versus H hysteresis loop that is exchange-biased by only a unidirectional anisotropy. (b) A hysteresis loop on a system that has both a unidirectional and a uniaxial anisotropy component. The switch fields H_{sw1} and H_{sw2} as well as the excess magnetization ΔM are defined in the figure. (Adapted from Levy, P.M. et al., Origin of anisotropy in transition metal spin glass alloys, *J. Appl. Phys.*, 53, 2168–2172, 1982, Fig. 2.)

3.2 Magnetic Properties of Dilute Magnetic Alloys

Figure 3.2 shows typical exchange-biased hysteresis loops of $Cu_{0.75}Mn_{0.25}$ measured at different temperatures. In fact, all low-temperature field-cooled (FC) hysteresis loops of $Cu_{1-x}Mn_x$ and $Ag_{1-x}Mn_x$ of concentration $0 < x < 0.3$ exhibit exchange bias. This concentration range includes the region where archetypal spin-glass behavior occurs. The concept magnetic-glass or spin-glass was first adopted in 1970 by Anderson, 1970 and Bancroft, 1970, inspired by discussions with B.R. Coles. The discovery of a cusp in the weak-field ac-susceptibility of dilute Au(Fe) alloys by Cannella and Mydosh (1972) promoted the idea that a unique low-temperature spin-glass phase exists in dilute magnetic alloys.

The archetypal spin glass is formed of dilute magnetic alloys at concentrations of the magnetic atoms below the percolation limit. In these systems, exchange interaction between the magnetic atoms is transmitted via RKKY interaction (Ruderman and Kittel, 1954, Kasuya, 1956, Yosida, 1957). The RKKY interaction is oscillatory in nature and decays with the cube of the distance between magnetic constituents, which in dilute alloys provides disorder (random distribution of the magnetic atoms on the lattice sites and thus different distances between the magnetic atoms) and frustration due to competition between ferromagnetic and antiferromagnetic exchange interaction. The simplest experimental identification of a spin-glass material is provided by zero-field-cooled (ZFC), FC, and thermoremanent magnetization (TRM) versus temperature curves recorded in low magnetic fields. Figure 3.3a shows such ZFC, FC, and TRM curves of a polycrystalline sample of Cu(13.5 at% Mn) in a weak applied magnetic field (Hudl et al., 2016).

There is a characteristic cusp in the ZFC curve at about 62 K. (A dynamic scaling analysis of ac-susceptibility data on this sample indicates a spin-glass transition temperature, T_g, of about 57 K.) The measured M(T)/H curves are quasi-static at temperatures below the maximum, T_{max}, in the ZFC curve; that is, if the heating (cooling) is stopped,

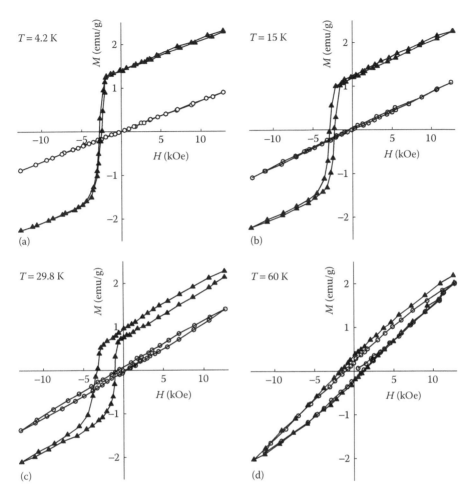

FIGURE 3.2 Field-cooled (solid triangles, H_{FC} = 12.7 kOe) and zero-field-cooled (open circles) hysteresis loops measured on a $Cu_{0.75}Mn_{0.25}$ alloy at different temperatures T in fields up to 12.7 kOe; T = a) 4.2 K, b) 15 K, c) 29.8 K, and d) 60 K. (Adapted from Beck, P.A., Properties of micto-magnets (spin glasses), *Prog. Mater. Sci.*, 23, 1–49, 1980, Figs. 13 and 16–18.)

the magnetization relaxes upward (ZFC) (or downward [TRM]). In addition, the system experiences magnetic aging at temperatures below T_{max} (Nordblad, 2013 and references therein). The behavior of the M(T)/H curves with increasing field is illustrated in Figure 3.3b; the susceptibility is significantly suppressed at temperatures above the point where the ZFC and FC curves merge. This point is pushed to lower temperatures with increasing field; in contrast, at lower temperatures, the susceptibility is enhanced with increasing field. Two reviews on spin-glass theory and experiments have recently been published (Kawamura and Taniguchi, 2015, Mydosh, 2015).

At constant low temperatures, the field dependence of the magnetization of a spin glass exhibits non-linearity and hysteresis. A plot of the remnant magnetization after

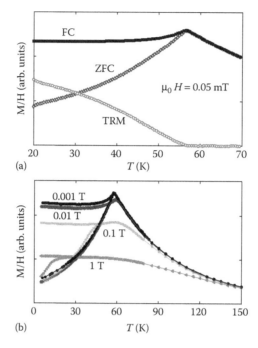

FIGURE 3.3   M/H versus temperature on the spin-glass Cu(13.5 at% Mn). (a) Low field (0.5 Oe) M/H versus T in zero-field-cooled (ZFC), field-cooled (FC), and thermoremanent (TRM) protocols. (b) M/H versus T measured in ZFC and FC protocols at different applied magnetic fields, as indicated in the figure.

field cooling (thermoremanent magnetization, TRM) and after a field pulse of certain duration (isothermal remanent magnetization: IRM of Cu(13.5 at% Mn) is shown in Figure 3.4a. The two curves saturate and asymptotically approach each other at high fields but follow separate tracks on decreasing the field to H = 0. (Such a field dependence of the low-temperature TRM and IRM of Cu(Mn) was first observed by Jacobs and Schmitt, 1959). Knowledge of this field dependence of the remanent magnetization is essential when investigating an exchange-biased hysteresis curve after FC and comparing it with the corresponding ZFC hysteresis behavior. In Figure 3.4b, a ZFC-hysteresis loop measured up to a maximum field of 3 T is shown, and in Figure 3.4c, the central part of this curve is shown together with the corresponding 3T FC loop. Both the FC- and the ZFC-hysteresis loops are exchange-biased. It should be noted that the behavior of both TRM(H) and IRM(H) is strongly temperature-dependent—the field required to reach the point where the two curves coalesce rapidly decreases with increasing temperature, Bouchiat and Monod, 1982. However, the overall shape of the curves remains similar if the magnetization—and field—axes are scaled to their values at saturation. It should again be emphasized that M versus H curves exhibiting exchange-biased FC loops without coercivity and linear reversible ZFC loops belong to the low-field region of TRM(H) and IRM(H) curves (cf. Figure 3.4a), whereas FC loops with significant coercivity and open ZFC loops (cf. Figure 3.4b and c) are recorded at fields approaching and above the

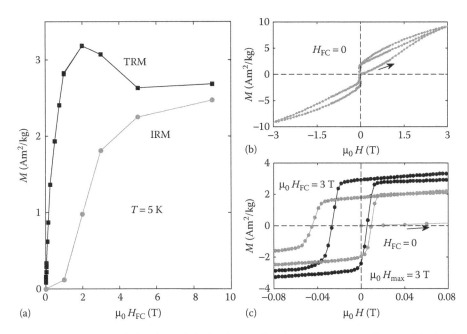

FIGURE 3.4 (a) TRM(H) and IRM(H) of Cu(13.5 at% Mn) versus H measured at 5 K. (b) A ZFC hysteresis loop measured in fields up to 3 T at 5 K. (c) An expanded view of the low-field part of the ZFC hysteresis loop in (b); the 3T FC loop is also included.

point where TRM(H) and IRM(H) coalesce. In a recent study, it is found that the width of the hysteresis loop (uniaxial part of the anisotropy) is controlled by the maximum field used in the experiment (Hudl et al., 2016). Figure 3.4 is inspired by corresponding results on a Cu(9 at% Mn) sample reported in Figures 3.3 and 3.4 of Knitter et al. (1977).

3.3 Exchange Bias

The hysteresis behavior of dilute magnetic alloys is strongly field-dependent, and on the time scale of standard M versus H measurements, any hysteresis vanishes at about the temperature for the maximum in the low-field ZFC-M(T)/H curve (cf. Figure 3.3a). Exchange-biased hysteresis loops first appear at temperatures well below this temperature. The characteristic parameters, coercivity $H_c = 1/2(H_{sw1} - H_{sw2})$ and exchange-bias field $H_e = 1/2(H_{sw1} + H_{sw2})$, are derived from the hysteresis loops. The switch fields, H_{sw1} and H_{sw2}, are indicated and defined in Figure 3.1. In addition, ΔM, the magnitude of the magnetization jump that occurs at the switch fields H_{sw1} and H_{sw2}, is indicated and defined in that figure. $\Delta M(H)$ depends strongly on temperature and vanishes well before T_{max}, and its size is controlled by the magnitude of the cooling field. At low temperatures, TRM(H) and $\Delta M(H)/2$ show similar field dependences. Figure 3.5 shows hysteresis loops measured on Cu(13.5 at% Mn) after field cooling. It is seen that at the investigated field strengths ($1 \leq \mu_0 H \leq 9$ T), H_{sw1} is near independent of the cooling field used, whereas H_{sw2} is controlled by the field (Hudl et al., 2016). Remarkably, hysteresis loops measured

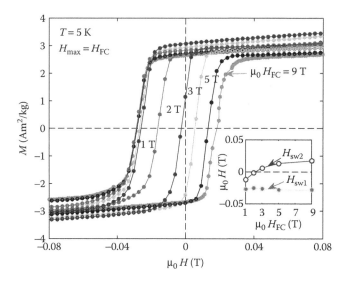

FIGURE 3.5 Field-cooled hysteresis loops of Cu(13.5 at% Mn) measured in different cooling fields (1, 2, 3, 5, and 9 T). The inset shows the field dependence of the fields H_{sw1} and H_{sw2}.

after FC in 5T and higher and ZFC loops measured to the same high fields yield quite similar exchange-biased hysteresis loops. The fact that exchange bias is observed in ZFC protocols and that it behaves quite similar to that observed under an FC protocol is surprising. However, the similarity of the FC and ZFC hysteresis loops when the maximum applied field is large enough was early demonstrated by Knitter et al. (1977).

Field-cooled exchange bias at low temperatures and well below the reversible region of the ZFC hysteresis loop yields $\Delta M(T, H)$ and $H_e(T, H)$ behaviors that are quite complex. Extensive studies and phenomenological modeling of the observations were, as mentioned earlier, reported by Kouvel, 1963 and a comprehensive summary on general magnetic properties of dilute magnetic alloys was given by Beck, 1980 in the review article "Properties of Micromagnets (Spin Glasses)." The early results on exchange anisotropy and shifted hysteresis loops in Mn-based dilute magnetic alloys were phenomenologically interpreted in models inspired by the discovery and modeling of exchange bias in compacts of Co nanoparticles covered by an antiferromagnetic shell of cobalt oxide (Meiklejohn and Bean, 1957). These nanostructured materials promoted an interpretation assigning the anisotropy to the interfaces between ferro- and antiferromagnetic substances, where the direction of the field-aligned magnetic moments of the ferromagnetic grains were locked to the direction of the rigidly confined antiferromagnetic phases. The ferromagnetic—antiferromagnetic (Kouvel, 1963) model for dilute Cu(Mn) and Ag(Mn) alloys thus prescribed shifted hysteresis loops. The model includes ensembles of antiferromagnetic and ferromagnetic domains that are fully compensated and yield zero magnetization at zero applied fields. When the system is field-cooled, an excess moment is introduced in the ensemble that becomes locked by the growth of a strong anisotropy in the antiferromagnetic domains. This locked magnetic moment gives rise to shifted hysteresis loops, where the switch field is higher when the field is applied antiparallel to the direction of the magnetic moment induced by the cooling field. A mictomagnetic model prescribing giant magnetic moments formed by clusters of Mn atoms was later developed (Chakravorty et al., 1971, Beck, 1980). Under field cooling, these giant moments become locked to the cooling field direction and produce the exchanged-biased loops.

An alternative approach to the exchange bias problem was pursued by several groups in France, assigning the observed exchange anisotropy to the DM interaction (Dzyaloshinsky, 1958, Moriya, 1960), defined by the Hamiltonian equation:

$$H_{DM} = \sum_{i,j} \boldsymbol{D}_{i,j} \cdot (\boldsymbol{S}_i \times \boldsymbol{S}_j)$$

$\boldsymbol{D}$ is a vector, whose components, $\boldsymbol{D}_{i,j}$, are controlled by the strength of the spin-orbit scattering of the atoms transmitting the interaction between the spins $\boldsymbol{S}_i$ and $\boldsymbol{S}_j$. Experiments on Cu(Mn) and Ag(Mn) alloys, with additional doping with a third element, indicated a direct relation between the scattering strength of the added dopants and the measured exchange anisotropy (Prejean et al., 1980). A summarizing article on DM-related experiments and models is given in the article "Origin of anisotropy in transition metal spin glass alloys" by Levy et al., 1982. One conclusion of this article is that the anisotropy can be described by a uniaxial and a unidirectional component (cf. Figure 3.1), where the unidirectional anisotropy originates from the DM interaction, whereas the uniaxial component is of unsettled origin. The interest and progress in the understanding of the shifted hysteresis loops and exchange bias in dilute magnetic alloys have declined since the early 1980s; however, some new work and reports appear quite regularly, for example, recently (Jiménez-Villacorta et al., 2012, Barnsley et al., 2013). A related finding that couples to the large activities on exchange bias in ferromagnetic–antiferromagnetic layer structures is discussed in the article "Exchange bias using a spin glass" by Ali et al., 2007, reporting exchange-biased hysteresis loops in a Co/Cu(6 at% Mn) bilayer system.

3.4 Tunable Exchange Bias

The article "Tunable exchange bias in dilute magnetic alloys—chiral spin glasses" by Hudl et al., 2016 reports results on exchange bias of a Cu(13.5 at% Mn) spin glass that suggest that the DM interaction accounts for both a unidirectional and a uniaxial anisotropy component. A key finding of this study is illustrated by the hysteresis loops shown in Figures 3.5 and 3.6.

The loops are measured at 5 K after field-cooling the sample in different applied magnetic fields and recording the loops on descending the field to $-H_{FC}$ and increasing the field back to H_{FC}. From these data, the parameters $\Delta M(H)$, $H_{sw1}(H)$, and $H_{sw2}(H)$ are derived. $\Delta M(H)$ has a field dependence that largely mimics that of TRM(H). As is seen in Figure 3.7a, $\Delta M(H)$ increases with increasing field, exhibits a weak maximum at about 2 T, and asymptotically approaches a constant level at higher fields. However, the corresponding anisotropy energy $E_{DM} \sim \mu_0 \Delta M H_{sw1}$ remains constant, as is seen in Figure 3.7b filled circles. Assuming that the DM interaction is responsible for the anisotropy, E_{DM} is a measure of the strength of this anisotropy. The switch field H_{sw2} almost coalesces

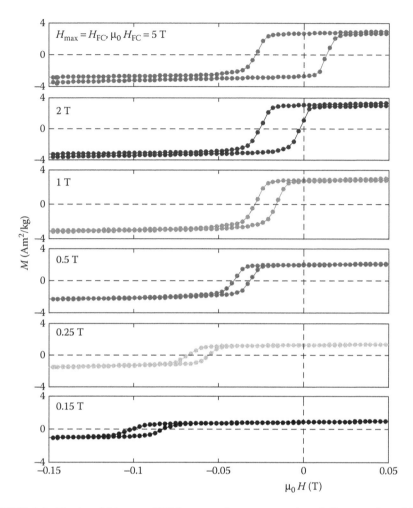

FIGURE 3.6 The low-field part of FC-hysteresis loops measured in different cooling fields, ranging from 0.15 to 5 T.

with H_{sw1} at lower fields but deviates more and more from H_{sw1} with increasing cooling field to, at the highest cooling field, approach $-H_{sw1}$. This behavior suggests that the anisotropy can be divided into one unidirectional part that is confined to the direction of the cooling field with respect to the sample, E_{UdS}, and one part confined to the direction of the excess magnetization, E_{UdM}, that is, a part that switches direction when ΔM switches direction. The energy related to these components can be derived by using $E_{UdS} - E_{UdM} = \mu_0 H_{sw2} \Delta M$ and $E_{UdS} + E_{UdM} = E_{DM} = \mu_0 H_{sw1} \Delta M$. In Figure 3.7b, the two energy terms, E_{UdS} (open triangles) and E_{UdM} (filled triangles), are plotted together with $\mu_0 H_{sw1} \Delta M$. The observed anisotropy may be assumed to originate from the DM interaction, and the direction of the anisotropy is governed by the chirality of the interacting spin pairs. The low-field behavior then reflects a chirality of the spin structure that is

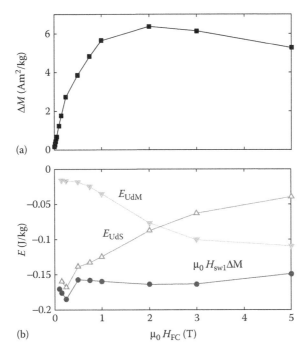

FIGURE 3.7 (a) $\Delta M(H)$ derived from Figure 3.6. (b) Anisotropy energy measures derived from Figure 3.6: $\mu_0 \Delta M H_{sw1}$ (filled circles), E_{UdS} (open triangles), and E_{UdM} (filled triangles) versus H.

confined to the direction induced by the cooling field, and it remains unchanged when the excess moment is flipped. However, with increasing field, parts of the chirality switch to become locked to the excess magnetization, thus yielding an apparent uniaxial anisotropy component. At high enough fields (>14 T at 5 K), the chirality of the spin structure becomes confined to the excess magnetization and the system essentially exhibits uniaxial anisotropy. On the other hand, at low temperatures and maximum fields that do not reach the saturation field, only a unidirectional anisotropy is observed and the chirality remains locked to the sample geometry and the original cooling field direction.

Coming back to the fact that the ZFC hysteresis loops are exchange-biased in the same way as the FC loops at high fields, where the IRM(H) and TRM(H) curves approach each other. An implication of this similarity is that the chirality of the system is aligned along the initial applied field's direction, and the division of chirality locked to the sample geometry and to the excess moment is, in both cases, governed by the maximum field used in the hysteresis measurement. However, after ZFC cooling, the starting configuration does not have a preferred direction for the chirality (averages to zero), and symmetric minor loops with weak coercivity and remanence governed by relaxation processes are obtained. Looking back at Figure 3.3a, this behavior is found at field strengths well below the broad maximum in TRM(H). It should again be noted that there is a strong temperature and concentration dependence of the TRM/IRM curves and that at low temperatures, the broad maximum in TRM(H) occurs at very high magnetic fields.

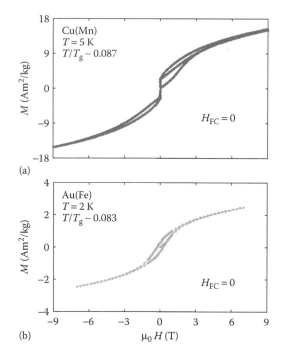

(a)

(b)

$\mu_0 H$ (T)

FIGURE 3.8 ZFC hysteresis loop of (a) Cu(13.5 at% Mn) at T = 5 K, T/T_g = 0.087, and (b) Au(6 at% Fe) at T = 2 K, T/T_g = 0.083.

The four panels in Figure 3.2 illustrate how the FC- and ZFC-hysteresis loops change with temperature for a Cu(25 at% Mn) sample.

Dilute magnetic alloys of Fe and Co atoms do not show exchange bias similar to that of the Mn-based alloys discussed here (Prejean et al., 1980). Figure 3.8 shows hysteresis loops measured on a Cu(Mn) spin glass and a Au(Fe) spin glass. The Mn-based system shows all characteristics of exchange-biased loops, whereas the Fe-based alloy exhibits a symmetric loop whose width is governed by the field sweep rate. The difference can be assigned to the Heisenberg nature of the Mn spins, allowing DM interaction and chirality-governed properties, whereas the Fe (and Co) spins, due to stronger single-ion anisotropy, have certain Ising character suppressing influences from the DM interaction.

3.5 Summary and Conclusions

Dilute magnetic alloys based on Mn atoms dissolved in nonmagnetic host metals exhibit exchange-biased hysteresis loops at low temperatures well below the temperature for the maximum in the low-field ZFC susceptibility versus temperature curve. The exchange bias is controlled by the direction and strength of the cooling field and the maximum field applied in the hysteresis loop measurement. Exchange-biased loops are also observed after ZFC. The field dependence of the width of the hysteresis loops shows that the chirality of the DM interaction originally points in random directions.

However, any finite cooling field through T_g confines the chirality to the direction of the applied field, giving rise to a unidirectional anisotropy and magnetization jumps of magnitude governed by the cooling field. Depending on the strength of the cooling field, a fraction of the system attains a chirality that, instead of being confined to the crystallographic structure, becomes locked to the direction of ΔM and thus switches direction on field reversal and appears as a uniaxial anisotropy. The experiments show that the anisotropy energy $EDM_a \sim \mu_0 \Delta M H_{sw1}$ is independent of the size of the cooling field, weakly temperature-dependent, and a measure of the strength of the DM interaction. On the other hand, the switch field on reversal, H_{sw2}, is strongly dependent on the maximum field used in the measurement of the hysteresis loop and directly controls the fraction of spin pairs with spin chirality that has become confined to ΔM. Experiments on single-crystal samples of $Cu_{1-x}Mn_x$, $0.14 < x < 0.23$ by Iwata et al., 1970 yielded similar exchange-biased hysteresis loops, independent of the crystallographic direction in which the magnetic field was aligned.

Current magnetism research includes many phenomena that are consequences of the DM interaction. This includes, for example, magnetoelectric coupling in multiferroics (Cheong and Mostovoy, 2007), skyrmions (Fert et al., 2013, Nagaosa and Tokura, 2013), and induced domain motion from spin transfer torque (Ryu et al., 2013, Yu et al., 2016). A method to directly measure the DM interaction is given by Gross et al., 2016. Extended studies of field and temperature dependence of exchange-biased hysteresis loops of dilute Mn-based alloys may provide further information on the DM interaction and the nature of the induced excess moment, ΔM. Figure 3.9 shows a ZFC-hysteresis loop and, in the inset, first-order reversal curves (FORC) of Ag(11 at% Mn $T_g = 33$ K. In a FORC experiment, the M versus H measurements are performed, starting from one and the same initial field, H_{max}, in a sequence where the reversal field (first field in the M vs. H measurement with increasing field) is continuously altered from H_{max} down

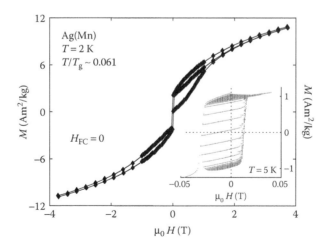

FIGURE 3.9 ZFC hysteresis curves measured on Ag(11 at% Mn) at 2 K ($T_g = 33$ K). Inset: First-order reversal curves on the same material measured at 5 K. The initial field in the FORC measurements is 4 T.

to $-H_{max}$, with a fixed field step in between measurements. The FORC curves indicate that the excess moment can be fractionally switched in any proportion, and thus, the effective strength of the DM interaction on reversal is set by the ratio of the chirality confined to the lattice to the chirality confined to the excess magnetization, both on a global and a local scale.

Acknowledgments

The FORC measurements on Ag(11 at% Mn) have been performed by Dr. Didier Hérisson. Financial support from the Swedish Research Council (VR) and the Göran Gustafsson Foundation is acknowledged.

References

Ali, M., Adie, P., Marrows, C.H., Greig, D., Hickey, B.J., and Stamps, R.L. Exchange bias using a spin glass. *Nature Mater.*, 2007, 6: 70–75.

Anderson, P.W. Localisation theory and the Cu-Mn problem—spin glasses. *Mat. Res. Bull.*, 1970, 5: 549–554.

Bancroft, M.H. Magnetic interactions in a weak moment system—Au-Co. *Phys. Rev B*, 1970, 2: 2597–2603.

Barnsley, L.C., MacA Gray, E., and Webb, C.J. Asymmetric reversal in aged high concentration CuMn alloy. *J. Phys.: Condens. Matter.*, 2013, 25: 086003.

Beck, P.A. Properties of mictomagnets (spin glasses). *Prog. Mater. Sci.*, 1980, 23: 1–49 and references therein.

Bouchiat, H., and Monod, P. Remanent magnetisation properties of the spin glass state. *J. Magn. Magn. Mater.*, 1982, 30: 175–191.

Cannella, V., and Mydosh, J.A. Magnetic ordering in gold-iron alloys. *Phys. Rev. B*, 1972, 6: 4220–4237.

Chakravorty, S., Panagrahy, P., and Beck, P.A. Mictomagnetism in Pd-Cr and V-Mn alloys. *J. Appl. Phys.*, 1971, 42: 1698–1699.

Cheong, S.-W., and Mostovoy, M. Multiferroics: a magnetic twist for ferroelectricity, *Nature Mater.*, 2007, 6: 13–20.

Dzyaloshinsky, I. A thermodynamic theory of "weak" ferromagnetism of antiferromagnets. *J. Phys. Chem. Solids*, 1958, 4: 241–255.

Fert, A., Cros, V., and Sampaio, J. Skyrmions on the track. *Nat. Nanotechnol.*, 2013, 8: 152–156.

Gross, I. et al. Direct measurement of interfacial Dzyaloshinskii-Moriya interaction in X|CoFeB|MgO heterostructures with scanning NV magnetometer (X = Ta, TaN, and W). *Phys. Rev. B*, 2016, 94: 064413.

Hudl, M., Mathieu, R., and Nordblad, P. Tunable exchange bias in dilute magnetic alloys—chiral spin glasses. *Sci. Rep.*, 2016, 6: 19964.

Iwata, T., Kai, K., and Nakamichi, T. Exchange anisotropy in single crystals of Cu-Mn alloys. *J. Phys. Soc. Jpn.*, 1970, 28: 582–589.

Jacobs, I.S., and Schmitt, R.W. Low-temperature electrical and magnetic behavior of dilute alloys: Mn in Cu and Co in Cu. *Phys. Rev.*, 1959, 113: 459–463.

Jiménez-Villacorta, F., Marion, J.L., Sepehrifar, T., and Lewis, L.H. Tuning exchange anisotropy in nanocomposite AgMn alloys. *J. Appl. Phys.*, 2012, 111: 07E141.

Kasuya, T. A theory of metallic ferro- and antferromagnetism on Zener's model. *Prog. Theor. Phys.*, 1956, 16: 45–57.

Kawamura, H., and Taniguchi, T. In *Handbook of Magnetic Materials 24* (Ed. Buschow K.H.J.) Ch. 1, 2015, Amsterdam, the Netherlands: Elsevier. pp. 1–137.

Knitter, R.W., Kouvel, J.S., and Klaus, H. Magnetic irreversibility and metastability in a Cu-Mn alloy. *J. Magn. Magn. Mater.*, 1977, 5: 356–359.

Kouvel, J.S., and Graham, Jr. C.D. Exchange anisotropy in disordered nickel-manganese alloys. *J. Phys. Chem. Solids*, 1959, 11: 220–235.

Kouvel, J.S. Exchange anisotropy in Cu-Mn and Ag-Mn alloys. *J. Appl. Phys.*, 1960, 31: 142S–147S.

Kouvel, J.S. A ferromagnetic-antiferromagnetic model for copper-manganese and related alloys. *J. Phys. Chem. Solids*, 1963, 24: 795–822 and references therein.

Levy, P.M., Morgan-Pond, C., and Fert, A. Origin of anisotropy in transition metal spin glass alloys. *J. Appl. Phys.*, 1982, 53: 2168–2172.

Meiklejohn, W.H., and Bean, C.P. New magnetic anisotropy. *Phys. Rev.*, 1956, 102: 1413–1414.

Meiklejohn, W.H., and Bean, C.P. New magnetic anisotropy. *Phys. Rev.*, 1957, 105: 904–913.

Moriya, T. New mechanism of anisotropic super exchange interaction. *Phys. Rev. Lett.*, 1960, 4: 228–230.

Mydosh, J.A. Spin glasses: redux: an updated experimental/materials survey. *Rep. Progr. Phys.*, 2015, 78: 052501.

Nordblad, P. Competing interaction in magnets: the root of ordered disorder or only frustration? *Phys. Scr.*, 2013, 88: 058301 and references therein.

Nagaosa, N., and Tokura, Y. Topological properties and dynamics of magnetic Skyrmions. *Nat. Nanotechnol.* 2013, 8: 899–911.

Prejean, J.J., Joliclerc, M.J., and Monod, P. Hysteresis in CuMn: the effect of spin-orbit scattering on the anisotropy in the spin glass state. *J. Physique*, 1980, 41: 427–435.

Ruderman, M.A., and Kittel, C. Indirect exchange coupling of nuclear magnetic moments by conduction electrons. *Phys. Rev.*, 1954, 96: 99–102.

Ryu, K.-S., Thomas, L., Yang, S.-H., and Parkin, S. Chiral spin torque at magnetic domain walls. *Nat. Nanotechnol.*, 2013, 8: 527–533.

Yosida, K. Magnetic properties of Cu-Mn alloys. *Phys. Rev.*, 1957, 106: 893–898.

Yu, J. et al. Spin orbit torques and Dzyaloshinskii-Moriya interaction in dual-interfaced Co-Ni multilayers. *Sci. Rep.*, 2016, 6: 32629.

4

Structural Complexity in Exchange-Coupled Core–Shell Nanoparticles

Xiao-Min Lin,
Quy Khac Ong, and
Alexander Wei

4.1   Introduction

The phenomenon of exchange-bias (EB) coupling between two different magnetic materials was first introduced and described 60 years ago, to account for a unidirectional shift in magnetic hysteresis of surface-oxidized Co nanoparticles cooled below a certain temperature (Meiklejohn and Bean, 1956). Although this seminal observation involved core–shell nanoparticles, EB coupling is now more widely studied by the magnetic thin-film community (Nogués and Schuller, 1999), largely because this effect can play an important role in magnetic-recording and data-storage technologies (Ching et al., 1994; Parkin et al., 1999). Recently, however, there is a renewed interest in the EB effects involving core–shell magnetic nanoparticles due to the possibility of utilizing these effects to address two critically important problems in magnetic nanoparticle applications:

(1) the size-dependent superparamagnetic limit, which has been a key bottleneck in using magnetic nanoparticles as memory bits in ultradense data-storage devices (Skumryev et al., 2003) and (2) new approaches to enhance energy product, which is a key parameter for applications of permanent magnet (Quesada et al., 2016).

The intuitive model of EB is that on cooling the antiferromagnetic (AFM) material under an applied field below the Néel temperature, the spontaneous formation of an ordered spin lattice results in a unidirectional magnetic anisotropy, which affects the switching behavior of the adjacent ferromagnetic (FM) layer through exchange coupling. Two manifestations of the EB effect are enhanced coercivity (H_C) and a magnetic loop shift (H_E) for samples that experience field cooling below a transition temperature, T_E, as shown in Figure 4.1. In the case of nanoparticles, thermal energy can overcome the magnetic anisotropy of individual particle, causing magnetic moments inside the particle to switch in unison. The critical temperature at which this behavior occurs is defined as the blocking temperature (T_B). In core–shell nanoparticles, T_B can be strongly influenced by the presence of exchange bias, which influences the overall magnetic anisotropy of the particle. This implies that the magnetic hysteresis loop for core–shell nanoparticles with EB coupling will depend on the temperature at which the sample is measured, relative to T_E and T_B (Figure 4.1).

However, this simple model does not completely capture the essence of EB in thin film and nanoparticle systems. Exchange bias is also highly dependent on AFM domain structure (Borchers et al., 1998), interface roughness, the thickness of AFM and FM layers (Feygenson et al., 2010), and particle's packing density (Nogués et al., 2005, 2006). Furthermore, variations in chemical synthesis can tune the microstructure of core–shell

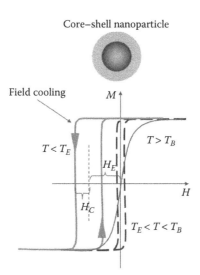

FIGURE 4.1 Magnetic behavior of core–shell nanoparticles in which the sample is field-cooled to temperature T. T_E is the characteristic temperature for materials-dependent exchange bias, and T_B is the blocking temperature above which the entire nanoparticle exhibits superparamagnetic behavior. H_E and H_C are exchange-bias field and coercivity, respectively.

nanoparticles (either rationally or inadvertently), with subsequent impact on magnetic properties. For example, closely related transition metal oxides may be classified as either AFM or ferrimagnetic (FiM), and their formation may depend on the method of oxidation used to produce the shell material.

Compared with thermally evaporated or sputtered thin films prepared in a pristine vacuum environment, most core–shell nanoparticles are prepared by *messy* solution-phase syntheses that involve multiple chemical precursors and stabilizing agents, with a wide range of temperatures used for particle nucleation and growth. Nanoparticle size, shape, internal structure, chemical composition, and degree of aggregation can vary dramatically between laboratories and even between batches within the same laboratory, leading to considerable uncertainty in the measured magnetic properties (Frey et al., 2009). This makes it difficult to compare results from different research groups, even for the same material system. In this chapter, we will illustrate how subtle variations in structural features can affect magnetic behavior, using several different EB-coupled core–shell nanoparticle systems as examples.

4.2 Ferromagnetic–Antiferromagnetic Core–Shell Nanoparticles: Co@CoO

Although EB was first discovered in Co@CoO core–shell nanoparticles, it was only recently that Co nanoparticles could be synthesized with narrow size dispersity (Puntes et al., 2001). Different phases of Co nanoparticles could be produced by using different types of precursors and surface ligands, including hexagonal close-packed (hcp), multi-twinned face-centered cubic (fcc), and a novel phase called ε-Co (Murray et al., 2001). Systematic studies of the EB effect in Co nanoparticles with different degrees of surface oxidation have been carried out (Gangopadhyay et al., 1993; Tracy et al., 2005; Feygenson et al., 2010). Overall, these measurements indicate that the EB field in Co@CoO core–shell nanoparticles has a strong dependence on CoO shell thickness, reaching a maximum value of 7 kOe at 30 K (Figure 4.2). On the other hand, slight variations in nanoparticle synthesis and sample preparation can introduce complications in the magnetic measurements. In some studies, a paramagnetic component was observed to be part of the overall magnetic response, attributed to either enhanced moment in the shell (Chen et al., 1995) or small clusters of Co in the diffusion layer between the core and the shell (Tracy et al., 2005). However, this paramagnetic component could also come from other sources, such as from Co^{2+} ions that are not completely reduced (Lin, 1999) or from clusters created by reductive etching of the CoO layer by excess ligand (Samia et al., 2005), thus becoming contaminants in the final product.

A relatively clean dataset on progressively oxidized Co@CoO nanoparticles (~11 nm) was presented by Aronson and coworkers (Feygenson et al., 2010). In Figure 4.2, field-cooled (FC) magnetic hysteresis loops for samples with different degrees of oxidation indicate that both the oxide thickness and internal core diameter have a significant impact on exchange bias. When the thickness of the oxide layer is comparable to the core diameter, both H_E and H_C are maximized. This non-monotonic behavior is intuitively easy to understand: when the shell is thin, it may not provide enough EB coupling to the FM core, whereas a

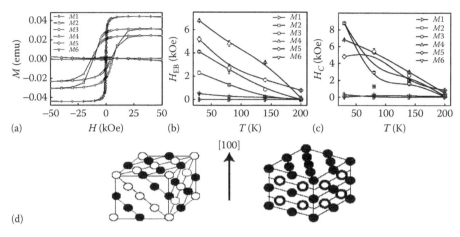

(d)

FIGURE 4.2 (a) Field-dependent magnetization of six samples of Co@CoO core–shell nanoparticles that have been oxidized to different degrees, measured at 30 K. The oxide thickness increases progressively from *M*1 to *M*6. (b) Temperature dependence of the exchange-bias field, H_E, for all six samples. Solid lines are guides for the eye. (c) Temperature dependence of the coercive field, H_C, for the same six samples. (From Feygenson, M. et al., *Phys. Rev. B*, 81, 195445, 2010.) (d) Magnetic modulation of CoO shell along two different directions [1/2,1/2,1/2] and [100] as determined by neutron scattering. The magnetic modulation along [100] direction becomes much stronger when the shell thickness is decreased, leading to uncompensated spins along the (100) interface. (From Inderhees, S.E. et al., *Phys. Rev. Lett.*, 101, 117202, 2008.)

completely oxidized particle should behave like an AFM material. However, the mechanism of the initial increase in H_E with increasing thickness is not as straightforward. There is some evidence that the Néel temperature of the AFM particle or layer depends on its size (Kodama, 1999; Sako et al., 2001), such that the strength of the EB effect decreases with thickness. However, it has also been argued that the tetragonal strain at the interface is a more dominant effect than layer thickness (Inderhees et al., 2008; Feygenson et al., 2010). This tetragonal strain can be much larger than bulk CoO (the thinner the CoO shell, the larger the tetragonal strain) and can force the canting of interface moments and decompensation of the (100) interface planes. Uncompensated spin moments at the AFM–FM interface can enhance exchange coupling, but a minimum thickness of the AFM layer is needed to produce an effective bias field against FM moment rotation under magnetic field switching. An AFM layer of intermediate thickness provides the best trade-off between these two effects and should thus produce the largest EB.

The T_E in AFM–FM coupled systems is not only largely dependent on the bulk Néel temperature (T_N) of the AFM material but is also influenced by AFM layer thickness and composition, such that the measured T_E can differ significantly from T_N. For samples shown in Figure 4.2, T_E is near 200 K, much lower than the T_N value of 291 K for bulk CoO. The lower T_E is likely due to its dependency on layer size or thickness. Furthermore, if the cobalt surface layer is not completely oxidized into CoO, a mixed Co_3O_4–CoO phase can form and push the measured T_E even lower, as bulk Co_3O_4 has a much lower Néel temperature ($T_N = 40$ K) than that of CoO.

4.3   Ferromagnetic–Ferrimagnetic Core–Shell Nanoparticles: Fe@Fe$_x$O$_y$

Two of the most common FM–FiM systems are represented by Fe@Fe$_3$O$_4$ and Fe@γ-Fe$_2$O$_3$ core–shell nanoparticles. These nanoparticles are often synthesized by the thermolysis of iron pentacarbonyl (Fe(CO)$_5$) in the presence of passivating ligands (Peng et al., 2006; Khurshid et al., 2010). Nanoparticle crystallinity and particle size apparently depend on the type of coordinating surfactant used and the decomposition temperature. Thermolysis of Fe(CO)$_5$ in the presence of oleylamine at 180°C, followed by exposure to air, produces core–shell Fe@Fe$_3$O$_4$ nanoparticles with amorphous metal cores (Figure 4.3) (Peng et al., 2006), whereas higher reaction temperatures, using oleic acid or trioctylphosphine as co-surfactants, promote core crystallization (Khurshid et al., 2010). Although there are some conflicting reports whether the shell is Fe$_3$O$_4$, γ-Fe$_2$O$_3$, or a mixture of two phases, the magnetic properties are not overly affected by subtle differences in chemical composition, as both Fe$_3$O$_4$ and γ-Fe$_2$O$_3$ are FiM materials.

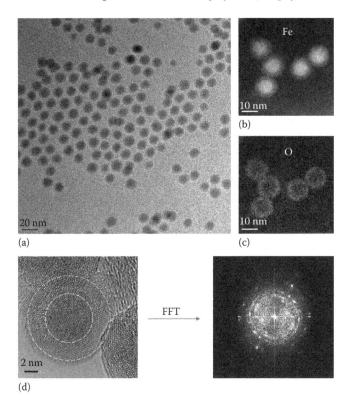

FIGURE 4.3   (a) Bright-field TEM image of Fe@Fe$_3$O$_4$ core–shell nanoparticles. (b, c) Energy-filtered TEM mapping of Fe and Co distribution of several core–shell nanoparticles. (d) High-resolution TEM image with fast Fourier transform (FFT) image of a single Fe@Fe$_3$O$_4$ nanoparticles, indicating an amorphous core and a polycrystalline shell. (From Ong, Q.K. et al., *Phys. Rev. B*, 80, 134418, 2009.)

Kirkendall effect

Fe-Fe$_3$O$_4$ core-shell nanoparticles $\xrightarrow[\text{2 h 240°C}]{\text{Me}_3\text{NO}}$ Fe$_3$O$_4$ hollow-shell nanoparticles

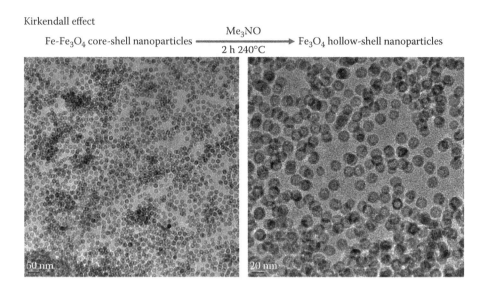

FIGURE 4.4 Hollow Fe$_3$O$_4$ nanoparticles, prepared by the controlled oxidation of Fe@Fe$_3$O$_4$ core–shell nanoparticles with trimethylamine N-oxide at 240°C. (From Ong, Q.K. et al., *Phys. Rev. B*, 80, 134418, 2009.)

If the Fe@Fe$_x$O$_y$ nanoparticles are subsequently exposed to oxygen or trimethylamine N-oxide, the core–shell nanoparticles will further evolve into hollow-shell nanoparticles (Figure 4.4). Void formation within core–shell nanoparticles is driven by the Kirkendall effect, which can be described as an imbalance in diffusion rates between metal atoms and oxidized species (Anderson and Tracy, 2014).

The controlled synthesis of both core–shell and hollow iron oxide nanoparticles have allowed us to compare their magnetic properties in the context of the EB effects below T_B ($T < 30$ K) (Ong et al., 2009, 2011). Magnetic measurements were conducted at both zero-field-cooled (ZFC) and FC conditions; the latter involved an applied field while cooling the sample to 5 K. The magnetic hysteresis loops of core–shell and hollow nanoparticles are shown in Figure 4.5. There are several interesting features that can be observed. One feature is a large EB-induced loop shift for FM–FiM nanoparticles under FC conditions, whereas the loop shift for hollow FiM nanoparticles is much smaller. In contrast, neither core–shell nor hollow nanoparticles exhibit a loop shift under ZFC conditions. Another observation is that the hysteresis loop of core–shell nanoparticles features an abrupt demagnetization or *jump* at low field (ΔM), which is evident in both positive- and negative-field sweep directions and under both FC and ZFC conditions. This sudden decrease in magnetic moment is more clearly distinguished by a sharp peak near zero field when dM/dH is plotted (Ong et al., 2009).

Low-field jumps are observed in both core–shell and hollow nanoparticles, which indicate that their origins lie on or within the FiM shell. High-resolution transmission electron microscopy (HRTEM) confirmed that the Fe$_3$O$_4$ shells are composed of randomly oriented nanocrystalline domains that span the thickness of the shell. This suggests that the low-field jump may be due to a sudden reorientation of moments along their

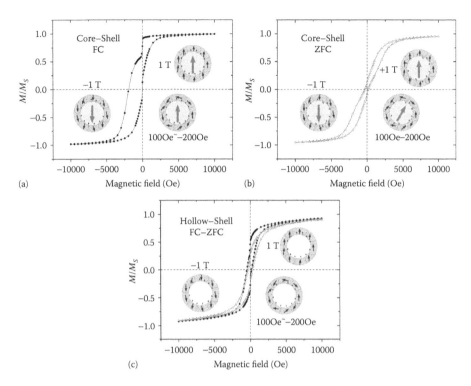

FIGURE 4.5 Magnetic hysteresis loops measured at 5K for (a) Fe@Fe$_3$O$_4$ core–shell nanoparticles with field-cooled (FC), (b) Fe@Fe$_3$O$_4$ core–shell nanoparticles with zero-field-cooled (ZFC), and (c) hollow Fe$_3$O$_4$ nanoparticles (FC and ZFC). Insets illustrate net orientations of spin moments within various domains of the core–shell nanoparticles. (From Ong, Q.K. et al., *Phys. Rev. B*, 80, 134418, 2009.)

respective easy axes, when the magnetocrystalline anisotropy becomes dominant at low field. For core–shell nanoparticles subject to FC, the strong coupling between the core and the shell layer is mediated by a uniaxially aligned interfacial spin layer that pins the magnetic moments along the applied direction above a threshold field strength (Figure 4.5a). When the applied field is sufficiently reduced (<200 Oe), a spontaneous randomization of shell moments can occur, accounting for up to 70% of the total shell moment. On the other hand, core–shell nanoparticles subject to ZFC or hollow nanoparticles in general lack a strong coupling between the core and shell moments. In these cases, the shell moments are not strongly pinned along the applied field's direction; only ~20% of the total moment participates in the spontaneous randomization at low field.

For hollow FiM shells, the local orientation of surface spins plays an important role in mediating the low-field magnetization jump. Surface spins of typical FiM spinel nanoparticles are known to be disordered, attributed to broken exchange bonds (Kodama et al., 1996). These spins are sensitive to local anisotropy influences and can be trapped at low temperatures. Thus, when the sample is cooled under a large field, the surface/interfacial spins tend to be aligned with the magnetic field and trapped with uniaxial orientation at low temperature, even after the magnetic field is removed, whereas those same spins are trapped in random orientations when the sample is cooled at zero field (Figure 4.5a and b, insets). This is evident in a comparison of magnetization–temperature (*M–T*) curves for FC versus ZFC, using a cooling field of 1 T and an applied field of 1 T for *M–T* measurements (Figure 4.6). A split between ZFC and FC curves is evident below 30 K, indicating that some of these surface/interfacial spins are trapped in random orientations under ZFC conditions and are unresponsive to the applied measuring field. At higher temperatures (*T* > 30K), these previously *frozen* spins can overcome local barriers and reorient along the applied field, allowing the ZFC and FC curves to merge. Similar divergences of ZFC–FC curves at high magnetic field and low temperature have also been observed with solid FiM nanoparticles and are attributed to similar arguments (Kodama et al., 1996). The power law in the *M–T* relationship above 30 K is nearly quadratic [$M = M_s(1-bT^{1.95})$], similar to the temperature dependency in bulk materials caused by spin-wave excitation (Figure 4.6, inset). Although the power law decay exponent of 1.95 deviates from the bulk value of 1.5, it is consistent with theoretical predictions based on the reduced density of states in nanoparticles (Hendriksen et al., 1993).

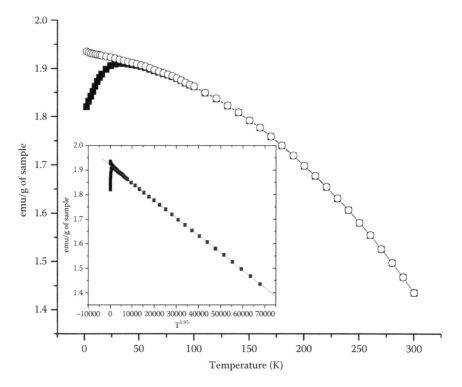

FIGURE 4.6 Magnetization versus temperature curve for Fe@Fe₃O₄ core–shell nanoparticles after FC (open circles) and after ZFC (filled squares), with a cooling and measuring field of 10 kOe. (From Ong, Q.K. et al., *Phys. Rev. B*, 80, 134418, 2009.)

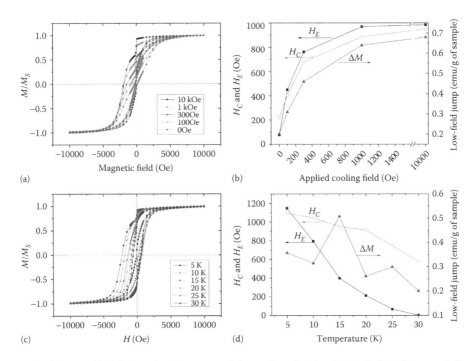

FIGURE 4.7 Modulating the alignment of frozen interfacial spins in Fe@Fe$_3$O$_4$ core–shell nanoparticles, as functions of (a, b) applied cooling field and (c, d) FC temperatures. In both cases, coercivity (H_C) and exchange-bias field (H_E) correlate with changes in the relative degree of interfacial spin alignment. (From Ong, Q.K. et al., *Phys. Rev. B*, 80, 134418, 2009.)

There is a strong correlation between the percentage of frozen interfacial spins aligned with the cooling field and the EB field in core–shell nanoparticles under FC conditions. This can be clearly observed in several experiments. First, different degrees of alignment can be established as a function of FC strength because of the large local anisotropy in interfacial spins (Figure 4.7a and b). As a consequence, both the coercivity (H_C) and the EB field (H_E) decrease in proportion to the cooling field. Second, samples that are field-cooled to different temperatures below 30 K show different degrees of interfacial spin alignment, with lower H_C and H_E values observed at higher temperatures (Figure 4.7c and d). This is consistent with the notion that the degree of alignment of interfacial spins with the applied field has a direct impact on the strength of EB coupling between the core and shell layers.

A quantitative approach for characterizing the relative number of surface/interfacial spins aligned with the cooling field is to measure the strength of the net moment that remain unchanged during magnetic switching, that is, $M_f = (M_+ + M_-)/2$, where M_+ and M_- are the saturation magnetizations in the positive (along the cooling field) and negative field directions, respectively. Switching the applied field back and forth between large positive and negative values ($\pm H_{\max}$) gradually decreases the number of frozen spins that are initially aligned in the FC direction, a phenomenon known as the training effect. Figure 4.8 shows that the population of aligned frozen spins decays exponentially

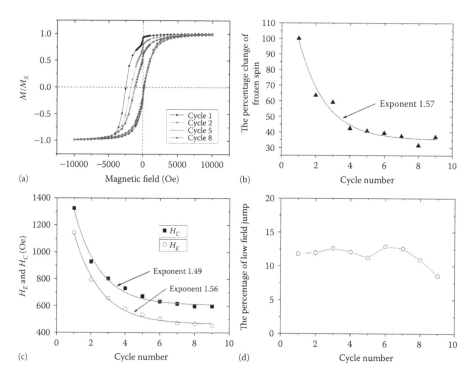

FIGURE 4.8 Training effect on exchange bias in Fe@Fe$_3$O$_4$ nanoparticles. (a) Changes in the magnetic hysteresis loop of dispersed nanoparticles over multiple measurement cycles. For clarity, only data from cycles 1, 2, 5, and 8 are shown. (b) Relative decrease in M_f as a function of cycle number (from an initial M_f/M_+ ratio of 0.33%). (c) Decrease in H_C and H_E as a function of cycle number. Exponential curve fits in (b) and (c) indicated by solid lines. (d) The relative intensity of the low-field demagnetization ($\Delta M/M_+$) is insensitive to the number of measurement cycles. (From Ong, Q.K. et al., *J. Phys. Chem. C*, 115, 2665–2672, 2011.)

with cycle number: the M_f/M_+ ratio decreases to 37% of its initial value after 9 cycles, corresponding to a relaxation constant of 1.57 (Figure 4.8b). The EB field, H_E, and coercivity, H_C, also decrease exponentially with the number of hysteresis cycles, with decay constants of 1.56 and 1.49, respectively (Figure 4.8c). The H_E relaxation has the same decay exponent as that associated with changes in frozen-spin population, supporting a strong correlation between these two parameters.

Further oxidation of Fe@Fe$_3$O$_4$ nanoparticles can induce a structural evolution from core–shell morphologies to partially hollow *yolk–shell* nanoparticles and eventually to entirely hollow nanoparticles. The morphology of nanoparticles at these intermediate oxidation stages can vary greatly, depending on oxidation conditions. For example, Fe@Fe$_3$O$_4$ nanoparticles stored in a matrix of ionic surfactant under ambient conditions for 1.5 years produced visible changes in their structure, based on TEM analysis (Figure 4.9). These aged nanoparticles exhibit higher contrast at the core–shell interface relative to freshly prepared Fe@Fe$_3$O$_4$ nanoparticles, initially suggestive of a detached core with a concentric shell (Figure 4.9a). However, careful profiling of the nanoparticle for

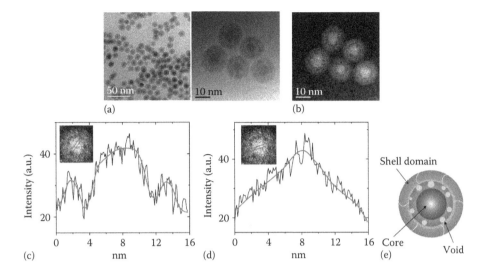

FIGURE 4.9 Fe@Fe$_3$O$_4$ nanoparticles after 1.5 years of storage in a didodecyldimethylammonium bromide matrix under ambient conditions. (a) TEM and HRTEM images and (b) EFTEM image on Fe distribution. (c, d) Cross section of intensity profile on a single aged Fe@Fe$_3$O$_4$ nanoparticle along slightly different directions. (e) Structural model of an aged core–shell nanoparticle. (From Ong, Q.K. et al., *J. Phys. Chem. C*, 115, 2665–2672, 2011.)

Fe distribution by energy-filtered TEM indicates that the core–shell interface is, in fact, mostly intact but with some small (≤ 2 nm) voids formed between layers (Figure 4.9b–d). This structural evolution is again attributable to the nanoscale Kirkendall effect, in which the oxidation-driven migration of metal atoms from the core to the shell produces vacancies at the interface that gradually coalesce into voids (Ong et al., 2011).

The subtlety of these structural changes belies their strong influence on the magnetic properties of the aged core–shell nanoparticles. Magnetic hysteresis loops measured at 5 K after 1 T field cooling is shown in Figure 4.10a for the same sample at the different stages of oxidation. After a 1.5-year storage period, the low-field jump ($\Delta M / M_+$) disappeared and was accompanied by a 25% loss in saturation magnetization; on the other hand, H_C and H_E increased slightly, and a dramatic increase in the relative percentage of M_f was observed. For freshly made Fe@Fe$_3$O$_4$ samples, roughly 1% of all moments are frozen after cooling the sample down to 5 K; however, this number increased to 6% after aging for 1.5 years (Figure 4.10b).

These effects were confirmed by the analysis of multiple Fe@Fe$_3$O$_4$ samples, all of which exhibit variable rates of change with age. The large increase of frozen interfacial spins can be attributed to the higher surface area at the interface, created by the growing number of voids over time. However, void formation also causes the interface between the FM core and the FiM shell to become highly corrugated, such that not all the interfacial spins are available to mediate exchange coupling between the core and the shell. The roughened and porous interface also means that H_C and H_E do not increase proportionally with the percentage of frozen interfacial spins, unlike the presumably seamless interface of freshly prepared core–shell nanoparticles. If, instead of air, the core–shell

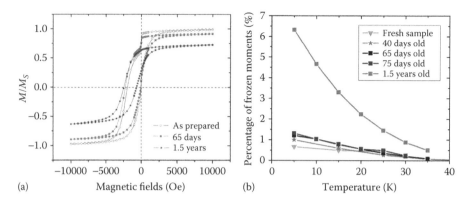

FIGURE 4.10 Changes in the magnetic properties of Fe@Fe$_3$O$_4$ core–shell nanoparticles at different stages of oxidation, during storage in a didodecyldimethylammonium bromide matrix under ambient conditions. (a) Magnetic hysteresis loop measured at 5 K after field-cooling at 1 T. (b) Changes in the percentage of frozen spin moments versus temperature. (From Ong, Q.K. et al., *J. Phys. Chem. C*, 115, 2665–2672, 2011.)

nanoparticles are exposed to a more aggressive oxidation agent such as trimethylamine N-oxide, the FM and FiM regions can become almost completely decoupled with sufficient oxidation. In these samples, the EB coupling will also be greatly diminished, proving that it is a short-range interaction that decays exponentially over distance.

4.4 Antiferromagnetic–Ferrimagnetic Core–Shell Nanoparticles: Fe$_x$O@CoFe$_2$O$_4$ and Fe$_x$O@Fe$_3$O$_4$

For FiM nanoparticles, the presence of frozen surface spins is enough to provide some EB coupling with bulk moments, as demonstrated with hollow Fe$_3$O$_4$ nanoparticles (Figure 4.5c). However, H_E is not large and quickly diminishes to zero for temperatures above the threshold *melting point* of frozen spins. The EB coupling can be significantly enhanced by an adjoining AFM layer: an excellent example is shown with monodisperse Fe$_x$O@CoFe$_2$O$_4$ core–shell nanoparticles ($x \sim 1$) prepared by the thermolysis of metal oleates (Bodnarchuk et al., 2009) (Figure 4.11). Magnetic measurements below the Néel temperature of FeO ($T_N \approx 200$ K) with field cooling reveal several unique features arising from the strong interfacial EB coupling between the AFM core and the FiM shell. Unlike solid FiM nanoparticles (Chalasani and Vasudevan, 2011), the Fe$_x$O@CoFe$_2$O$_4$ nanoparticles exhibit anomalously large vertical shifts in their hysteresis loops and large H_E values that persist up to 200 K (Figure 4.12). This suggests that EB is dictated by spins on the AFM core rather than by interfacial spins on the FiM layer. Furthermore, most spins in the FiM layer are strongly exchange-coupled with the AFM core and exhibit great reluctance to switch when the magnetic field is reversed, resulting in a large vertical shift in the hysteresis loop and a substantial H_E (Figure 4.12). It is worth noting that it is possible to raise T_B above room temperature by changing nanoparticle size or shape, meaning that these core–shell nanoparticles have finite coercivity, an important feature for potential applications in magnetic storage.

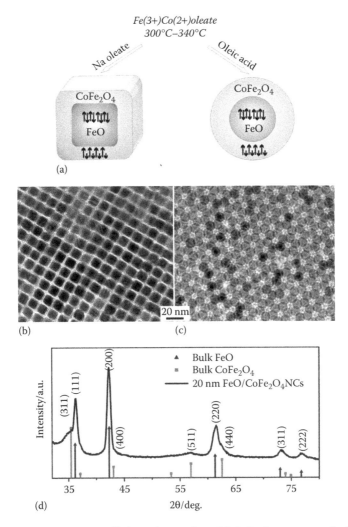

FIGURE 4.11 (a) Shape-controlled synthesis of $Fe_xO@CoFe_2O_4$ nanocrystals (NCs) with AFM–FiM core–shell structures. (b, c) TEM images of superlattice films formed from 11-nm cubic and 11-nm spherical $Fe_xO@CoFe_2O_4$ NCs, respectively. (d) Powder X-ray diffraction (XRD) pattern of 20 nm $Fe_xO@CoFe_2O_4$ NCs. (From Bodnarchuk, M.I. et al., *Small*, 5, 2247–2252, 2009.)

In an independent study, $Co_{0.3}Fe_{0.7}O@Co_{0.6}Fe_{2.4}O_4$ core–shell nanoparticles were synthesized by a one-pot thermolysis of iron and cobalt precursors, with sizes ranging from 6 to 18 nm (Lottini et al., 2016). The core has the AFM character due to the low stoichiometry of cobalt, and the EB coupling is very large ($H_E = 8.6$ kOe at 10 K), similar to the $FeO@CoFe_2O_4$ system.

The oxidation of Fe_xO nanoparticles into $Fe_xO@Fe_3O_4$ core–shell nanoparticles also gives rise to an AFM–FiM system; however, the EB effects reported by different research groups are variable (Kavich et al., 2008; Sun et al., 2012; Lak et al., 2013).

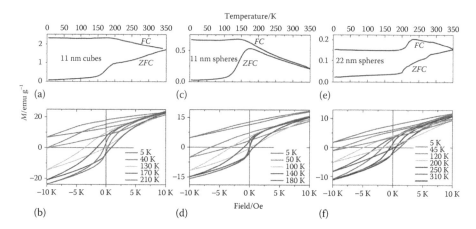

FIGURE 4.12 ZFC–FC magnetization curves for $Fe_xO@CoFe_2O_4$ NCs, measured with 100 Oe: (a) 11 nm cubes, (c) 11 nm spheres; and (e) 22 nm spheres. Hysteresis curves at various temperatures for the same samples, after cooling in a 10 kOe field: (b) 11 nm cubes, (d) 11 nm spheres; and (f) 22 nm spheres. Vertical loop shifts can be observed in all samples at temperatures below 100 K. (From Bodnarchuk, M.I. et al., *Small*, 5, 2247–2252, 2009.)

Recent experiments comparing two different sizes of $Fe_xO@Fe_3O_4$ nanoparticles demonstrate the influence of several experimental factors, including particle size, the stoichiometry of oxygen in the Fe_xO core, and the internal strain related to the core–shell structure (Figure 4.13) (Estrader et al., 2015). In particular, the iron content in small (~9 nm) spherical Fe_xO nanoparticles is sub-stoichiometric, as probed by neutron scattering ($x \sim 0.80$), causing the Fe_xO core to be nonmagnetic. Therefore, the EB effect in smaller $Fe_xO@Fe_3O_4$ nanoparticles is similar to that observed in Fe_3O_4 nanoparticles and appears to be dominated by interfacial frozen spins that dissipate at temperatures above 30 K. On the other hand, the stoichiometry of Fe_xO in larger (45 nm) core–shell

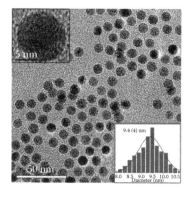
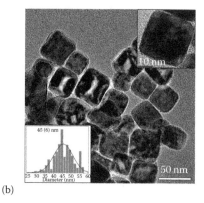

(a)   (b)

FIGURE 4.13 TEM images of (a) $Fe_xO@Fe_3O_4$ spherical core–shell nanoparticles (9 nm), and (b) $Fe_xO@Fe_3O_4$ nanocubes (45 nm).   (*Continued*)

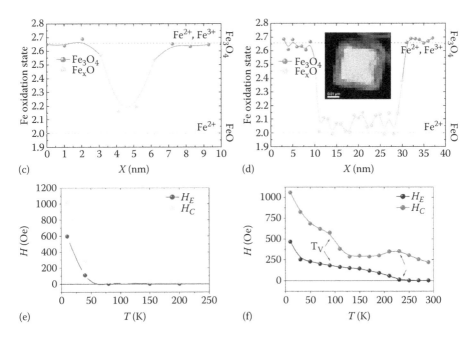

FIGURE 4.13 (Continued) TEM images of (c, d) Fe oxidation states along the particle cross-section of $Fe_xO@Fe_3O_4$ nanospheres (9 nm) and nanocubes (45 nm), respectively. (e, f) Temperature dependence of coercivity (H_C) and loop shift (H_E) for $Fe_xO@Fe_3O_4$ nanospheres (9 nm) and nanocubes (45 nm), respectively. (From Estrader, M. et al., *Nanoscale*, 7, 3002–3015, 2015.)

nanocubes is much closer to unity ($x \sim 0.95$), permitting the core to exhibit AFM behavior. In this case, the EB effect is dominated by the AFM core and persists until the temperature approaches 240 K, which is slightly higher than the Néel temperature for bulk FeO. A similar observation was made in a second, independent study by using smaller (9-nm) $Fe_xO@Fe_3O_4$ nanocubes, with a large EB effect, up to 200 K (Hai et al., 2010). This shows that slight variations in the preparation of experimental samples can affect the stoichiometry of the Fe_xO layer, with significant consequences on its magnetic structure.

4.5 Conclusions

In summary, a wide variety of core–shell magnetic nanoparticles exhibit EB effects. The structural complexities of the core–shell interface have significant ramifications on the magnetic coupling between these two layers, including but not limited to variations in chemical composition, interfacial strain effects, and local structural defects. The examples presented earlier represent just the tip of an iceberg of phenomena that are affected by magnetic exchange bias. Many unresolved issues remain concerning the character of the nanoparticle shell layer, which is essentially a 3D surface: for instance, local strains induced by polycrystallinity and lattice mismatch with the core material can lead to heterogeneous shell growth (Liu et al., 2015). Variations in intrinsic parameters (e.g., oxidation state) and extrinsic parameters (e.g., cooling field and temperature) can reveal novel manifestations of the EB phenomena, such as AFM coupling across the core–shell interface and a positive EB field (Nogues et al., 2000; Estrader et al., 2013). Microstructural changes at the core–shell interface are often subtle, yet they have a critical influence on the overall magnetic properties of the nanoparticle. Thus, there is fertile ground for the future development of synthetic techniques and analytical methods that can be applied toward the elucidation of such structure–property relationships.

Acknowledgments

Much of this work was performed at the Center of Nanoscale Materials, a U.S. Department of Energy, Office of Science, Office of Basic Energy Sciences User Facility under Contract No. DE-AC02-06CH11357. We also gratefully acknowledge the support from the National Science Foundation (CHE-0957738) and the National Institutes of Health (RC1 CA-147096).

References

Anderson BD, Tracy JB (2014) Nanoparticle conversion chemistry: Kirkendall effect, galvanic exchange, and anion exchange. *Nanoscale*, 6:12195–12216.

Bodnarchuk MI, Kovalenko MV, Groiss H, Resel R, Reissner M, Hesser G, Lechner RT, Steiner W, Schaffler F, Heiss W (2009) Exchange-coupled bimagnetic wustite/metal ferrite core/shell nanocrystals: Size, shape, and compositional control. *Small*, 5:2247–2252.

Borchers JA, Ijiri Y, Lee SH, Majkrzak CF, Felcher GP, Takano K, Kodama RH, Berkowitz AE (1998) Spin-flop tendencies in exchange-biased Co/CoO thin films. *J Appl Phys*, 83:7219–7221.

Chalasani R, Vasudevan S (2011) Form, content, and magnetism in iron oxide nanocrystals. *J Phys Chem C*, 115:18088–18093.

Chen JP, Sorensen CM, Klabunde KJ, Hadjipanayis GC (1995) Enhanced magnetization of nanoscale colloidal cobalt particles. *Phys Rev B*, 51:11527–11532.

Ching T, Fontana RE, Tsann L, Heim DE, Speriosu VS, Gurney BA, Williams ML (1994) Design, fabrication and testing of spin-valve read heads for high density recording. *IEEE Trans Magn*, 30:3801–3806.

Estrader M et al. (2013) Robust antiferromagnetic coupling in hard-soft bi-magnetic core/shell nanoparticles. *Nat Commun*, 4:2960

Estrader M et al. (2015) Origin of the large dispersion of magnetic properties in nanostructured oxides: $Fe_xO/Fe3O4$ nanoparticles as a case study. *Nanoscale*, 7:3002–3015.

Feygenson M, Yiu Y, Kou A, Kim K-S, Aronson MC (2010) Controlling the exchange bias field in Co core/CoO shell nanoparticles. *Phys Rev B*, 81:195445.

Frey NA, Peng S, Cheng K, Sun S (2009) Magnetic nanoparticles: Synthesis, functionalization, and applications in bioimaging and magnetic energy storage. *Chem Soc Rev*, 38:2532–2542.

Gangopadhyay S, Hadjipanayis GC, Sorensen CM, Klabunde KJ (1993) Exchange anisotropy in oxide passivated Co fine particles. *J Appl Phys*, 73:6964–6966.

Hai HT, Yang HT, Kura H, Hasegawa D, Ogata Y, Takahashi M, Ogawa T (2010) Size control and characterization of wustite (core)/spinel (shell) nanocubes obtained by decomposition of iron oleate complex. *J Colloid Interf Sci*, 346:37–42.

Hendriksen PV, Linderoth S, Lindgard PA (1993) Magnetic-properties of Heisenberg clusters. *J Phys-Condens Matter*, 5:5675–5684.

Inderhees SE, Borchers JA, Green KS, Kim MS, Sun K, Strycker GL, Aronson MC (2008) Manipulating the magnetic structure of Co Core/CoO shell nanoparticles: Implications for controlling the exchange bias. *Phys Rev Lett*, 101:117202.

Kavich DW, Dickerson JH, Mahajan SV, Hasan SA, Park JH (2008) Exchange bias of singly inverted FeOFe$_3$O$_4$ core-shell nanocrystals. *Phys Rev B*, 78:174414.

Khurshid H, Tzitzios V, Li W, Hadjipanayis CG, Hadjipanayis GC (2010) Size and composition control of core-shell structured iron/iron-oxide nanoparticles. *J Appl Phys*, 107:09A333.

Kodama RH (1999) Magnetic nanoparticles. *J Magn Magn Mater*, 200:359–372.

Kodama RH, Berkowitz AE, McNiff JEJ, Foner S (1996) Surface spin disorder in NiFe$_2$O$_4$ nanoparticles. *Phys Rev Lett*, 77:394–397.

Lak A, Kraken M, Ludwig F, Kornowski A, Eberbeck D, Sievers S, Litterst FJ, Weller H, Schilling M (2013) Size dependent structural and magnetic properties of FeO-Fe$_3$O$_4$ nanoparticles. *Nanoscale*, 5:12286–12295.

Lin XM (1999) unpublished results.

Liu XJ, Pichon BP, Ulhaq C, Lefevre C, Greneche JM, Begin D, Begin-Colin S (2015) Systematic study of exchange coupling in core shell Fe$_3$O$_4$@CoO nanoparticles. *Chem Mater*, 27:4073–4081.

Lottini E, Lopez-Ortega A, Bertoni G, Turner S, Meledina M, Van Tendeloo G, Fernandez CD, Sangregorio C (2016) Strongly exchange coupled ore/shell nanoparticles with high magnetic anisotropy: A strategy toward rare-earth-free permanent magnets. *Chem Mater*, 28:4214–4222.

Meiklejohn WH, Bean CP (1956) New magnetic anisotropy. *Phys Rev*, 102:1413–1414.

Murray CB, Sun SH, Gaschler W, Doyle H, Betley TA, Kagan CR (2001) Colloidal synthesis of nanocrystals and nanocrystal superlattices. *IBM J Res Dev*, 45:47–56.

Nogues J, Leighton C, Schuller IK (2000) Correlation between antiferromagnetic interface coupling and positive exchange bias. *Phys Rev B*, 61:1315–1317.

Nogués J, Schuller IK (1999) Exchange bias. *J Magn Magn Mater*, 192:203–232.

Nogués J, Skumryev V, Sort J, Stoyanov S, Givord D (2006) Shell-driven magnetic stability in core-shell nanoparticles. *Phys Rev Lett*, 97:157203.

Nogués J, Sort J, Langlais V, Skumryev V, Suriñach S, Muñoz JS, Baró MD (2005) Exchange bias in nanostructures. *Phys Rep*, 422:65–117.

Ong QK, Wei A, Lin X-M (2009) Exchange bias in Fe/Fe$_3$O$_4$ core-shell magnetic nanoparticles mediated by frozen interfacial spins. *Phys Rev B*, 80:134418.

Ong QK, Lin X-M, Wei A (2011) The role of frozen spins in the exchange anisotropy of core–shell Fe@Fe$_3$O$_4$ nanoparticles. *J Phys Chem C*, 115:2665–2672.

Parkin SSP et al. (1999) Exchange-biased magnetic tunnel junctions and application to nonvolatile magnetic random access memory (invited). *J Appl Phys*, 85:5828–5833.

Peng S, Wang C, Xie J, Sun S (2006) Synthesis and stabilization of monodisperse Fe nanoparticles. *J Am Chem Soc*, 128:10676–10677.

Puntes VF, Krishnan KM, Alivisatos AP (2001) Colloidal nanocrystal shape and size control: The case of cobalt. *Science*, 291:2115–2117.

Quesada A et al. (2016) Energy product enhancement in imperfectly exchange-coupled nanocomposite magnets. *Adv Electron Mater*, 2:1500365.

Sako S, Ohshima K, Sakai M (2001) Magnetic property of oxide passivated Co nanosized particles dispersed in two dimensional plane. *J Phys Soc Japan*, 70:2134–2138.

Samia ACS, Hyzer K, Schlueter JA, Qin C-J, Jiang JS, Bader SD, Lin X-M (2005) Ligand effect on the growth and the digestion of Co nanocrystals. *J Am Chem Soc*, 127:4126–4127.

Skumryev V, Stoyanov S, Zhang Y, Hadjipanayis G, Givord D, Nogues J (2003) Beating the superparamagnetic limit with exchange bias. *Nature*, 423:850–853.

Sun XL, Huls NF, Sigdel A, Sun SH (2012) Tuning exchange bias in core/shell FeO/Fe$_3$O$_4$ nanoparticles. *Nano Lett*, 12:246–251.

Tracy JB, Weiss DN, Dinega DP, Bawendi MG (2005) Exchange biasing and magnetic properties of partially and fully oxidized colloidal cobalt nanoparticles. *Phys Rev B*, 72:064404.

5

Exchange-Bias Effect in Manganite Nanostructures

A. Wisniewski,
I. Fita, R. Puzniak,
and V. Markovich

5.1 Properties of Manganites: Nanoparticles versus Bulk Samples

Magnetic nanoparticles (NPs) of manganites were recently intensively studied due to both the involved physics and the potential technological applications. When the size of magnetic NPs is reduced to the nanometer scale, many of their basic magnetic properties, for example, spontaneous magnetization, the magnetic transition temperature, and coercivity, differ significantly from the bulk values and become strongly dependent on the particle size. Effect of defects, broken bonds, fluctuations in the number of atomic neighbors and interatomic distances, causing topological and magnetic disorder at particle surfaces, becomes more and more pronounced with downsizing and results in changes of the magnetic properties of the system.

It is well established in bulk samples that hydrostatic pressure affects saturation magnetization, transition temperatures, and unit cell volume. Sarkar et al. (2008a) suggested that surface pressure acts on the nanocrystals like high hydrostatic pressure. They estimated that induced pressure is equivalent to about 6 GPa hydrostatic pressure for manganite NPs with 15 nm in diameter and falls below 1 GPa for particles larger than 100 nm. Martinelli et al. (2013) studied the crystal and magnetic structures of 10- and

20-nm-sized $La_{1-x}Ca_xMnO_3$ ($x = 0.37, 0.50$, and 0.75) NPs by means of Rietveld refinement of neutron powder diffraction data, coupled with transmission electron microscope (TEM) observation and magnetization measurements. Their TEM observation revealed that TEM analyzes evidence invalidity of the core–shell picture in the studied samples: NPs are strongly affected by strain fields, probably originating from surface pressure, that distort the crystal lattice and produce about 3- to 4-nm-sized strained regions. According to the authors, these strain fields promote phase separation, which takes place on cooling.

Usually, magnetic particles with reduced size are of single domain and exhibit superparamagnetic behavior. However, interparticle interactions that induce collective behavior may strongly modify magnetic properties of NPs' assemble. Interparticle interactions are strongly dependent on volume concentration of particles. Pramanik and Banerjee (2010) have showed that these interactions also depend on NPs' size. They studied three samples consisting of NPs of half-doped $Pr_{0.5}Sr_{0.5}MnO_3$ with different average particle size (ranging from 15.7 to 26.6 nm). Alternating current (AC) and direct current (DC) magnetic measurements have shown the superparamagnetic behavior of these NPs, and it was noticed that the blocking temperature increases with the increasing particle size. The presence of interparticle interaction was confirmed by the temperature variation in coercive field and the analysis of frequency-dependent ac susceptibility. The authors proved that this interaction is of dipolar type and showed that its strength decreases with the increasing particle size. The interparticle interaction give rise to the unidirectional anisotropy, which was confirmed by the presence of the exchange bias (EB) effect.

Electron-doped $RE_{1-x}Ca_xMnO_3$ (RE—rare earth, $x > 0.5$) manganites are a good example of the material with new magnetic properties that emerge as a consequence of a reduction from bulk to nanometric size. Since charge ordering (CO) and the subsequent antiferromagnetic (AFM) arrangements are long-range-order phenomena, the reduction of crystallite size dimension causes changes in the magnetic and electronic ordering. The suppression of the AFM state and the suppression/disappearance of CO occur for basically AFM electron-doped materials at some critical size, while at the surface, short-range ferromagnetic (FM) clusters appear. The uncompensated surface spins favor the FM coupling, leading to the formation of FM clusters and their growth with reducing particle size. This causes a creation of natural AFM/FM interfaces and leads to the EB effect. This behavior is commonly described in the framework of the core–shell model, where FM clusters generally form close to the surface of the nanograins, and this allows for an occurrence of the EB interface with the AFM core.

In fact, it was found that a reduction of NPs' size to about 20 nm results in the suppression of the charge-ordered (AFM/CO) state (e.g., Lu et al., 2007; Rozenberg et al., 2008; Markovich et al., 2010a, b). Phenomenological models and Monte Carlo studies of AFM nanomanganites predicted an enhancement of the surface charge density and confirmed a suppression of AFM/CO phase and an emergence of the FM order with spin-glass (SG)-like behavior near the surface (Dong et al., 2007, 2008).

Studies of suppression of charge and orbital ordering in manganites were also carried out for half-doped ($x = 0.5$) regimes (e.g., Sarkar et al., 2008a, b; Rozenberg et al., 2009; Wang and Fan, 2011), for low-bandwidth $Pr_{1-x}Ca_xMnO_3$ ($0.35 \leq x \leq 0.5$) (Biswas and Das, 2006; Chai et al., 2009; Zhang and Dressel, 2009; Jirak et al., 2010; Zhang et al., 2011;

Jammalamadaka et al., 2011), and for half-doped $Nd_{0.5}Ca_{0.5}MnO_3$ (Rao and Bhat, 2009, 2010; Zhou et al., 2011) and $Sm_{0.5}Ca_{0.5}MnO_3$ materials (Giri et al., 2011).

The doped manganites were synthesized in various nanostructures such as NPs, nanowires, and nanorods (Jiang et al., 2004; Tian et al., 2006; Carretero-Genevrier et al., 2011; Zhou et al., 2011; Zhi et al., 2012). The doped rare earth manganite nanostructures were so far synthesized by the sol–gel method (to obtain NPs [Sarkar et al., 2007; Chai et al., 2009] and nanowires [Rao and Bhat, 2010]), the hydrothermal technique (Rao et al., 2005), the molten-salt route (Tian et al., 2006), the electrospinning method (Zhi et al., 2012), ball milling (Levstik et al., 2009), microwave irradiation (Sadhu et al., 2012), solid-state calcination of oxides (Frontera et al., 2008), and pulsed laser deposition (Jiang et al., 2004). One should keep in mind that for a given compound, due to such variety of methods used for fabrication of manganite nanostructures, the properties may differ due to variation in chemical composition, number of vacancies, strain at the surface, and so on.

Similarly, as in the chapter "Exchange-Bias Effect in Bulk Perovskite Manganites," we draw attention of the readers to the recent review devoted to the EB effect, with particular interest in the EB effect in perovskite oxides and some of their nanostructures published by Giri et al. (2011). Hence, in this chapter, mainly the papers that were published during last 5–6 years are reviewed. The chapter content is not limited to only NPs, since during last few years, quite a few important papers concerning, for example, manganites' heterostructures and nanowires exhibiting the EB effect were published.

5.2 Training Effect

One of the characteristic features, usually observed in the EB systems, is the so-called *training effect* (TE). The TE is manifested as a difference between subsequent recorded hysteresis loops. The effect is due to the magnetic-field-cycling-caused modification of the magnetic state of metastable interfaces, present in the phase-separated sample. In quantity terms, the TE depends critically on the strength of exchange interaction between the FM and AFM domains and, of course, on the thermal energy. In the phase-separated systems, under the influence of the external magnetic field, the AFM domains are transformed into metastable FM domains, aligned along the direction of the magnetic field. These metastable FM domains are transformed back to stable AFM domains just after the external magnetic field is removed. These magnetic-field-cycling-induced magnetic transitions lead to the local change of the magnetic state of the sample. The training effect, as exemplified below, may depend on various factors (size of NPs, magnetic field applied, and temperature), and in some cases, the physics behind the effect is not fully understood.

Das and Das (2015) have studied properties of the charge-ordered AFM $La_{0.46}Ca_{0.54}MnO_3$ manganite, with average grain size of 100, 45, 21, and 15 nm. In particular, they found significant impact of the reduction of the particle sizes on TE, which disappears for the smallest NPs (15 nm). Reducing the particle size leads to systematical increase of the FM counterpart and charged ordering is gradually suppressed and finally disappears; hence, according to the authors, an apparent correlation between CO and TE is observed.

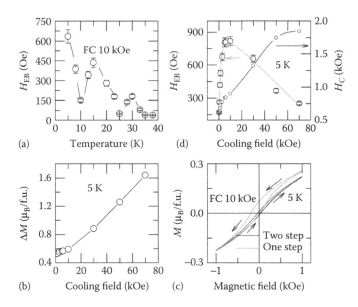

FIGURE 5.1 (a) Temperature variation of oscillatory exchange-bias field. (b) Variation of magnetization shift, ΔM, with field-cooled (FC) strength. (c) Magnetization M–H loops resulted from one-step and two-step FC process. Symmetric M–H loops are evident after two-step process. (d) Variation of exchange-bias field, H_{EB}, and coercivity, H_C, with FC. (Reprinted with permission from Jammalamadaka, S.N. et al., *AIP Adv.*, 2, 012169. Copyright 2012, American Institute of Physics.)

The TE was also studied for 10-nm particles of $Pr_{0.5}Ca_{0.5}MnO_3$ by Jammalamadaka et al. (2012). They observed that temperature dependence of the EB field (H_{EB}) is not a monotonic function but exhibits oscillatory character and is tunable as a function of cooling field strength below the SG freezing temperature (Figure 5.1).

The authors attributed the effect of oscillatory EB to the presence of charge/spin density wave in the AFM core of the NPs. However, despite the small size of the studied NPs, the pronounced TE is noticed at 5 K—the reduction in the EB field, H_{EB}, is observed after cycling the system through several consecutive hysteresis loops, and the dependence of H_{EB} on the number of field cycles (n) on the field-cooled (FC) process has well been described by power-law behavior for $n > 1$.

The TE is also discussed in the subsequent paragraph, which is devoted to size-dependent EB.

5.3 Size-Dependent Exchange Bias in Nanoparticles

As a detailed example of magnetic properties of half-doped manganites, one may refer to studies of $Sm_{0.5}Ca_{0.5}MnO_3$ NPs performed by Giri et al. (2011). They found that the charge/orbital order state characteristic for bulk AFM $Sm_{0.5}Ca_{0.5}MnO_3$ compound is suppressed in NPs, and weak FM appears below 65 K, followed by SG-like transition at 41 K. The destabilization of AFM/CO state, attributed mainly to the surface reconstruction of the electronic states and the consequent renormalization of magnetic exchange state, enhances the double-exchange interactions and, in consequence, the FM interaction. The observed EB effect can be tuned by the strength of the cooling magnetic field (H_{cool}). The values of the exchange field (H_{EB}), coercivity (H_C), remanence asymmetry (M_{EB}), and magnetic coercivity (M_C) were found to be strongly depend on H_{cool} and temperature (Figure 5.2).

The value of H_{EB} initially increases with increasing H_{cool}, but for larger H_{cool} values, it decreases due to the growth of FM cluster size. The EB effect reveals a magnetic training effect, which can be described very well by the spin-relaxation model proposed by Binek (2004). The temperature dependence of EB parameters suggests that the pinned or frozen disordered FM/SG spins give rise to the unidirectional shift of the magnetic hysteresis loops. A schematic picture showing destabilization of CO and surface disorder at the shell of NPs is presented in Figure 5.3.

In later studies, Giri et al. (2014) have investigated the effect of grain size variation on EB effect in $Sm_{0.5}Ca_{0.5}MnO_3$. When particle size is reduced (from 150 to 17 nm, in

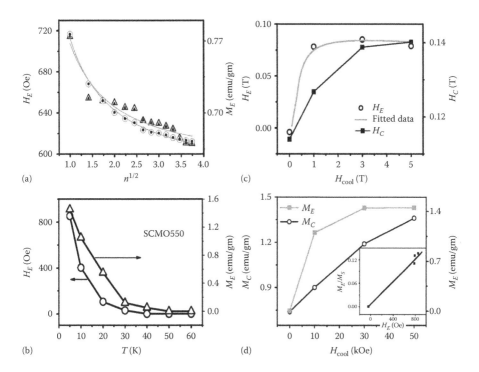

FIGURE 5.2 (a) The number of field cycles n dependences of H_{EB} (H_E in figure) and M_{EB} (M_E in figure) at 5 K after FC in 7 kOe magnetic field. (b) The variation of exchange bias parameters H_{EB} and M_{EB} with temperature after FC with magnetic field of 7 kOe. (c) H_{EB} and H_C versus cooling field (H_{cool}) plot for 17-nm $Sm_{0.5}Ca_{0.5}MnO_3$ nanoparticles. (d) The M_{EB} (M_E in figure) and M_C versus cooling field plot for 17-nm $Sm_{0.5}Ca_{0.5}MnO_3$ nanoparticles. Inset shows the plot of scaled vertical shift with horizontal shift. (Reprinted with permission from Giri, S.K. et al., *AIP Adv.*, 1, 032110. Copyright 2011, American Institute of Physics.)

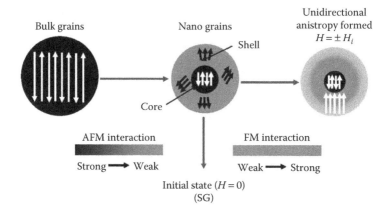

FIGURE 5.3 The schematic representation of the phenomenological model for CO/AFM bulk manganites and the corresponding nanograins. (Reprinted with permission from Giri, S.K. et al., *AIP Adv.*, 1, 032110. Copyright 2011, American Institute of Physics.)

this case), surface effects become dominating and suppress the CO states due to the modification of surface spin configuration. At the same time, the enhancement of FM clusters increases the FM fraction. The observed EB effect was found to be tuned with the reduction of particle sizes—the values of H_{EB}, H_C, M_{EB}, and M_C depend strongly on the particle size. In particular, the variation of H_{EB} with downsizing is non-monotonic and exhibits maximum value (1205 Oe) for NP size of 150 nm at 5 K. The linear relationship was found between H_{EB} and M_{EB}. Both these values decrease exponentially with increasing the temperature below the SG-like or cluster-glass-like freezing temperature. Strong magnetic training effect was explained in a frame of spin-relaxation model (Binek, 2004).

Zhang et al. (2011) have studied the evolution of magnetic properties of the bulk and 100-nm $Pr_{0.67}Ca_{0.33}MnO_3$ NPs. They found that for the bulk $Pr_{0.67}Ca_{0.33}MnO_3$, the AFM phase transforms with increasing field to the FM phase and gradually disappears at 50 kOe below the FM transition temperature, T_C. In magnetic field of about 40 kOe, the system reaches the percolation of the metallic FM phase. For nanosized particles, although the CO transition still exists, the percolation of the metallic FM phase and the complete FM transition do not occur up to 60 kOe and the AFM transition disappears. For bulk sample, a small EB effect with $H_{EB} = 36$ Oe is observed at 10 K. The coexistence of the FM and AFM phases leads to the natural FM/AFM interfaces. The spins of the FM cluster align parallel to the external magnetic field. The interfacial FM spins on the exterior surface of the AFM or canted AFM inner core tend to be coupled with AFM spins at the interface below the Néel temperature, T_N, which causes the EB behavior. For NPs, the EB field significantly increases compared with the one for the bulk: $H_{EB} = 312$ Oe (Figure 5.4). According to the authors, this behavior is caused by the surface phase separation. For NPs, the surface spins act as an FM on the AFM core. Thus, due to the coupling between surface FM and AFM at the surface, the FM spins drag AFM spins to the external fields and produce additional EB fields, besides that in the core, which results in a larger H_{EB}.

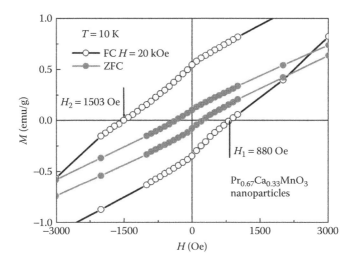

FIGURE 5.4 Zero-field-cooled (ZFC) and FC magnetic hysteresis loop measured at 10 K for $Pr_{0.67}Ca_{0.33}MnO_3$ nanoparticles. (Reprinted with permission from Zhang, T. et al., *J. Phys. Chem. C*, 115, 19482. Copyright 2011 American Chemical Society.)

This behavior can be attributed to a change of the spin configuration from collinear AFM to the coexistence of disordered surface spins, canted AFM, and collinear AFM arrangement, which are induced by the surface effect and the coupling among the shell and interface spins, and the EB field increases compared with that of the bulk. All these features can be ascribed to the surface pressure and uncompensated surface spins.

Markovich et al. (2012a) have studied magnetic properties of electron-doped $La_{0.23}Ca_{0.77}MnO_3$ manganite NPs, with average size of 12 and 60 nm. They showed that weak FM moment, attributed to the shell of NPs, appears at temperature as high as 270 K, while additional FM component emerges at $T \sim 50$ K, and it may be attributed to the presence of domains of orbitally disordered phase within the AFM core. For smaller, 12-nm $La_{0.23}Ca_{0.77}MnO_3$ particles, the double fit of the $T^{3/2}$ Bloch law to spontaneous magnetization shows that it may be well approximated by two independent FM contributions. The appearance of low-temperature FM component results in the appearance of cluster-glass-like features, such as a difference between the zero-field-cooled (ZFC) and FC magnetization and significant frequency dependence of ac susceptibility. On field cooling, the particles display both vertical and horizontal shifts of the hysteresis loop (Figure 5.5). The EB effect is strongly size-dependent. The changes in temperature variation of H_{EB}, remanent magnetization, and spontaneous magnetic moment with changing particle size were explained in terms of the magnetic coupling of the anisotropic basically AFM core with FM-like shell and regions of orbitally disordered phase inside the core.

The FM components, which are more pronounced in smaller particles, occupy only a small fraction of the NP volume, and the AFM ground state remains stable. It is found that the magnetic hysteresis loops display size-dependent horizontal and vertical shifts in FC processes, exhibiting the EB effect.

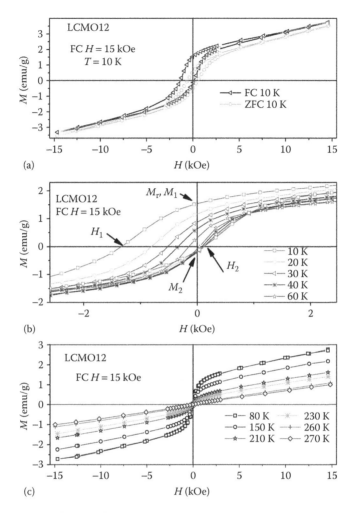

FIGURE 5.5 (a) Field dependence of magnetization of $La_{0.23}Ca_{0.77}MnO_3$ nanoparticles with average size of 12 nm (LCMO12) at 10 K after ZFC and FC under $H = 15$ kOe. (b, c) Field dependence of magnetization of LCMO12 sample at various temperatures after FC. (With kind permission from Springer: *J. Nanopart. Res.*, Size dependent magnetic properties of electron doped $La_{0.23}Ca_{0.77}MnO_3$ nanoparticles, 14, 2012a, 1119, Markovich, V. et al.)

Iniama et al. (2014) studied the physical properties of half-doped manganite $La_{0.5}Ca_{0.5}MnO_3$, with crystallite sizes ranging from 15 to 40 nm. They noticed that, as expected, the FM order at expense of AFM one is reduced in 15-nm NPs. On the other hand, surprisingly, an enhancement of saturation magnetization, M_S, is observed as crystallite size increases from 15 to 22 nm. This is accompanied by a decrease in the unit cell volume. The authors have attributed the enhancement of M_S to two contributions: an FM-like behavior of the SG at crystallite boundaries and an effective induced hydrostatic pressure (due to small size of the NPs) that promotes FM behavior inside the AFM cores. Owing to a core-shell structure with AFM/FM coupling for crystallite sizes as small as 15 nm, FC hysteresis loops show EB effect and coercivity enhancement for increasing cooling fields.

Markovich et al. (2013, 2014) have studied size-dependent magnetism and the EB effect in $Sm_{1-x}Ca_xMnO_3$ NPs. Highly distorted perovskite $GdFeO_3$-type structure of the narrow-bandwidth $Sm_{1-x}Ca_xMnO_3$ is favorable for charge localization but detrimental for double-exchange (DE) interactions (Hejtmánek et al., 1999; Martin et al., 1999). Charge ordering appears in the doping range $0.4 < x < 0.85$, and different CO configurations are stabilized depending on the Mn^{3+}/Mn^{4+} content. Charge ordering in the bulk $Sm_{0.43}Ca_{0.57}MnO_3$ appears at CO temperature, $T_{CO} \approx 283$ K, the highest T_{CO} in the SmCaMnO system, and is characterized by a huge difference between T_{CO} and $T_N \approx 135$ K.

Size-dependent magnetism and the EB effect in $Sm_{0.27}Ca_{0.73}MnO_3$ NPs, with average particle size ranging from 20 to 80 nm, were discussed in the paper of Markovich et al. (2013). The CO transition was found to shift gradually to lower temperatures with downsizing, and it almost disappeared for 20-nm particles. At the same time, the relative volume of the FM phase increases monotonously. Field-induced transition from AFM to FM state in 80-nm particles is observed at the same magnetic field as in the bulk. In small, 20-nm particles, the transition is strongly suppressed due to surface spin disorder. Magnetic hysteresis loops show size-dependent EB effect with exchange field, remanence asymmetry, and magnetic coercivity that depend on cooling magnetic field and temperature. At different cooling fields, various frozen magnetic configurations of the system are formed, thus allowing for wide range tuning of the values of H_{EB}, H_C, M_{EB}, and M_C. Decrease in both H_{EB} and M_{EB} is observed after cycling of smaller, 20-nm NPs through several consecutive hysteresis loops, manifesting a TE. The observed TE can be described within the relaxation model used for other classical EB systems, indicating a unique mechanism for exchange anisotropy. The analysis of the thermoremanent and isothermoremanent magnetization curves showed that the inner core of the AFM 20-nm NPs behaves like a two-dimensional diluted AFM.

Results for $Sm_{0.43}Ca_{0.57}MnO_3$ NPs with average size of 15–60 nm were presented in the paper of Markovich et al. (2014). It was found that CO is gradually suppressed with decreasing particle size and fully disappears in 15-nm NPs. Onset of FM contribution to the magnetization appear below 90 K, independently on the particle size, while spontaneous magnetization at 10 K increases with decreasing particle size. Magnetic hysteresis loops exhibit size-dependent EB effect. Interestingly, variation of H_{EB} and M_{EB} shows opposite size dependences; H_{EB} decreases while M_{EB} increases with the decreasing particle size. For large, 60-nm particles, H_{EB} decreases monotonously with increasing temperature, while for the smallest, 15-nm particles, it shows nonmonotonic temperature dependence (Figure 5.6). Size and temperature dependences were discussed in terms of magnetic coupling between AFM cores and shells containing FM clusters in a frustrated-spin configuration.

Markovich et al. (2012b) studied the EB effect in $Sm_{0.1}Ca_{0.9}MnO_3$ NPs with average particle size of 25 and 60 nm. It should be noted that ground state of bulk $Sm_{0.1}Ca_{0.9}MnO_3$ differs significantly from the one for electron-doped bulk AFM CO

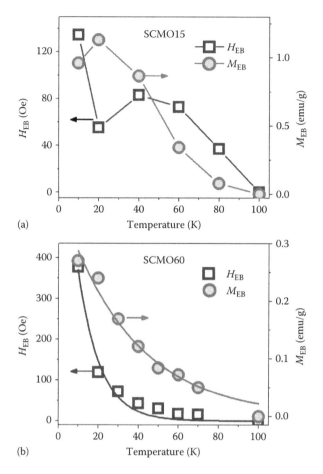

FIGURE 5.6 Temperature dependence of H_{EB} and M_{EB} for $Sm_{0.43}Ca_{0.57}MnO_3$ nanoparticles with average size of (a) 15 nm (SCMO15) and (b) 60 nm (SCMO60), after FC in 15 kOe. The lines in Figure 5.6a are guides to the eyes. Lines in Figure 5.6b are the best fits with equations: $H_{EB} = H_{EB}(0)$ $\exp(-T/T_1)$ and $M_{EB} = M_{EB}(0)\exp(-T/T_2)$, where $H_{EB}(0)$ and $M_{EB}(0)$ are the values of H_{EB} and M_{EB} extrapolated to $T = 0$ K, and T_1 and T_2 are constants. (Reprinted with permission from Markovich, V. et al., *J. Phys. Chem. C.*, 118, 7721. Copyright 2014 American Chemical Society.)

manganites, since relatively large FM clusters are distributed in the AFM matrix and the magnetization at low temperatures reaches the value of about 1 μ_B/f.u. (Martin et al., 1999). Furthermore, on field cooling, $Sm_{0.1}Ca_{0.9}MnO_3$ NPs display size-dependent EB effect, manifested by both horizontal and vertical shifts of the hysteresis loop, which is relatively small in 60-nm and much larger in 25-nm NPs (~ 900 Oe at $T = 10$ K). For 25-nm NPs, the values of H_{EB} and H_C exhibit non-monotonic variation with H_{cool}, while values of M_{EB} and M_C increase monotonously with H_{cool}. It was suggested that observed size dependence of the EB is due to the suppression of the FM phase with decreasing particle size, that is, due to the decrease in the size of the FM clusters embedded in the AFM matrix of the cores of NPs.

Rostamnejadi et al. (2014) have studied the magnetic properties of 16-nm $La_{0.45}Sr_{0.55}$ MnO_3 NPs. In this system, local disorder and surface effects change the competition between DE and superexchange (SE) interactions, and randomly oriented canted FM regions form in an AFM matrix. Hence, the EB effect occurs due to exchange interaction at the interface between the canted FM and AFM regions within the NPs. It was found that the hysteresis loop shift, coercivity, and remanence asymmetry decrease strongly with increasing cooling field above 1 T (Figure 5.7), unlike in a conventional FM/AFM EB system. This behavior may be caused by a magnetization process involving coalescence of canted FM clusters with increasing field, which reduces the interface area with the AFM matrix.

Manna et al. (2011) have studied multiferroic $BiFe_{0.8}Mn_{0.2}O_3$ NPs. From neutron diffraction, thermoremanent magnetization, and isothermoremanent magnetization measurements, the NPs were found to be core-shell in nature, consisting of an AFM core and a two-dimensional diluted AFM shell with a net magnetization under a field. The values of the EB parameters and their temperature dependence are shown in Figure 5.8a. The analysis of the TE data using the Binek's model showed (Figure 5.8b) that the observed loop shift arises entirely due to an interface exchange coupling between the core and the shell. It is worthwhile to stress that a significantly high value of the EB field (200 Oe) has been observed at room temperature.

Chauhan et al. (2013) have studied the EB properties of h-$YMnO_3$ NPs. They showed that large EB can be obtained in these NPs, even after zero-field cooling from an unmagnetized

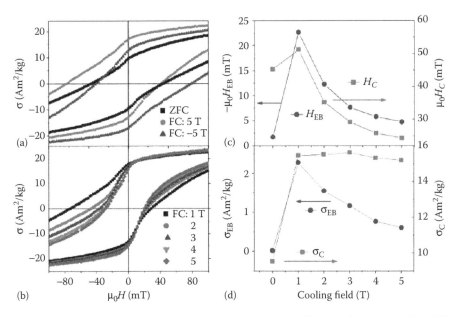

FIGURE 5.7 (a) ZFC and FC magnetization hysteresis loops of the sample at 5 K. Cooling field dependence of (b) the hysteresis loop, (c) $-\mu_0 H_{EB}$ and $\mu_0 H_C$, and (d) σ_{EB} and σ_C at 5 K (in Figure 5.7d, σ_{EB} and σ_C mean M_{EB} and M_C, respectively). (Reprinted with permission from Rostamnejadi, A. et al., *J. Appl. Phys.*, 116, 043913. Copyright 2014, American Institute of Physics.)

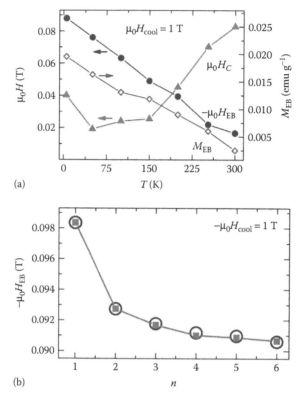

(a)

(b)

FIGURE 5.8 (a) Temperature dependence of $-\mu_0 H_{EB}$ ($-\mu_0 H_{EB}$ in figure), $\mu_0 H_C$, and M_{EB} (M_{EB} in figure). (b) EB field ($-\mu_0 H_{EB}$: open circle) dependence on the number of field cycles (n). The solid squares represent the calculated data points using Binek's recursive formula, and the solid line is a guide to the eye. (Reprinted with permission from Manna, P.K. et al., *Phys. Rev. B,* 83, 184412, 2011. Copyright 2011 by the American Physical Society.)

state without any remanent magnetization. The H_{EB} and M_{EB} values were found to vary nonmonotonically with both particle size change (ranging from 45 to 90 nm) and maximum applied-field variation. Large H_{EB} value of 1024 Oe at 2 K was obtained for the 55-nm particles. The authors attribute the EB effect to the exchange interaction between the compensated AFM spins and uncompensated surface spins of the NPs.

Liu et al. (2014) have studied double perovskite $La_{1.6}Sr_{0.4}NiMnO_6$ NPs of different sizes (18–150 nm). For double-perovskite oxides, fully ordered arrangement of the transition metal (TM) ions is difficult to obtain because of the easy formation of anti-site defect (creation of imperfections by interchanging the TM ionic positions), which leads to the AFM interaction based on SE mechanism between the TM ions. The existence of anti-site defects results in the so-called antiphase boundaries, which can introduce similar AFM interactions, even for an ordered double perovskite (Wang et al., 2009). Most of the magnetic properties of double perovskites are strongly influenced by the ordering of TM ions at the *B* site. Liu et al. (2014) have found that there is an FM transition at

$T_C \sim 245$ K for the particle with diameter $D \sim 150$ nm. With downsizing, an SG transition appears at ~ 60 K, the T_C value lowers, and the FM transition becomes blurry. The size-dependent saturation magnetization shows a non-monotonic variation, with a maximum of 1.27 μ_B/f.u. at the particle diameter ~ 42 nm. The EB effect is clearly observed for $D > 32$ nm and becomes indistinct at $D \leq 32$ nm. As the particle size decreases, the H_{EB} decreases and almost vanishes for ~ 42-nm NPs and the H_C first increases to a maximum and then decreases for ~ 42-nm NPs. The magnetic properties of the NPs can be understood within a complex core-shell model; that is, both AFM in antiphase boundaries and FM domains coexist in the core and the SG phase mainly resides at the surface of each particle.

It should be mentioned that occurrence of the EB effect is not *obligatory* in manganite NPs. Singh and Bhat (2012) studied the magnetic properties of 30-nm NPs of $Bi_{0.25}Ca_{0.75}MnO_3$ (BCMO) and compared them with the results for bulk samples. While the bulk samples exhibit a CO transition at about 230 K and an AFM transition at about 130 K, in the NPs, the CO phase disappears and a transition to a FM state is observed at 120 K. However, the EB effect is absent in these NPs.

5.4 Exchange-Bias Effect in Nanostructures of Manganites

In the few last years, several relevant papers on EB effect in manganite heterostructures, nanowires, and nanorods were published. For practical applications, it is favorable that significant EB coupling occurs at zero or minimum applied cooling fields. Nanowires or nanorods with 1D morphology lead to shape anisotropy, useful to increase the overall magnetic anisotropy of the system. The 2D nanosheets or thin films exhibit enhanced anisotropy along the micrometer-length dimensions, which makes their physical properties different in comparison with those of the conventional NPs and bulk material. In general, low-dimensional manganite structures have the advantage of direct implementation as potential building blocks for next-generation nanodevices.

In particular, EB, as a prototypical interfacial magnetic interaction between different spin orders, was reported in some manganite-based heterostructures and superlattices; see, for example, Kobrinskii et al. (2009), Wu et al. (2010), Ziese et al. (2010), Gibert et al. (2012), Wu et al. (2013), and Ding et al. (2013). Below, we discuss in more details some selection of the papers devoted to EB effect in various manganite nanostructures.

Sadhu and Bhattacharyya (2013) have found spontaneous EB in stacked nanosheets of $Pr_{1-x}Ca_xMnO_3$ ($x = 0.3$ and 0.49). Stacked 10- to 14-nm-thick nanosheets were obtained by decomposition of carbon-coated $CaCO_3/MnCO_3$ microsheets and $Pr_2O_2CO_3$ aggregates; the latter was synthesized under pressure at 500°C−800°C. The occurrence of FM domains within the AFM matrix results in FM moments at 5 K and spontaneous exchange bias (SEB) coupling between the AFM/FM spins observed under zero-field cooling. For $Pr_{0.7}Ca_{0.3}MnO_3$ and $Pr_{0.51}Ca_{0.49}MnO_3$, at 5 K, hysteresis loop shift of 22.7 and 6.2 mT, respectively, in the negative-field direction, was observed in the absence of an applied cooling field. When the divalent alkaline earth Ca^{2+} ions replace the trivalent Pr^{3+} ions, the concentration of Mn^{3+} ions decreases at the cost of Mn^{4+} ions to maintain charge balance. According to the authors, large FM moments and SEB are possible by the long-range magnetic interactions in the stacked 2D arrangement of Mn^{3+}/Mn^{4+} spins, where CO is completely suppressed. The nanosheets exhibit flatter Mn–O–Mn tilt angles in the MnO_6 octahedra. Ealier Sadhu et al. (2012) showed that in lightly doped $Pr_{1-x}Ca_xMnO_3$ ($x = 0.023$ and 0.036) NPs, the flattening of the Mn–O–Mn bond angles and reduced orthorhombic strain promoted enhancement of FM moments at 5 K. It was found that the SEB behavior observed for $Pr_{0.7}Ca_{0.3}MnO_3$ and $Pr_{0.51}Ca_{0.49}MnO_3$ depends on the initial magnetization process and the direction of rotation of the random spins at the AFM/FM interface.

Ding et al. (2013) have studied the magnetic properties of manganite bilayers composed of the *G*-type AFM $SrMnO_3$ (SMO) and double-exchange FM $La_{0.7}Sr_{0.3}MnO_3$ (also called LSMO). An SG state was observed due to competing magnetic orders and spin frustration at the $La_{0.7}Sr_{0.3}MnO_3/SrMnO_3$ interface. An ideal *G*-type AFM $SrMnO_3$ characterizes a compensated spin configuration, but the bilayers exhibit the EB effect (Figure 5.9) below the SG freezing temperature, which is much lower than the T_N of $SrMnO_3$. The results indicate that the spin frustration that originates from the competition between the AFM SE and the FM DE interactions can induce a strong magnetic anisotropy at the $La_{0.7}Sr_{0.3}MnO_3/SrMnO_3$ interface.

Tian et al. (2014) have studied interfacial magnetic coupling in ultrathin $La_{0.7}Sr_{0.3}MnO_3/TbMnO_3$ superlattices. In this system, rare earth manganite $TbMnO_3$ exhibits multiferroic properties and coupling between ferroelectric and magnetic orders. As a result of the competing exchange interactions, in $TbMnO_3$, the following orderings are observed: (1) sinusoidal AFM Mn^{3+} spin ordering at $T_N = 41$ K; (2) spiral Mn^{3+} spin ordering at the ferroelectric transition temperature at 28 K; and (3) long-range FM Tb^{3+} spin ordering below 7 K. Orthorhombic structure of $TbMnO_3$ is compatible with other perovskites; hence, various $TbMnO_3$-based heterostructures may be fabricated. The correlation effects involving multiple degrees of freedom are expected to be enhanced at interfaces in ultrathin heterostructures. Tian et al. have observed that in a series of epitaxial manganite superlattices composed of ultrathin $La_{0.7}Sr_{0.3}MnO_3/TbMnO_3$ films, low-temperature magnetism with hysteresis is stabilized down to a $La_{0.7}Sr_{0.3}MnO_3$ layer thickness of two unit cells, which is accompanied by drastically enhanced coercive field of more than 3000 Oe (Figure 5.10a). The EB effect of about 170 Oe was observed at 10 K (Figure 5.10b) and was attributed to the strong magnetic coupling across the $La_{0.7}Sr_{0.3}MnO_3/TbMnO_3$ interface.

Cui et al. (2013) have reported that the EB effect can emerge in a $La_{2/3}Sr_{1/3}MnO_3$ thin film when large compressive stress is caused by a lattice-mismatched substrate. To enlarge the compressive strain, 45 unit cells of $La_{2/3}Sr_{1/3}MnO_3$ thin film ($a = 3.870$ Å for the bulk) were grown on $LaSrAlO_4$ (001) substrates, with a comparatively small lattice constant ($a = 3.756$ Å). The interfacial reconstruction of structure and composition was directly observed, resulting in two magnetic sublattices and intrinsic EB in manganite (Figure 5.11). The intrinsic EB behavior originates from the exchange coupling between the FM $La_{2/3}Sr_{1/3}MnO_3$ and a unique $LaSrMnO_4$-based SG, formed under a large interfacial strain.

Schumacher et al. (2013) have studied EB in $La_{0.67}Sr_{0.33}MnO_{3-\delta}/SrTiO_3$ thin films induced by strain and oxygen deficiency. The samples were prepared by pulsed laser deposition and high-pressure sputter deposition in oxygen atmosphere at different

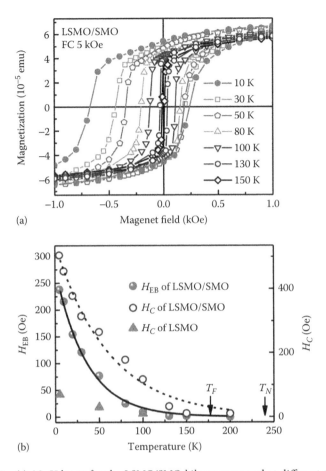

FIGURE 5.9 (a) *M−H* loops for the LSMO/SMO bilayer measured at different temperatures. For clarity, only the data between −1 and 1 kOe are shown in the figures, while the actual measurements took place between −5 and 5 kOe. (b) Temperature dependence of H_{EB} and H_C for the LSMO/SMO bilayer. Also shown are the H_C data measured for the LSMO single layer. The two arrows mark the Néel temperature T_N of bulk SMO and the freezing temperature T_F, respectively. (Reprinted with permission from Ding, J.F. et al., *Phys. Rev. B*, 87, 054428, 2013. Copyright 2013 by the American Physical Society.)

oxygen pressures. The reduced Curie temperatures indicated oxygen deficiencies in the samples grown at lower oxygen pressures. It was found that the EB and coercive fields increase with decreasing oxygen pressure, while the growth at high pressures does not lead to an EB effect. In the samples exhibiting the EB effect, the depth-sensitive method of polarized neutron reflectometry revealed a region with drastically reduced net magnetization in the $La_{0.67}Sr_{0.33}MnO_{3−\delta}$ layer at the interface with $SrTiO_3$, but such reduction was not observed in the samples that do not exhibit EB. The authors suggested that an AFM structure is formed in this part of the $La_{0.67}Sr_{0.33}MnO_{3−\delta}$ thin film, which causes the EB effect in this system.

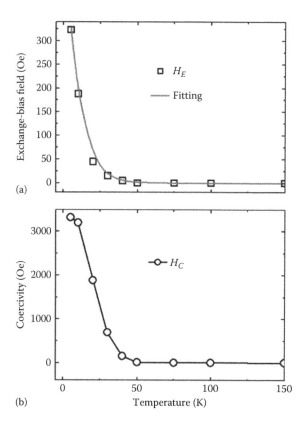

FIGURE 5.10 Temperature dependence of (a) the exchange-bias field, H_{EB} (H_E in the figure) and (b) the coercivity, H_C, measured on the epitaxial manganite superlattices composed of ultrathin $La_{0.7}Sr_{0.3}MnO_3/TbMnO_3$ films. The solid line in (a) presents fitting with expression: $H_{EB} = H_{EB}(0)$ $\exp(-T/T_0)$. (Reprinted with permission from Tian, Y.F. et al., *Appl. Phys. Lett.*, 104, 152404. Copyright 2014, American Institute of Physics.)

Vafaee et al. (2016) have studied the effect of interface roughness on EB in $La_{0.7}Sr_{0.3}MnO_3/BiFeO_3$ heterostructures. They characterized the interfaces of heterostructures with different stack sequences of $La_{0.7}Sr_{0.3}MnO_3/BiFeO_3$ by using TEM. Magnetometry and magnetoresistance measurements did not show any EB coupling for the heterostructures with sharp interface. In contrast, the heterostructures with rough and chemically intermixed interfaces exhibited the EB coupling, with $H_{EB} = 340$ Oe at 5 K. They found an exponential decay of coercive and EB fields with temperature, suggesting a possible SG-like state at the interface of both stacks.

He et al. (2012) have found interfacial ferromagnetism and EB in epitaxially grown $CaRuO_3$ (CRO)/$CaMnO_3$ (CMO) superlattices. The $CaRuO_3$ is a paramagnetic metal, and $CaMnO_3$ is an AFM insulator. The FM in these superlattices is attributed to the leakage of itinerant electrons from $CaRuO_3$ to $CaMnO_3$. Scanning transmission electron microscopy and electron energy-loss spectroscopy indicated that the difference in the magnitude of the Mn valence states between the center of the $CaMnO_3$ layer and

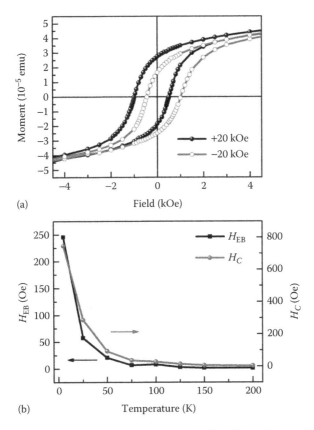

(a)

(b)

FIGURE 5.11 Intrinsic exchange bias in an LSMO *single* film. (a) *M–H* loops of the 45 unit cells of $La_{2/3}Sr_{1/3}MnO_3$ thin film measured at 5 K after field cooling from room temperature in +20 kOe (solid spheres) and −20 kOe (open circles). For clarity, only the data between −4.5 and 4.5 kOe are shown in the figures, while the actual measurements were carried out between −20 and 20 kOe. (b) Temperature dependence of H_{EB} and H_C for the sample. The solid lines through the data points are guides to the eye. (Reprinted by permission from Macmillan Publishers Ltd. *Sci Rep*, Cui, B. et al., Strain engineering induced interfacial self-assembly and intrinsic exchange bias in a manganite perovskite film, 3, 2542, copyright 2013.)

the interface region points toward DE interaction among the Mn ions at the interface. Polarized neutron reflectivity and the $CaMnO_3$ thickness dependence of the EB field (Figure 5.12) indicate that the interfacial FM is only limited to one unit cell of $CaMnO_3$ at each interface.

Peng et al. (2014) found the existence of the EB effect in a single $LaMnO_3$ thin film, manifested by the asymmetry in coercive field, induced by vertical electronic-phase separation. Using combined X-ray absorption spectroscopy for microstructure characterizations and magnetic measurements, they found evidence that the occurrence of Mn^{2+} ions allows for DE (Mn^{2+}–O–Mn^{3+}) that creates robust ferromagnetism in the upper part of the film. That part is exchange-coupled with the AFM's bottom part that is dominated by Mn^{3+} ions. In the studied system, the Mn ions are not distributed

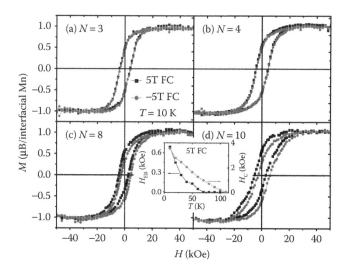

FIGURE 5.12 Field dependence of magnetization of (CRO)$_3$/(CMO)$_N$ superlattices at $T = 10$ K for (a) $N = 3$, (b) $N = 4$, (c) $N = 8$, and (d) $N = 10$. The measurements are done after 5 T (black squares) and −5 T (gray circles) field cooling, respectively, from 300 K. Inset of (c) and (d) shows T dependence of H_C and H_{EB} for (CRO)$_3$/(CMO)$_8$. (Reprinted with permission from He, C. et al., *Phys. Rev. Lett.*, 109, 197202, 2012. Copyright 2012 by the American Physical Society.)

uniformly in the depth profile, which causes vertical electronic-phase separation, instead of commonly assumed lateral-phase separation in manganites.

Peng et al. (2015) have fabricated and studied strain-engineered (La,Sr)MnO$_3$/LaNiO$_3$ heterostructures. They showed that charge transfer from Mn^{3+} ions to Ni^{3+} ions is accompanied by the formation of hybridized Mn/Ni $3z^2-r^2$ orbits at the interface, instead of strain stabilized Mn and Ni x^2-y^2 orbits that are formed in the bulk films. This orbital reconstruction induced by charge transfer results in magnetization frustration of (La,Sr)MnO$_3$ at the interface. It was also shown that the strain effect exerted on the top LaNiO$_3$ layer leads to larger magnetization frustration and to the EB effect at the interface. Peng et al. (2015) found in studied heterostructures the $H_{EB} \approx 50$ Oe at 5 K. The spins in the frustrated SG regions, partly frozen collinearly with the field direction, serve as the uncompensated spins and induce the EB effect. Interestingly, EB is absent in the bilayers with LaNiO$_3$ thickness less than 36 unit cells, while the coercivity enhancement is still present. Hence, the authors concluded that both H_{EB} and H_C enhancements originate from the exchange coupling at the interface. The coercivity enhancement occurs as long as there is exchange coupling, while the EB effect arises only when there is enough AFM phase or SG to pin the FM part.

Chandra et al. (2013) have found the EB effect in single-crystalline La$_{0.5}$Sr$_{0.5}$MnO$_3$ nanowires. Single-crystalline La$_{0.5}$Sr$_{0.5}$MnO$_3$ nanowires were obtained by using hydrothermal method. The nanowires undergo a paramagnetic to FM transition at $T_C \approx 310$ K, followed by an FM/AFM transition at $T_N \approx 210$ K. Both magnetic phases coexist, which results in completion of DE and SE interactions arising from the FM and AFM domains. With decreasing temperature, the FM volume fraction increases and the DE overcomes the AFM interactions at about 75 K. At about 42 K, the surface spins freeze. At this temperature, onset of the EB effect is observed and the H_{EB} is equal to about 70 Oe at 5 K.

Cui et al. (2014) have studied the effect of strain-induced intrinsic H_{EB} on the magnetization rotation process in a nominally $La_{2/3}Sr_{1/3}MnO_3$ film. The intrinsic EB appears when the film is grown on $LaAlO_3$ substrate. It was shown that the magnetization rotation in two dimensions can be manipulated by the EB field. Hence, the H_{EB} is an effective tool for tuning the in-plane magnetization rotation, for example, producing a 360° instead of 180° periodicity in the anisotropic magnetoresistance curves measured in a low external magnetic field.

Zhang et al. (2015) have fabricated and studied self-assembled $La_{0.7}Sr_{0.3}MnO_3$:NiO vertically aligned nanocomposite films. The films were grown by using pulsed laser deposition with a uniform morphology of highly ordered, ultrafine NiO nanopillars embedded in the $La_{0.7}Sr_{0.3}MnO_3$ matrix. It was noticed that after cooling in 1 T to 10 K, the $M(H)$ curve of the film exhibits a horizontal shift, with H_{EB} of −77 Oe and a coercive field of 450 Oe. In contrast, due to the lack of exchange coupling, pure $La_{0.7}Sr_{0.3}MnO_3$ film shows no EB shift and a smaller value of $H_C = 396$ Oe. The authors stressed that such a vertical interface coupling enables a dynamic and reversible control of magnetotransport properties in epitaxial oxide nanocomposite films.

Very recently, Fan et al. (2016) have reported strong perpendicular EB in epitaxial $La_{0.7}Sr_{0.3}MnO_3$:$LaFeO_3$ nanocomposite thin films. Most EB effects reported in thin films are based on multilayer structures in which the magnetization of FM layer is pinned in the in-plane direction, which is parallel to the coupling interfaces. However, perpendicular EB is more advantageous for high-density data storage. Fan et al. have grown thin films of $(La_{0.7}Sr_{0.3}MnO_3)_{1-x}$:$(LaFeO_3)_x$, with $x = 0.33$, 0.5, and 0.67, using pulsed laser deposition. The authors stressed that systematic property modulation can be achieved by strain tuning effect with composition control. Under 1 T magnetic field cooling, high EB field of about 800 Oe was achieved at 5 K in the film with $x = 0.33$. The authors attributed the EB effects in perpendicular direction to the high-quality epitaxial co-growth of vertically aligned FM $La_{0.7}Sr_{0.3}MnO_3$ and AFM $LaFeO_3$ phases, as well as to the vertical interface coupling associated with a disordered SG state.

5.5 Concluding Remarks

For NPs of electron-doped manganites, the suppression of both AFM state and the suppression/disappearance of CO occur at some critical size for basically AFM electron-doped materials, while at the surface, short-range FM clusters appear. The uncompensated surface spins favor the FM coupling, leading to the formation of FM clusters and their growth with reducing particle size. This causes a creation of natural AFM/FM interfaces and the EB effect. Owing to surface pressure, NPs are strongly affected by strain fields that distort the crystal lattice and produce about 3- to 4-nm-sized strained regions at the surface. These strain fields promote phase separation, which takes place on cooling, and hence also lead to EB effect.

The EB effect significantly varies with downsizing—the values of H_{EB}, H_C, M_{EB}, and M_C depend strongly on the particle size. However, variation of these parameters with reduction of particle size may be non-monotonic. The EB effect can also be tuned by the strength of the cooling magnetic field. A variation of, for example, H_{EB} on H_{cool} also may be non-monotonic. The TE observed for NPs depends on various factors (size of NPs, magnetic field applied, and temperature), and in some cases, the physics behind the effect is not fully understood.

Exchange bias was observed in some manganite-based heterostructures and superlattices. The spin frustration that originates from the competition between the AFM SE and the FM DE interactions can induce a strong magnetic anisotropy at the interface in such a system and, in consequence, may lead to the EB effect. In the heterostructures, roughness, chemically intermixed interfaces, and interfacial strain also promote the EB coupling. In some systems, the Mn ions are not distributed uniformly in the depth profile, which causes vertical electronic-phase separation, instead of commonly assumed lateral-phase separation in manganites. The EB effect is observed in single-crystalline nanowires of manganites.

Acknowledgment

This work was partly supported by the Polish NCN grant 2014/15/B/ST3/03898.

References

Binek C (2004) Training of the exchange-bias effect: A simple analytic approach. *Phys Rev B*, 70:014421.

Biswas A, Das I (2006) Experimental observation of charge ordering in nanocrystalline $Pr_{0.65}Ca_{0.35}MnO_3$. *Phys Rev B*, 74:172405.

Carretero-Genevrier A et al. (2011) Single crystalline $La_{0.7}Sr_{0.3}MnO_3$ molecular sieve nanowires with high temperature ferromagnetism. *J Am Chem Soc*, 133:4053–4061.

Chai P et al. (2009) Particle size-dependent charge ordering and magnetic properties in $Pr_{0.55}Ca_{0.45}MnO_3$. *J Phys Chem C*, 113:15817–15823.

Chandra S, Biswas A, Datta S, Ghosh B, Raychaudhuri AK, Srikanth H (2013) Inverse magnetocaloric and EB effects in single crystalline $La_{0.5}Sr_{0.5}MnO_3$ nanowires. *Nanotechnology*, 24:505712.

Chauhan S, Srivastava SK, Chandra R (2013) Zero-field cooled exchange bias in hexagonal $YMnO_3$ nanoparticles. *Appl Phys Lett*, 103:042416.

Cui B, Song C, Wang GY, Mao HJ, Zeng F, Pan F (2013) Strain engineering induced interfacial self-assembly and intrinsic exchange bias in a manganite perovskite film. *Sci Rep*, 3:2542.

Cui B et al. (2014) Exchange bias field induced symmetry-breaking of magnetization rotation in two-dimension. *Appl Phys Lett*, 105:152402.

Das K, Das I (2015) A comparative study of magnetic training effect in bulk and nanocrystalline $La_{0.46}Ca_{0.54}MnO_3$ compound. *J Appl Phys*, 118:084302.

Ding JF et al. (2013) Interfacial spin glass state and exchange bias in manganite bilayers with competing magnetic orders. *Phys Rev B*, 87:054428.

Dong S, Gao F, Wang ZQ, Liu JM, Ren ZF (2007) Surface phase separation in nanosized charge-ordered manganites. *Appl Phys Lett*, 90:082508.

Dong S, Yu R, Yunoki S, Liu JM, Dagotto E (2008) Ferromagnetic tendency at the surface of CE-type charge ordered manganites. *Phys Rev B*, 78:064414.

Fan M, Zhang W, Jian J, Huang J, Wang H (2016) Strong perpendicular exchange bias in epitaxial $La_{0.7}Sr_{0.3}MnO_3$:$LaFeO_3$ nanocomposite thin films. *APL Materials*, 4:076105.

Frontera C, García-Muñoz JL, Beran P, Bellido N, Margiolaki I, Ritter C (2008) Short- and long-range orbital order in phase separated $Pr_{0.50}Ca_{0.50}Mn_{0.99}Ti_{0.01}O_3$: Its role in thermal hysteresis. *Chem Mater*, 20:3068–3075.

Gibert M, Zubko P, Scherwitzl R, Iniguez J, Triscone JM (2012) Exchange bias in $LaNiO_3$–$LaMnO_3$ superlattices. *Nature Mater*, 11:195–198.

Giri SK, Poddar A, Nath TK (2011) Surface spin glass and exchange bias effect in $Sm_{0.5}Ca_{0.5}MnO_3$ manganites nano particles. *AIP Adv*, 1:032110.

Giri SK, Yusuf SM, Mukadam MD, Nath TK (2014) Enhanced exchange bias effect in size modulated $Sm_{0.5}Ca_{0.5}MnO_3$ phase separated manganite. *J Appl Phys*, 115:093906.

He C et al. (2012) Interfacial ferromagnetism and exchange bias in $CaRuO_3$/$CaMnO_3$ Superlattices. *Phys Rev Lett*, 109:197202.

Hejtmánek J et al. (1999) Interplay between transport, magnetic, and ordering phenomena in $Sm_{1-x}Ca_xMnO_3$. *Phys Rev B*, 60:14057–14065.

Iniama G et al. (2014) Unexpected ferromagnetic ordering enhancement with crystallite size growth observed in $La_{0.5}Ca_{0.5}MnO_3$ nanoparticles. *J Appl Phys*, 116:113901.

Jammalamadaka SN, Rao SS, Bhat SV, Vanacken J, Moshchalkov VV (2012) Oscillatory exchange bias and training effects in nanocrystalline $Pr_{0.5}Ca_{0.5}MnO_3$. *AIP Adv*, 2:012169.

Jammalamadaka SN, Rao SS, Vanacken J, Stesmans A, Bhat SV, Moshchalkov VV (2011) Martensite-like transition and spin-glass behavior in nanocrystalline $Pr_{0.5}Ca_{0.5}MnO_3$. *AIP Adv*, 1:042151.

Jiang J, Henry LL, Gnanasekhar KI, Chen C, Meletis EI (2004) Self-assembly of highly epitaxial (La,Sr)MnO_3 nanorods on (001) $LaAlO_3$. *Nano Lett*, 4:741–745.

Jirak Z et al. (2010) Ferromagnetism versus charge ordering in the $Pr_{0.5}Ca_{0.5}MnO_3$ and $La_{0.5}Ca_{0.5}MnO_3$ nanocrystals. *Phys Rev B*, 81:024403.

Kobrinskii AL, Goldman AM, Varela M, Pennycook SJ (2009) Thickness dependence of the exchange bias in epitaxial manganite bilayers. *Phys Rev B*, 79:094405.

Levstik A et al. (2009) Ordering of polarons in the charge-disordered phase of $Pr_{0.7}Ca_{0.3}MnO_3$. *Phys Rev B*, 79:153110.

Liu W, Shi L, Zhou S, Zhao J, Li Y, Guo Y (2014) Size-dependent multiple magnetic phases and exchange bias effect in hole-doped double perovskite $La_{1.6}Sr_{0.4}NiMnO_6$. *J Phys D: Appl Phys*, 47:485003.

Lu CL et al. (2007) Charge-order breaking and ferromagnetism in $La_{0.4}Ca_{0.6}MnO_3$. *Appl Phys Lett*, 91:032502.

Manna PK, Yusuf SM, Shukla R, Tyagi AK (2011) Exchange bias in $BiFe_{0.8}Mn_{0.2}O_3$ nanoparticles with an antiferromagnetic core and a diluted antiferromagnetic shell. *Phys Rev B*, 83:184412.

Markovich V et al. (2010a) Size effect on the magnetic properties of antiferromagnetic $La_{0.2}Ca_{0.8}MnO_3$ nanoparticles. *Phys Rev B*, 81:094428.

Markovich V et al. (2010b) Size-driven magnetic transitions in $La_{1/3}Ca_{2/3}MnO_3$ nanoparticles. *J Appl Phys*, 108:063918.

Markovich V et al. (2012a) Size dependent magnetic properties of electron doped $La_{0.23}Ca_{0.77}MnO_3$ nanoparticles. *J Nanopart Res*, 14:1119.

Markovich V et al. (2012b) Magnetic properties of $Sm_{0.1}Ca_{0.9}MnO_3$ nanoparticles. *J Appl Phys*, 112:063921.

Markovich V et al. (2013) Size-dependent magnetism and exchange bias effect in $Sm_{0.27}Ca_{0.73}MnO_3$ nanoparticles. *J Nanopart Res*, 15:1862.

Markovich V et al. (2014) Particle size effects on charge ordering and exchange bias in nanosized $Sm_{0.43}Ca_{0.57}MnO_3$. *J Phys Chem C*, 118:7721–7729.

Martin C, Maignan A, Hervieu M, Raveau B (1999) Magnetic phase diagrams of $L_{1-x}A_xMnO_3$ manganites (L = Pr, Sm; A = Ca, Sr). *Phys Rev B*, 60:12191–12199.

Martinelli A et al. (2013) Structural, microstructural and magnetic properties of $(La_{1-x}Ca_x)MnO_3$ nanoparticles. *J Phys: Condens Matter*, 25:176003.

Peng JJ et al. (2014) Exchange bias in a single $LaMnO_3$ film induced by vertical electronic phase separation. *Phys Rev B*, 89:165129.

Peng J et al. (2015) Charge transfer and orbital reconstruction in strain-engineered $(La,Sr)MnO_3/LaNiO_3$ heterostructures. *ACS Appl Mater Interfaces*, 7:17700–17706.

Pramanik AK, Banerjee A (2010) Interparticle interaction and crossover in critical lines on field-temperature plane in $Pr_{0.5}Sr_{0.5}MnO_3$ nanoparticles. *Phys Rev B*, 82:094402.

Rao SS, Anuradha KN, Sarangi S, Bhat SV (2005) Weakening of charge order and antiferromagnetic to ferromagnetic switch over in $Pr_{0.5}Ca_{0.5}MnO_3$ nanowires. *Appl Phys Lett*, 87:182503.

Rao SS, Bhat SV (2009) Realizing the 'hindered charge ordered phase' in nanoscale charge ordered manganites: Magnetization, magneto-transport and EPR investigations. *J Phys Condens Matter*, 21:196005.

Rao SS, Bhat SV (2010) Probing the existing magnetic phases in $Pr_{0.5}Ca_{0.5}MnO_3$ (PCMO) nanowires and nanoparticles: Magnetization and magneto-transport investigations. *J Phys Condens Matter*, 22:116004.

Rostamnejadi A, Venkatesan M, Kameli P, Salamati H, Coey JMD (2014) Cooling-field dependence of exchange bias effect in $La_{0.45}Sr_{0.55}MnO_3$ nanoparticles. *J Appl Phys*, 116:043913.

Rozenberg E et al. (2008) Nanometer size effect on magnetic order in $La_{0.4}Ca_{0.6}MnO_3$: Predominant influence of doped electron localization. *Phys Rev B*, 78:052405.

Rozenberg E et al. (2009) Size and nonstoichiometry effects on magnetic properties of $La_{0.5}Ca_{0.5}MnO_3$ manganite. *IEEE Trans Magn*, 45:2576–2579.

Sadhu A et al. (2012) Ferromagnetism in lightly doped $Pr_{1-x}Ca_xMnO_3$ (x = 0.023, 0.036) nanoparticles synthesized by microwave irradiation. *Chem Mater*, 24:3758–3764.

Sadhu A, Bhattacharyya S (2013) Stacked Nanosheets of $Pr_{1-x}Ca_xMnO_3$ (x = 0.3 and 0.49): A ferromagnetic two-dimensional material with spontaneous exchange bias. *J Phys Chem C*, 117:26351–26360.

Sarkar T, Mukhopadhyay PK, Raychaudhuri AK, Banerjee S (2007) Structural, magnetic, and transport properties of nanoparticles of the manganite $Pr_{0.5}Ca_{0.5}MnO_3$. *J Appl Phys*, 101:124307.

Sarkar T, Ghosh B, Raychaudhuri AK, Chatterji T (2008a) Crystal structure and physical properties of half-doped manganite nanocrystals of less than 100 nm size. *Phys Rev B*, 77:235112.

Sarkar T, Raychaudhuri AK, Chatterji T (2008b) Size induced arrest of the room temperature crystallographic structure in nanoparticles of $La_{0.5}Ca_{0.5}MnO_3$. *Appl Phys Lett*, 92:123104.

Schumacher D et al. (2013) Inducing exchange bias in $La_{0.67}Sr_{0.33}MnO_{3-\delta}/SrTiO_3$ thin films by strain and oxygen deficiency. *Phys Rev B*, 88:144427.

Singh G, Bhat SV (2012) Charge order suppression, emergence of ferromagnetism and absence of exchange bias effect in $Bi_{0.25}Ca_{0.75}MnO_3$ nanoparticles: Electron paramagnetic resonance and magnetization studies. *J Appl Phys*, 111:123913.

Tian Y, Chen D, Jiao X (2006) $La_{1-x}Sr_xMnO_3$ ($x = 0$, 0.3, 0.5, 0.7) Nanoparticles nearly freestanding in water: Preparation and magnetic properties. *Chem Mater*, 18: 6088–6090.

Tian YF et al. (2014) Interfacial magnetic coupling in ultrathin all-manganite $La_{0.7}Sr_{0.3}MnO_3/TbMnO_3$ superlattices. *Appl Phys Lett*, 104:152404.

Vafaee M et al. (2016) The effect of interface roughness on exchange bias in $La_{0.7}Sr_{0.3}MnO_3/BiFeO_3$ heterostructures. *Appl Phys Lett*, 108:072401.

Wang X et al. (2009) The influence of the antiferromagnetic boundary on the magnetic property of La_2NiMnO_6. *Appl Phys Lett*, 95:252502.

Wang Y, Fan HJ (2011) Magnetic phase diagram and critical behavior of electron-doped $La_xCa_{1-x}MnO_3$ ($0 \leq x \leq 0.25$) nanoparticles. *Phys Rev B*, 83:224409.

Wu SM et al. (2010) Reversible electric control of exchange bias in a multiferroic field-effect device. *Nature Mater*, 9:756–761.

Wu SM, Cybart SA, Yi D, Parker JM, Ramesh R, Dynes RC (2013) Full electric control of exchange bias. *Phys Rev Lett*, 110:067202.

Zhang T, Dressel M (2009) Grain-size effects on the charge ordering and exchange bias in $Pr_{0.5}Ca_{0.5}MnO_3$: The role of spin configuration. *Phys Rev B*, 80:014435.

Zhang T, Wang XP, Fang QF (2011) Evolution of the electronic phase separation with magnetic field in bulk and nanometer $Pr_{0.67}Ca_{0.33}MnO_3$ particles. *J Phys Chem C*, 115:19482–19487.

Zhang W et al. (2015) Perpendicular exchange-biased magnetotransport at the vertical heterointerfaces in $La_{0.7}Sr_{0.3}MnO_3$: NiO nanocomposites. *ACS Appl Mater Interfaces*, 7:21646–21651.

Zhi M, Koneru A, Yang F, Manivannan A, Li J, Wu N (2012) Electrospun $La_{0.8}Sr_{0.2}MnO_3$ nanofibers for a high-temperature electrochemical carbon monoxide sensor. *Nanotechnology*, 23:305501.

Zhou S, Guo Y, Zhao J, He L, Wang C, Shi L (2011) Particle size effects on charge and spin correlations in $Nd_{0.5}Ca_{0.5}MnO_3$ nanoparticles. *J Phys Chem C*, 115:11500–11506.

Ziese M et al. (2010) Tailoring magnetic interlayer coupling in $La_{0.7}Sr_{0.3}MnO_3/SrRuO_3$ superlattices. *Phys Rev Lett*, 104:167203.

6

Monte Carlo Study of the Exchange Bias Effects in Magnetic Nanoparticles with Core–Shell Morphology

M. Vasilakaki,
G. Margaris,
E. Eftaxias, and
K. N. Trohidou

6.1 Introduction

Bi-magnetic nanoparticles (NPs) with core–shell morphology are gaining considerable interest both in research and technological applications [1]. The tuning of the magnetic properties through the interfacial exchange coupling of the two distinct components has been widely studied, initially in thin films [2]. Significantly, the recent advances in chemical synthesis [3,4] or in technical production of NPs (e.g., gas-phase NPs [5], super-fluid helium droplets [6]) have allowed the delicate control of their structural parameters (e.g., size, shape, and composition) and the possibility of designing new bi-magnetic NPs suitable for the development of permanent magnets [7], magnetic recording media [8], spin-valve sensors [9], microwave absorption [1], and biomedical applications (magnetic resonance imaging, drug delivery, or hyperthermia [10–12]).

Nowadays, the technological demand for thermal stability of very small particles (a few nm) at room temperature, namely the increase in their blocking temperature [8], has led to the search of a great variety of bi-magnetic core–shell NPs, whose core and shell are consisted of a soft and a hard magnetic phase of either ferromagnetic (FM), ferrimagnetic (FiM), antiferromagnetic (AFM), or even spin-glass like structure [1,13,14]. The synergetic combination of the properties of these two constituents and their interaction [1,15] can improve and tune the single-phase properties. In addition, the interplay between surface and interface (IF) effects can make the bi-magnetic NPs advantageous over single magnetic NPs in applications bringing about new functionalities of these two-phase systems [16,17].

The origin of the enhanced properties of the bi-magnetic NPs is considered to be the IF exchange coupling between the two different magnetic phases that results to the *exchange bias effect* [2,18]. This effect gives shifted hysteresis loop, with the exchange bias field (H_E) defined as the half of the loop width, and increased coercive field (H_C) under a cooling field procedure in a static magnetic field (H_{cool}).

Meiklejohn and Bean [18] first discovered the exchange bias effect in Co (FM core)/CoO (AFM shell) NPs, and they attributed it to an extra unidirectional anisotropy induced by the exchange coupling at the IF between the soft and the hard material [19,20]. In the subsequent years, the study of the exchange bias effect was developed in bilayered magnetic structures. Various models have been developed to explain the exchange bias mechanism in the layered structures, but still there is no definitive theory to account for the observed effects. This is attributed to the diversity of the studied magnetic nanostructured materials [21]. In the approach introduced by Malozemoff [22], the exchange bias effect does not appear in the perfect FM–AFM IF of a spin valve due to the lack of the IF's roughness. In other models, the exchange anisotropy is attributed to additional mechanisms such as uncompensated spins at the IF of the AFM [23] or to the motion of magnetic domains that are created in a diluted AFM [24]. Though these models and theories were developed for the explanation of the shifted hysteresis loops in layered systems, the same physical arguments are expected to hold for the NPs. Indeed, numerical simulations on core–shell NPs, that is, soft FM core–hard AFM [25,26] or FiM shell NPs [27–31] show that the exchange bias effect depends on the structure of the IF (uncompensated spins). Interestingly, even in the absence of additional roughness or lattice vacancies, the spins of the AFM shell which are coupled to the spins of the FM core can create an imperfect IF when they are not equally distributed on the FM–AFM IF of the NP. Also Monte Carlo (MC) simulations show [32] that the AFM shell itself induces the exchange anisotropy along the IF that is in turn responsible for the fact that the smaller NPs have higher coercive fields than the bigger ones at low temperatures. In addition, the vertical shift of the hysteresis loops usually observed in FM–AFM core–shell NPs has been attributed to the uncompensated spins of the AFM shell [33,34]. Thus, the explanation of the exchange bias effect in the case of core–shell NPs has similarities with the layered systems, but it can appear in all bi-magnetic NPs due to the intrinsic inhomogeneity caused by their shape and the finite-size effects that are associated with the IF and surface effects.

The FM core–AFM (or FiM) shell NPs have been extensively studied. More recently, there is an increasing interest in, the so-called, inverted and doubly inverted structures, in which the core is AFM, and the shell is FM or FiM, for example, MnO/Mn_3O_4, [35–41], FeO/Fe_3O_4, $Fe_{1-x}O/Fe_{3-\delta}O_4$ [42–54], $CoO/CoFe_2O_4$, CoO/Fe_3O_4 [55–61], Cr_2O_3/Fe_3O_4, [62–64], $FePt/\gamma-Fe_2O_3$ [65], or even multiferroic, for example, $BiFeO_3/\gamma-Fe_2O_3$ [66,67]. These inverted structures can overcome some of the limitations of FM–AFM NP systems [61–65], as the AFM structures can be much better controlled in the core than in the shell (where usually they are forced to grow in nonideal conditions). It has been demonstrated, experimentally and theoretically, that the poor crystallinity of the AFM counterpart can result in considerably inferior exchange bias properties [68,69]. In fact, inverted structures have already demonstrated very large coercivities and loop shifts, tunable blocking temperatures, enhanced Néel temperatures or proximity effects [35–69], and have been proposed as potential magneto electric random access memories [66]. However, despite their potential, systematic studies of size effects (i.e., core diameter or shell thickness) are still rather scarce [35,39,45,48,58–60]. Remarkably, similar effects related to the role of the contribution of the different magnetic phases (core vs. shell) also arise in other types of bi-magnetic core–shell NPs such as hard-FiM–soft-FiM versus soft-FiM–hard-FiM NPs known as *exchange-spring magnets* [7], in which systems with the hard core material can have enhanced or different properties from the ones with soft-FiM core [70,71]. Thus, understanding the role of the position of the diverse magnetic phases is of major importance in the development of novel applications of bi-magnetic core–shell NPs.

Magnetic NPs are commonly formed in random or ordered assemblies. In these assemblies, the crucial role of interparticle interactions in determining their magnetic properties have been recognized long ago [72].

The study of NP assemblies was based on the single spin model of Stoner–Wohlfarth [73–76]. However, as the NP size decreases, it was necessary to take into account its internal structure, namely core versus surface contribution. So numerical and experimental studies have been focused on the surface effects going beyond the Stoner–Wohlfarth model [77–80]. Currently, the exploitation of the exchange bias effect in assemblies of core–shell NPs has attracted a lot of interest [68,69,81–86]. Despite the research effort on the microscopic mechanism of this effect in individual NPs, much less attention has been paid so far to the modification of the magnetic hysteresis behavior due to interparticle interactions in these assemblies. In this direction, Co NPs embedded in Mn matrix [82] were shown to freeze below a temperature owing to the competition between the exchange anisotropy at the core–shell IF and the interparticle dipolar interactions, and an increase in the exchange bias field was reported. Similarly, increase of the exchange bias field due to magnetostatic interparticle coupling was found in stripes of Co/CoO NPs [87] and interdot magnetostatic interactions were shown to produce asymmetric anomalies in the magnetization reversal mechanism of Co/CoO dot arrays [88]. MC simulations on Co NPs embedded in Mn matrix show that the intraparticles characteristics could enhance the exchange bias effect, whereas Margaris et al. [69] have shown that in random assemblies of FM–AFM NPs the exchange interparticle interactions play a major role causing the increase in the H_C and the H_E with the concentration of the NPs. Therefore, the modification of the coercive and exchange-bias fields in cluster assembled NPs with core–shell morphology that results from the competition between the intra-particle exchange anisotropy and interparticle interactions is a challenging issue. It is evident that the basic understanding of the magnetic properties of bi-magnetic NPs in ordered and disordered structurally assemblies is of crucial importance for the next generation of high performance applications of magnetic nanomaterials.

6.1.1 Metropolis Monte Carlo Simulations for the Exchange Bias Behavior of Core–Shell Magnetic Nanoparticles

The MC simulation technique with the implementation of the Metropolis algorithm [89] has been proved a very powerful tool for the systematic study of the magnetic behavior of NPs and NP assemblies at finite temperature. Especially in the case of bi-magnetic NPs with core–shell morphology, the technique is advantageous because it gives the pos-sibility for an atomic scale treatment of the NPs, so the details of their microstructure along the IF can be explicitly studied. The appropriate choice of a model Hamiltonian is the starting point of MC simulations, and then the use of random number genera-tor to simulate statistical fluctuations to generate the correct thermodynamical prob-ability distribution according to a canonical ensemble [90]. In this way, one can obtain microscopic information about complex systems that cannot be studied analytically or that might not be accessible in a real system. Unlike the Landau-Lifshitz or Langevin equations, the MC scheme with the Metropolis algorithm provides the straightforward implementation of the temperature. To simulate the magnetic NPs and the NP assem-blies and to derive thermodynamic averages, the elementary physical quantity that we use is the spin. In the case of the single NPs, we consider a classical spin at each atomic site, and we simulate using the MC technique the stochastic movement of the system in the phase space. In the case of assemblies, we consider an effective spin to represent the magnetic state of each domain of the NP namely the core, the IF of the core, the IF of the shell, the shell, and the surface [91] unlike the single spin treatment of assemblies in the context of the SW model [73] used for assemblies. We must notice here that the number of effective spins we are using depends on the characteristics of the system we study, for example, the shell thickness or the magnetic structure of the NP [69].

Although the obtained dynamics in the MC simulations is intrinsic and the time evolution of the system does not come from any deterministic equation for the magneti-zation, the results of the MC simulations reproduce qualitatively the trends of the exper-imental data [91–93]. Actually this good qualitative agreement between the simulation results and the experimental data enable us to have a better insight into the nanoscale phenomena, though some of them stem from nonequilibrium processes [94].

During MC simulation, several MC steps (MCS) are performed where spins are suc-cessively chosen at random from a system of N spins, and a new orientation of each spin is generated by a small random deviation from its initial orientation. The attempted new direction is chosen in a spherical segment around the initial orientation. Then the energy difference ΔE between the new and the initial orientation is calculated. In the Metropolis MC algorithm, if $\Delta E \leq 0$, the new orientation is accepted, if $\Delta E > 0$, the attempted new orientation is accepted with a certain probability, the Boltzmann probability $\exp(-\Delta E/k_B T)$, the decision taken by a new random number [95]. In addition, a limit of the attempted deviation is set at 50% to detect confinement in metastable states that are responsible for the hysteresis and to achieve true relaxation in different temperatures [96–100]. In this algorithm, states are generated with a certain probability (Importance Sampling) by rejecting a certain number of the first MCS (thermalization process). In this way, the thermal averages of the thermodynamic quantities turn to arithmetical averages. Note here that special care has been taken of the time and ensemble averaging of the magnetization of the system by properly choosing the number of MCS, and a rather big number of different samples with independent random number sequences, corresponding to different realizations of thermal fluctuations.

In what follows, we review our MC simulations results for atomic scale modeling on FM–AFM, FM–FiM NPs, exchange-spring magnets FiM–FiM, and inverted and doubly inverted AFM–FiM NPs. The exchange bias field dependence on external parameters (temperature and applied cooling field) and the intrinsic particle properties (size of shell and core, size and type of anisotropy, and magnetic structure) are discussed. Then our MC results of the effect of the interparticle interactions on the exchange bias behavior of ordered and disordered assemblies in a mesoscopic scale approach are discussed. The effect of the exchange IF coupling on the macroscopic magnetic behavior of these NP assemblies and its interplay with the interparticle interactions is analyzed. The characteristics of the hysteresis loop and the temperature-dependent magnetization (field cooled [FC], zero-field cooled [ZFC]) are also studied.

A discussion on potential applications and a comparison with experimental findings are given in all cases.

6.2 Atomic Scale Modeling of the Exchange Bias Behavior of Core–Shell Nanoparticles

Modeling of the exchange bias behavior of core–shell NPs is a very challenging issue because of the variety of core–shell structures. The broad variation of particles' structural characteristics, such as composition, defects, shapes, or sizes, does not allow the formation of a uniform model. Thus, this modeling is based on experimental evidence, for example transmission electron microscope (TEM) and scanning transmission electron microscope (STEM) imaging, electron energy loss spectroscopy mapping and magnetization measurements (hysteresis loops and ZFC–FC magnetization curves). For this reason, in Section 6.2.1, we will introduce the general characteristics of a core–shell NP model, and we will continue in Section 6.2.2 with the description of the details of specific structures of single bi-magnetic NPs and the associated exchange bias behavior.

6.2.1 The Model

The most common shape of core–shell bi-magnetic NPs is the spherical one. In Figure 6.1, we give the 2D schematic representation of a spherical NP with N classical spins placed

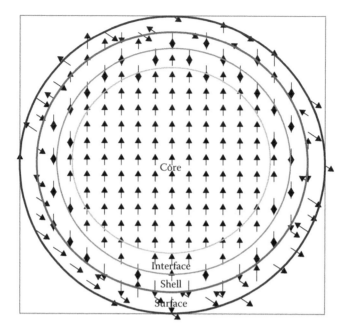

FIGURE 6.1 2D schematic representation of a spherical core–shell nanoparticle with spins placed at the nodes of a simple cubic lattice.

on the sites of a three-dimensional simple cubic (SC) lattice, within a radius of R lattice spacings from the central site. The NP consists of a FM, AFM or FiM core, and an AFM or FiM shell, surrounding the core (Figure 6.1). The IF between the core and the shell is defined by the spins at the outer layer of the core (core IF) and the first shell layer (shell IF). The surface of the particle is defined by the spins in the outer layer of the particle. In our simulations, as we will see in Section 6.2.2, we also consider NPs, of cubic shape or different other shapes corresponding to the experimental situation.

The classical spins in the particles interact with nearest neighbors Heisenberg exchange interaction, and at each crystal site, they experience a uniaxial anisotropy. We take into account explicitly the exchange interaction between the spins in the core, at the IF, in the shell, and at the surface [25,28,33]. In the presence of an external magnetic field, the energy of the system is as follows:

$$E = -J_{core} \sum_{\langle i,j \in core \rangle} \vec{S}_i \cdot \vec{S}_j - J_{IF} \sum_{\langle i \in core, j \in shell \rangle} \vec{S}_i \cdot \vec{S}_j - J_{shell} \sum_{\langle i,j \in shell \rangle} \vec{S}_i \cdot \vec{S}_j$$

$$- \sum_{i \in core} K_{core} (\vec{S}_i \cdot \hat{e}_i)^2 - \sum_{i \in IF} K_{IF} (\vec{S}_i \cdot \hat{e}_i)^2 - \sum_{i \in shell} K_{shell} (\vec{S}_i \cdot \hat{e}_i)^2 - \vec{H} \cdot \sum_i \vec{S}_i$$

where:

S_i is the atomic spin at site i

$\hat{e}_i$ is the unit vector in the direction of the easy axis of anisotropy at site i

The first term gives the exchange interaction between the spins in the core with exchange coupling constant J_{core}. The second term gives the exchange interaction at the IF between the core and the shell (with exchange coupling constant J_{IF}) and the third term gives the exchange interaction in the shell (with exchange coupling constant J_{shell}). The angular brackets denote a summation over the nearest neighbors only. The fourth and fifth terms give the anisotropy energy of the core and the IF with anisotropy coupling constants K_{core} and K_{IF}, respectively. The sixth term gives the anisotropy energy of the shell. Many studies have shown that, due to the reduced symmetry of the surface, the surface crystal anisotropy is stronger than the bulk [101,102] and of different type. So usually there are two anisotropies constants K_{shell} and K_S for the anisotropy inside the shell and at the surface, respectively. The last term gives the Zeeman energy in the presence of an external magnetic field.

In order to obtain the coercive and exchange bias field, we have calculated the complete hysteresis loop. A FC procedure was performed initially: we started at a temperature between the critical temperatures of the two constitutes of the core–shell NP [2], and we cooled the NP at a constant rate in the presence of a magnetic field H_{cool} along the z-axis. Once the desired temperature was reached, we slowly varied a magnetic field starting from a maximum value along the +z-direction reducing it by very small constant steps. At each field step, several MCS per spin (MCSS) were executed after thermalization, then for the subsequent MCSS, the magnetization was calculated, the field was changing again, and so on. The resulting hysteresis loops had a horizontal asymmetry. The value of the loop shift along the field axis was expressed by the exchange bias field $H_E = -(H_{right}+H_{left})/2$, and the coercive field was defined as $H_C = (H_{right}-H_{left})/2$, H_{right} and H_{left} being the points at which the loop intersected the field axis. Magnetization was normalized to the magnetization at saturation (M_s).

In our results, the temperature T is measured in units of J_{FM}/k_B, the magnetic field H in units of $J_{FM}/g\mu_B$, and the anisotropy coupling constants in units of J_{FM} that is the exchange coupling constant of a pure ferromagnet $J_{FM} = 1$ as a reference value. The estimation of the values of the chosen parameters is made using the mean field theory arguments and taking into account the size effects of the bulk values and the structural characteristics of each specific experimental system.

The number of the MCSS has been carefully chosen depending on the fluctuations of the magnetization and the exchange bias field values. Usually, we take MCSS of the order of 18×10^3 up to 4×10^4. Results have been checked by calculating the magnetization and the coercivity for different sequences of random numbers (10–40 runs). The statistical error is very small even at high temperatures. Including the corresponding error bars in our figures would not affect the information obtained from them, so they have been omitted.

6.2.2 Exchange Bias Behavior of Core–Shell Nanoparticles: Study of Intraparticle Characteristics

6.2.2.1 FM–AFM (or FiM) Core–Shell Nanoparticles

We have investigated the exchange bias mechanism in the spherical FM core–AFM shell NPs with a FM interfacial coupling. We studied the influence of the IF structure, the exchange IF coupling strength, and the shell thickness on their magnetic behavior [25,32]. Also we have studied the vertical shift and training effect that in some cases accompanies the exchange bias phenomenon.

We present our results for Co/CoO, FM core–AFM shell NPs. In this case, we set $J_{core} = 1$, $J_{shell} = -J_{core}/2$, because the Néel temperature of the AFM oxide ($T_N = 1.5$ [103]) is lower than the Curie temperature of the FM Co (for the SC lattice, $T_C = 2.9$). The IF exchange coupling constant J_{IF} is taken equal to the J_{shell} in size, in agreement with theoretical studies in layered systems [20,24], and the IF interaction is considered FM. The anisotropy constants are set as $K_C = 0.05 J_{FM}$, $K_{IF} = 0.5 J_{FM}$, $K_{shell} = 0.5 J_{FM}$, and $K_S = 1.0 J_{FM}$. For the NPs, there is some evidence that the easy axis is along one of the crystallographic directions, even though in cubic bulk materials, the easy axis is not uniaxial but along the three cubic axes [104]. The core and the IF anisotropies are therefore considered uniaxial along the z-axis. The surface is random.

We study a single FM–AFM NP with a shell thickness equal to four lattice spacings and a surface layer thickness of one lattice spacings. In Figure 6.2, we show our results for the exchange bias field and the coercive field as a function of temperature for two pairs of particles with radii of 10.0 and 11.0 (circles), 12.0 and 12.35 (squares). Assuming that the exchange interaction along the IF is FM, the bond energy for the spins across the FM–AFM IF is minimum when they are aligned as parallel and maximum when they aligned as antiparallel. The opposite would happen in the case of an AFM interaction along the IF. During the field-cooling procedure, the spins are aligned in such a way that the energy of the system is minimized. Along these lines, for a FM IF interaction the parallel spin alignment is favorable. This alignment together with the strong IF anisotropy makes it hard for the spins to turn when the field goes from H to −H, and it results in a high coercive field. When the spins align along the negative direction, by changing the field again from −H to H, they need less energy to turn. Developing this picture, we call up bonds, the pairs of spins along the FM–AFM IF, which are parallel and down bonds, the antiparallel ones. So according to our calculations, a NP with a radius of $R = 10.0$ has up bonds = 360 and down bonds = 318, and the NP with a radius

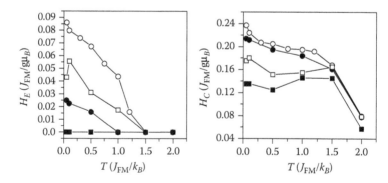

FIGURE 6.2 Temperature dependence of the exchange bias field (H_E) and coercive field (H_C) for two sets of FM–AFM nanoparticles with similar sizes but very different proportions of up and down bonds at the interface $R = 10$ (closed circles) and $R = 11.0$ (open circles), $R = 12.0$ (closed squares) and $R = 12.35$ (open squares). (Modified from Eftaxias, E. and Trohidou, K. N., *Phys. Rev. B*, 71, 134406, 2005.)

of $R = 11.0$ has up bonds $= 606$ and down bonds $= 288$. We can see that though these two particles are very close in size, they have very different numbers of up and down bonds.

Significantly, comparing each pair of particles of similar size in Figure 6.2, the ones with a bigger proportion of up bonds have higher coercive fields. Notably, the difference is more pronounced in the pair of particles with sizes 12.0 and 12.35 having the biggest difference in the proportion of up bonds. The exchange bias field decreases as the temperature increases due to thermal fluctuations that cancel the IF effects. Moreover, if we compare the pair of particles, the size dependence of the coercive field as a function of temperature has no difference in behavior from the one observed previously [35]. The smaller particles have a higher coercive field at low temperatures than the bigger ones, and this behavior is reversed at higher temperatures. However, the exchange bias field values depend on the number of bonds and not so much on the actual size of the particle. When the difference of the up and down bonds is big, such as $R = 11.0, 12.35$, the particle magnetization turns easily by going from the $-H$ to the H field. The H_E is stronger, and it follows the temperature dependence of the H_C, whereas in the case of small up or down bonds, the difference in the behavior of the two branches of the hysteresis loop is similar. This results in a reduced H_E with strong temperature dependence. Thus, our simulations show that it is the number of uncompensated bonds, namely the difference between FM and AFM bonds, along the IF that plays the key role in the appearance of H_E and not the number of uncompensated spins (difference between up and down spins). Note here that the temperature dependence of the exchange field is in good agreement with the experimental findings of Co/CoO NPs [105].

We have demonstrated that the strength of the exchange coupling constant of the AFM shell also affects the exchange bias behavior. In Figure 6.3, we show the temperature dependence of the coercive and the exchange bias fields for a particle with size $R = 11.0$ and shell thickness of 4 lattice spacings for three cases: (a) IF coupling constant $J_{IF} = -J_{shell} = J_{FM}/2$ (squares), (b) $J_{IF} = -2 J_{shell} = J_{FM}$ (circles), and (c) $J_{IF} = -J_{shell} = J_{FM}$ (triangles). Notably, an increase in the strength of the IF exchange coupling (case b) results in a reduction of the coercive field at low temperatures and an enhancement of

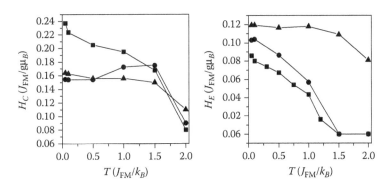

FIGURE 6.3 Coercive field (left) and exchange bias field (right) versus temperatures for a FM–AFM nanoparticle of radius $R = 11.0$ with (a) $J_{IF} = -J_{shell} = J_{FM}/2$ (squares), (b) $J_{IF} = -2 J_{shell} = J_{FM}$ (circles), (c) $J_{IF} = -J_{shell} = J_{FM}$ (triangles). (Modified from Eftaxias, E. and Trohidou, K. N., *Phys. Rev. B*, 71, 134406, 2005.)

the exchange bias field. This is due to the fact that the stronger IF-exchange coupling results in a faster reversal of the ferromagnetically aligned spins with the AFM shell. At temperatures higher than T_N, the behavior is similar to that of $J_{IF} = J_{FM}/2$, because at these temperatures the shell becomes paramagnetic (PM) and does not influence the FM core. Having the IF-exchange coupling constant enhanced as previously and by increasing also the exchange coupling constant of the AFM shell (case c), the coercive field decreases. The exchange bias field is further increased and also persists at high temperatures as expected because now T_N is higher than it was in the previous cases.

We must note that changing the type of the surface anisotropy from random to radial [106] affects very little our results. This is expected since, as we have demonstrated above, in the core–shell NPs the major contribution to the exchange bias effect comes from the IF.

It is well known that in the case of bilayers, the thickness of the AFM shell influences the exchange bias properties [20]. The same happens in the case of core–shell NPs in which the AFM shell thickness also affects the exchange bias behavior.

We considered a particle with core thickness seven lattice spacings, and we started to add AFM layers. In Figure 6.4, we have plotted the exchange bias field as a function of the shell thickness at a low temperature $T = 0.05\ J_{FM}/k_B$ and at a high $T = 1.0$ J_{FM}/k_B. At the low temperature, we observe that H_E is approximately constant after the second layer in agreement with the experimental findings of Reference [107] in which they observed very fast stabilization of the H_E with the oxygen dose in Co/CoO NPs. However, more AFM layers are needed to increase and stabilize the exchange bias field at the higher temperature ($T = 1.0\ J_{FM}/k_B$), because of the thermal fluctuations at the IF we need a thicker shell to stabilize the IF contribution. Also in Figure 6.4, we observe that after a certain number of AFM layers, roughly when the shell size becomes equal to the core size and then further increases, the exchange bias field is decreasing because of the enhancement of the AFM contribution that masks the IF role.

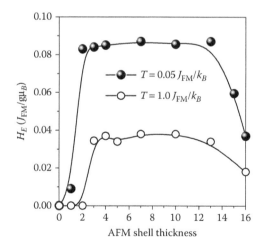

FIGURE 6.4 Exchange bias field as a function of the AFM shell thickness, starting from a particle with ferromagnetic core radius $R_C = 7.0$ at $T = 0.05\ J_{FM}/k_B$ and $T = 1.0\ J_{FM}/k_B$. Solid lines are guides to the eye. (Modified from Eftaxias, E. et al., *Mod. Phys. Lett. B*, 21, 1169–1177, 2007.)

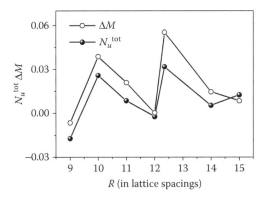

FIGURE 6.5 The vertical shift ΔM (open circles) and the normalized total number of uncompensated spins in the nanoparticle N_u^{tot} (closed circles) as a function of the particle size for a constant shell thickness of 4 lattice spacings at $T = 0.05\ J_{FM}/k_B$. Solid lines are guides to the eye. (Modified from Vasilakaki, M. et al., *Phys. Stat. Solid. A*, 205, 1865–1871, 2008.)

One effect related to the exchange bias is the vertical shift namely the shift in the magnetization remanence in the hysteresis loop. We have calculated the vertical shift for FM–AFM NPs with constant shell thickness 4 lattice spacings. The vertical shift arises from the total number of uncompensated spins in the shell and not only from the IF spins as it is the case for bilayers [2]. In Figure 6.5, we have plotted the vertical shift (open circles) and the normalized total number of uncompensated spins (closed circles) as a function of the particle size. Indeed, the behavior of the vertical shift curve follows the total number of uncompensated spin curve. This behavior is in agreement with the experimental work on Co/CoO NPs [34] in which the vertical shift in M was also attributed to the total number of uncompensated spins in the NP.

Another effect related to the exchange bias is the training effect, namely the reduction of the exchange bias field, the coercive field, and the magnetization remanence by the hysteresis loops cycling. Our studies have shown that in order to have a measurable training effect, we need a strong IF and shell anisotropy and medium size core, because only the IF and the shell contribute to the effect [30,33]. In Figure 6.6, we have plotted the training effect for a NP of radius $R = 7$ with AFM shell thickness 6, 10, and 15 lattice spacings. We consider strong shell and IF anisotropy $K_{IF} = 1.5\ J_{FM}$, $K_{shell} = 1.5\ J_{FM}$ with uniaxial along the z-axis core and IF anisotropies and random shell anisotropy. We observe that the training effect increases with the shell thickness. As the shell thickness increases, more shell layers contribute to the training effect, and at a thickness of 15 layers, the exchange bias field and consequently the unidirectional exchange anisotropy at the IF decrease so the spins are allowed to move with the field. The increase of the number of the random shell spins with the shell thickness affects the IF AFM spins, and they are forced more rapidly to follow the applied field. We also observe a smaller reduction of the H_C with the loop cycling than that of the H_E. This is due to the fact that the coercive field depends not only on the IF but also on the details of the whole NP.

We next present our studies on the exchange bias mechanism in spherical FM core–FiM shell NPs. We study the influence of the shell thickness, of the applied cooling field,

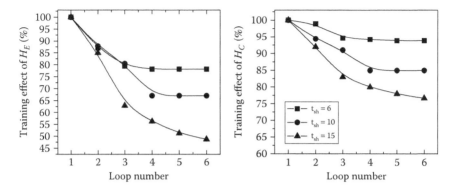

FIGURE 6.6 Percentage of reduction of the exchange bias field H_E (left) and coercive field H_C (right), with the loop cycling for a FM–AFM nanoparticle with size of the FM core 7 lattice spacings and AFM shell size 6,10,15 lattice spacings at temperature $T = 0.01\ J_{FM}/k_B$.

of the shape of the NP on their exchange bias behavior together with the vertical shift, the training, and the aging effect.

In the case of FM–FiM NPs, we considered the magnitude of the atomic spins in the two FiM sublattices of the shell equal to 1 and 1.5, respectively. The exchange coupling constant of the core is $1\ J_{FM}$, of the IF $0.5\ J_{FM}$ and that of the shell $-0.5\ J_{FM}$. The anisotropy constants are set as $K_C = 0.05\ J_{FM}$, $K_{IF} = 0.5\ J_{FM}$ for the FM IF and $K_{IF} = 1.5\ J_{FM}$ for the FiM IF, $K_{shell} = 1.5\ J_{FM}$, and $K_S = 1.5\ J_{FM}$ [28]. We have introduced the strong random anisotropy in the shell and at surface in order to simulate a spin-glass like phase.

We first investigated the role of the disordered FiM shell thickness in the exchange bias behavior of the NP. For a spherical NP with core size 5 lattice spacings, we plotted in Figure 6.7 the shell thickness dependence for the H_C, the H_E, and ΔM at a low temperature $T = 0.01\ J_{FM}/k_B$. We observe that the H_C initially increases continuously and then remains constant for a broad range of shell thicknesses. The increase in H_C with the increase in the disordered layer is expected because of the fact that the thicker shell

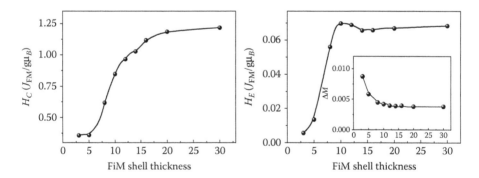

FIGURE 6.7 Shell thickness dependence of the coercive field H_C and the exchange bias field H_E. Inset, the vertical shift (ΔM) for a FM–FiM nanoparticle with core radius $R_C = 5$ lattice spacings. (Modified from Vasilakaki, M. and Trohidou, K. N., *Phys. Rev. B*, 79, 144402, 2009.)

has bigger disorder. H_E also increases more abruptly at the beginning, it reaches a maximum value and then decreases slowly, and it remains constant with the shell thickness. This behavior is different from the AFM shell case in which we have decrease of H_E after a shell thickness size (see Figure 6.4). We expect that in the case of the FiM disordered shell the reduction will also occur but for much bigger shell thickness than in the AFM ordered shell case. Our simulations also show that the remanent magnetization has an exponential decay with the shell thickness [28]. This exponential reduction of the remanent magnetization is attributed to the enhancement of the disorder with the increase of the FiM shell. The magnetic contribution is diminished as it has also been observed experimentally [108]. The vertical shift also decreases with the shell size and vanishes after a critical size of around 10 lattice spacings (inset of Figure 6.7). Our MC results agree with the experimental findings for Fe/Fe oxide core–shell NPs by Baker et al. [108].

The behavior of the vertical shift confirms the fact that ΔM originates from the uncompensated spins in the shell as in the AFM shell case [30]. As the shell thickness increases, the number of the uncompensated spins relatively to the total number of the spins in the NP decreases; subsequently ΔM decreases.

In Figure 6.8, we show the coercive field and the exchange bias field, as a function of temperature for a NP with FM core 5 lattice spacings and FiM shell thickness 9 lattice spacings. We observe that H_E decreases rapidly with increasing temperature, because the IF interaction is masked by the thermal fluctuations. We also observe that the decrease in H_E is exponential as it is observed experimentally in Reference [108]. The coercive field also exhibits an exponential decay with temperature in agreement with the experimental findings in Reference [109]. The observed temperature dependence of H_C and H_E is different from that reported in FM–AFM bilayers [110] and FM–AFM NPs [32,33,111].

There are many studies [110,112,113] mainly on FM–AFM bilayers that show that H_E and H_C can be tuned by the strength of the cooling field. For low cooling field values, the exchange bias and the coercive field increase although for higher cooling fields the magnetic behavior depends on the type of the interfacial couplings. Fiorani et al. [114] in their study of granular systems where Fe particles are embedded in a Fe oxide matrix with a spin-glass like phase showed that H_E and H_C depend on the cooling field and the

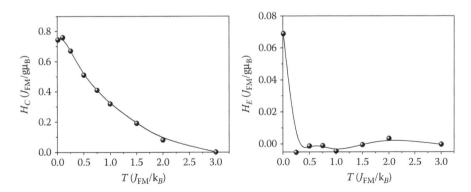

FIGURE 6.8 H_C and H_E as a function of temperature for FM–FiM nanoparticles with core $R = 5$ lattice spacings and shell thickness 9 lattice spacings.

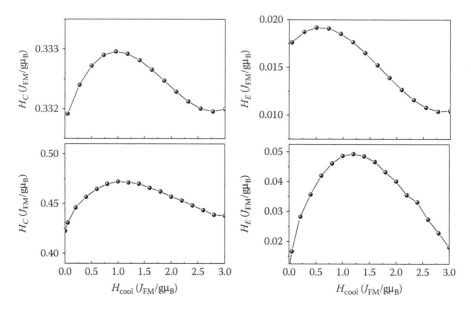

FIGURE 6.9 Cooling field dependence of the coercive field H_C and the exchange bias field H_E for a FM–FiM nanoparticle with core size five lattice spacings and shell thickness of 4 (upper panel) and 7 (lower panel) lattice spacings. (Modified from Vasilakaki, M. and Trohidou, K. N., *Phys. Rev. B*, 79, 144402, 2009.)

temperature from which the cooling field process initiates. They attributed this behavior to the fact that depending on the cooling field value, an energetically favorable spin configuration at the spin-glass-like shell is selected through the exchange interaction between the frozen FiM spins and the FM spins at the IF.

We have simulated the Cooling Field procedure for various applied cooling fields H_{cool} for a NP of FM core size 5 lattice spacings and FiM shell of 4 and 7 lattice spacings thickness. In Figure 6.9, we show our MC results of the H_{cool} dependence of H_C and H_E. We observe that the increase of H_{cool} causes an increase in both H_E and H_C. The gradual increase in H_{cool} tends to align a certain amount of FI spins at the IF along the field direction. Further increase in H_{cool} results in a decrease of these two quantities. For these higher cooling field values, the Zeeman coupling between the field and the FI spins dominates the magnetic interactions inside the system. So the FI spins follow the applied field and as a result the exchange bias field and the coercive field decrease. As the shell thickness increases (Figure 6.9), we observe that H_E decreases faster indicating that the FI shell plays the major role to the effect. We note here that studies on the cooling field dependence of the H_E for FM–AFM ultrafine particles [115] have shown that the H_E increases with the increasing cooling field and then saturates for field values above 20 kOe. This cooling field behavior is attributed to the polarization of uncompensated spins as the cooling field is increased. However, in the case of FM–FiM NPs with disordered shell, the existence of the spin-glass like phase causes the decrease of the H_E for large cooling fields. This is in agreement with the behavior of FM–AFM NP systems with spin disorder [116].

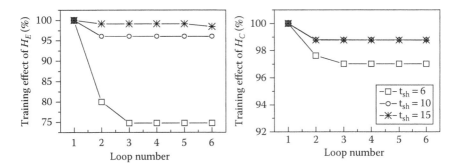

FIGURE 6.10 Percentage of the training effect of coercive H_C and exchange bias field H_E for FM core–FiMshell nanoparticles with core size of 7 lattice spacings and shell thicknesses of 6, 10, and 15 lattice spacings, respectively.

Moreover, we have investigated the training effect. In Figure 6.10, we have plotted the training of H_E and H_C for three NPs with core radius of seven lattice spacings and shell thicknesses of 6, 10, and 15 lattice spacings. The hysteresis loop was calculated six consecutive times at temperature $T = 0.01\ J_{FM}/k_B$ with cooling field $H_{cool} = 0.4\ J_{FM}/g\mu_B$. In Figure 6.10, we observe that in the NP with shell thickness seven lattice spacings after the first loop, there is a big reduction of H_E, whereas exchange bias field for the other two NPs has smaller reduction with the loop cycling. These two NPs have the same size of the exchange bias field as our simulations show. H_C has a bigger reduction with the loop cycling for the smaller shell NP than in the other two shell sizes. This behavior indicates that the IF has major contribution in the training effect for the chosen NPs and that the bigger the shell thickness is the harder the shell spins rotate so the training effect is reduced. Our simulations showed that as the shell thickness decreases, we have contribution from the shell and the core to the training effect. Actually this behavior is quite opposite to the AFM shell as we showed previously indicating that the training effect depends not only on the IF properties but also on the particle size and the shell properties. Indeed, it has been found experimentally that the training effect has a small reduction ~12% in the case of FM–AFM Co/CoO [117] NP system whereas it has a large 50% reduction in FM/FI NPs [92,118] with small shell thickness.

When the FiM shell is disordered (in a spin-glass like phase), collective nonequilibrium dynamics is present leading to the aging effect, that is, the slowing down of the spin dynamics with increasing the time t_w (waiting time) spent in the frozen state, before any field variation. We have numerically investigated the exchange bias properties of a single NP with a FM core surrounded by a magnetically disordered FiM shell. The anisotropy constants are set as $K_C = 0.05\ J_{FM}$, $K_{IF} = 0.5\ J_{FM}$, $K_{shell} = 1.5\ J_{FM}$, and $K_S = 1.0\ J_{FM}$. We introduced the strong random anisotropy at the IF, in the shell, and at surface in order to simulate a spin-glass like phase [92]. We simulate the field cooling procedure starting with the NP at temperature $T = 3.0\ J_{FM}/k_B$ (for the sc lattice the Curie temperature $T_C = 2.9\ J_{FM}/k_B$). The NP is cooled down to the temperature of $0.75\ J_{FM}/k_B$ in zero field. At $T_i = 0.75\ J_{FM}/k_B$, we apply h_L along the z-axis for different waiting times t_w (1, 10, 50, 100, 1000), expressed in MCSS). Then, we continue the field cooling at a constant rate, down to $T_f = 0.01\ J_{FM}/k_B$, in which the loop is calculated [92]. The aging of the system through an

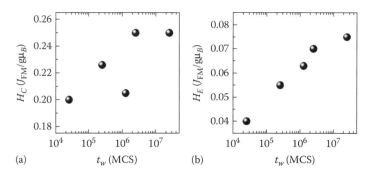

FIGURE 6.11 MC simulations of the coercive field (H_C) (a) and exchange bias field (H_E), (b) as a function of the waiting time t_w. (Modified from Trohidou, K. N. et al., *J. Magn. Magn. Mater.*, 316, e82–e85, 2007.)

increase of t_w at T_i, enhances the exchange bias effect at temperature $T_f = 0.01\ J_{FM}/k_B$ as we can see in Figure 6.11. As t_w elapses at T_i, the shell spins relax toward a more stable energy state, in which the IF exchange interaction energy with the moments of the FM component is minimized. Then, with reducing the temperature from T_i to T_f, the system remains trapped in this energy minimum, as it is more difficult to overcome the energy barriers separating different states. Therefore, by properly aging the sample, a final spin configuration at T_f is selected, corresponding to a stronger exchange coupling at the IF, which is responsible for the increase in H_E. The relaxation process taking place during t_w results in a net increase of the shell magnetization along the cooling field direction, accounting for a small increase in M_r versus t_w at T_f [92]. In the exchange bias phenomenon, a change in H_E is often accompanied by a change in H_C. In our case, an enhanced H_C is found after waiting for a long t_w, but fluctuating values are measured at shorter t_w (Figure 6.11) because intrinsic sources of anisotropy other than exchange anisotropy contribute to its value.

These results are in agreement with the experimental ones [92,114], for a system of iron particles (mean size ~6 nm) dispersed in a structurally and magnetically disordered iron oxide matrix (a mixture of Fe_3O_4 and γ-Fe_2O_3; mean grain size ~2 nm).

Shape effects influence the exchange bias behavior of the NPs, because they modify the shape of the IF and the surface and consequently the number of uncompensated spins. We study the shape effect on the exchange bias and the coercive field of FMcore–FiM shell NPs for four different shapes: sphere, cube, octahedron, and truncated cuboctahedron (Figure 6.12). The core and shell anisotropy were assumed to be uniaxial along the z axis ($K_{core} = 0.05\ J_{FM}$, $K_{IF} = K_{shell} = 0.5\ J_{FM}$) and the surface anisotropy randomly oriented $K_S = 1.5\ J_{FM}$ according to experimental indications for NPs produced with the ion-beam technique [119].

In Figure 6.13, the temperature dependence of the coercive and the exchange bias field are plotted for the FM–FiM NPs of cubic, octahedral, and truncated cuboctahedral shape together with the spherical ones for comparison. At low temperatures, the exchange bias field is significant only for the spherical NP due to the large number of the uncompensated IF bonds. The other shapes are almost symmetrical at the IF so the exchange bias effect of these NPs is negligible. As the temperature increases up to $T = 0.5\ J_{FM}/k_B$ the exchange bias field of the spherical NP goes to zero because the interfacial spins thermally

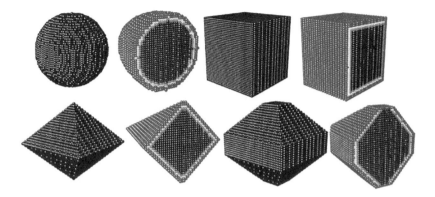

FIGURE 6.12 3D shapes of spherical, cubic, octahedral, truncated cuboctahedral complex FM–FiM nanoparticles. In each of them the internal structure (core, interface, shell, and surface) is also shown.

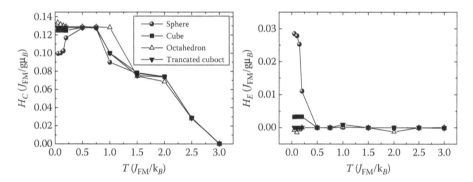

FIGURE 6.13 Temperature dependence of H_C and H_E for FM–FiM nanoparticle with core size $R = 16$ lattice constants $t_{SH} = 3$ lattice constants for the different shapes.

fluctuate breaking the uncompensated bonds. Despite this breaking they cannot follow the applied field because its competition with the hard shell does not let them rotate freely so the coercive field increases until $T = 0.75\,J_{FM}/k_B$. Then as the temperature increases further, the unfrozen shell spins start to rotate along with the PM core spins leading to the reduction of the coercive field. In cubic, octahedral, and truncated cuboctahedral NPs at low temperatures the shell spins remain blocked whereas at temperature $T = 1.0\,J_{FM}/k_B$ the blocked surface spins start to rotate dragging the PM core causing the reduction of H_C.

6.2.2.2 Exchange Spring Nanomagnets: FiM Soft–FiM Hard Core–Shell Nanoparticles

Though after the discovery of the exchange bias effect in Co/CoO NPs the studies had been focused on FM core–AFM (or FiM) shell NPs in the recent years, there is a vast amount of research on experimental systems of bi-magnetic NPs with a variety of combinations in the core and shell magnetic constituents. In what follows, we will present our studies of the exchange bias behavior of these systems discussing the underline mechanisms.

First, we present our results for the study of exchange bias effects in the core–shell NPs: FiM soft–FiM hard and FiM hard–FiM soft, known as exchange-spring magnets, simulating the experimental system of iron oxide—manganese oxide based NPs of Reference [27].

In the case of FiM soft–FiM hard (Fe_3O_4/Mn_3O_4), we consider the magnitude of the atomic spins in the two FiM sublattices of the core to be equal to 1.0 and 1.4 and of the shell to be equal to 1 and 0.5, respectively. The anisotropy constants are set $K_C = 0.01\,J_{FM}$, $K_{IF} = 0.01\,J_{FM}$ for the FM IF and $K_{IF} = 0.05\,J_{FM}$ for the FiM IF, $K_{shell} = 0.05\,J_{FM}$ and $K_S = 1.2\,J_{FM}$. The critical temperature of Fe_3O_4 is 20 times larger than the critical temperature of Mn_3O_4; thus, we set the exchange coupling constant of the soft counterpart $J_{core} = -30.0\,J_{FM}$ and of the hard material $J_{shell} = -1.5\,J_{FM}$. We set $J_{IF} = J_{core}$. We introduce strong radial anisotropy at the surface and uniaxial along z axis in the rest of the particle. We also examine an inverted spherical particle FiM hard–FiM soft (Mn_3O_4/Fe_3O_4) with analogous parameters and $K_S = 0.05\,J_{FM}$ due to smaller particle size.

It is demonstrated by Reference [27] that these NPs have an interfacial AFM coupling that leads to the presence of positive exchange bias at high enough cooling fields, which has never been observed neither in core–shell NPs nor in hard-soft bilayers. Indeed our MC simulations confirm the experimental observations of the cooling field effect on the H_C and H_E of two different NPs: a FiM soft–FiM hard with core radius 10 lattice constants and shell thickness 4 lattice constant and a special notched morphology and a spherical FiM hard–FiM soft with core radius 4 lattice spacings and shell thickness 5 lattice spacings. As we can see from Figure 6.14a as the cooling field increases, the H_E decreases and, for large cooling fields, changes sign becoming positive. This effect is ascribed to an AFM coupling between IF layers. In contrast to H_E, the cooling field only affects weakly the coercivity H_C that is mainly governed by the intrinsic characteristics rather than the AFM character of the coupling. The simulations of the spherical inverted NP show the same trends in H_C and H_E (Figure 6.14b) though the sign reversal of H_E occurs at considerably higher cooling fields as observed experimentally. Importantly, the core size or the core–shell morphology affects the strength of the AFM coupling but not its existence, and it results in an easy control of the sign of the exchange bias by means of the cooling field.

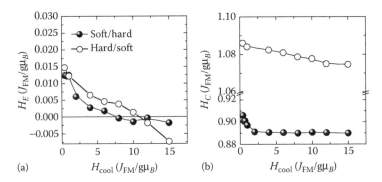

FIGURE 6.14 Monte Carlo simulations of the exchange bias H_E (a) and the coercive H_C (b) fields as a function of the applied cooling field H_{cool} for the soft–hard with notched morphology and the spherical hard–soft nanoparticles. (Modified from Estrader, A. et al., *Nat. Commun.*, 4, 2960, 2013.)

6.2.2.3 Inverted AFM Core–FiM Shell Nanoparticles

At the nanoscale the microstructural differences due to the oxidation state of the transition metals may lead to novel magnetic properties [120]. The rather critical structural alterations that often occur as the size is reduced down to few nanometers have to be carefully studied and properly analyzed. In this regard, we have studied AFM soft–FiM hard core–shell NPs that show a broad dispersion of magnetic properties going from small $R = 4.5$ lattice spacings to large $R = 9$ lattice spacings NP size [121]. The experimental results reveal that although the Fe_xO core in the large NPs is AFM and the shell has a Verwey transition, the Fe_xO core in the small particles is highly nonstoichiometric and strained, displaying no significant antiferromagnetism. We have simulated the large cubic NPs on a SC lattice, consisting of an AFM core with an edge length ($L_c = 12$) and a FiM shell ($t_{SH} = 3$), surrounding the core and a surface layer surrounding the shell. We set $J_{core} = -0.5 J_{FM}$, $J_{shell} = -1.5 J_{FM}$, $K_C = 0.05 J_{FM}$, $K_{IF} = K_{shell} = 0.3 J_{FM}$, and $K_S = 3.0 J_{FM}$. Based on the neutron diffraction results, K_C was assumed to be along the [1 1 0] direction and constant in the whole temperature range studied. To account for the strong nonmonotonic temperature dependence of the K of Fe_3O_4 around the Verwey transition [122], we have set a different K_{shell} value at each simulated temperature (Figure 6.15). The surface anisotropy is taken to be random and smoothly changing with T from $K_S = 3.0 J_{FM}$ ($T = 0.01 J_{FM}/k_B$) to $1.6 J_{FM}$ ($T = 0.5 J_{FM}/k_B$). To account for the experimental random distribution of NPs, the cooling fields have been applied in different directions defined by spherical coordinates (θ, φ), where $\theta = 0, 15,..., 180$ and $\varphi = 0, 15,..., 345$. The final hysteresis loop was calculated by averaging the hysteresis loops for each magnetic field direction.

Assuming the known strong temperature dependence of the anisotropy of Fe_3O_4, [122] for the shell anisotropy, K_{shell} (see the inset in Figure 6.15), and a reorientation of the easy axis from [1 1 1] to [1 0 0] direction around T_V, MC simulations clearly showed that

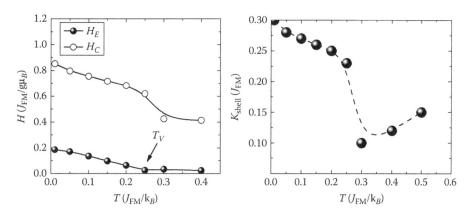

FIGURE 6.15 Monte Carlo simulation of the temperature dependence of H_C and H_E (left panel) for the large AFM soft–FiM hard core–shell nanoparticles of $R = 9.0$ lattice spacings calculated using the temperature dependence of the Fe_3O_4-shell anisotropy (right panel) and the change of the anisotropy orientation around the temperature T_V corresponding to the Verwey transition of the FiM shell. (Modified from Estrader, M. et al., *Nanoscale*, 7, 3002–3015, 2015.)

although H_C has a rather steep change at T_V, H_E changes more smoothly (Figure 6.15), in concordance with the experimental results [121].

On the other hand, the small NPs are simulated using a spherical morphology with a FiM shell ($t_{SH} = 3$ lattice constants) and a core size of three lattice spacings in diameter. The material parameters for the shell are the ones used above for the large particles at low T as there is no indication of the presence of a Verwey.

The small NPs were simulated using a spherical morphology with a FiM shell ($t_{SH} = 3$ lattice constants) and a core size of three lattice spacings in diameter. The material parameters for the shell were the ones used above for the large particles at low T, as there is no indication of the presence of a Verwey or Néel transitions due to the small particle size. Significantly, the large H_E and H_C observed for the small NPs indicated that, even if the AFM-core is nonmagnetic, the uncompensated spins of the FiM-shell at the AFM–FiM IF may also contribute to the exchange bias properties. In fact, MC simulations have demonstrated that a core–shell NP with a PM core ($J_{core} = 0.00$, $K_C = 0.00$) can have a rather large exchange bias (Figure 6.16). Notably, the presence of a sizable exchange bias can be explained without the need of an AFM or FiM (with the same parameters as those of the shell) counterpart. Although it is known that surface effects can give rise to an exchange bias [123], MC simulations have shown that the presence of a core–shell IF (even when the core is PM) can contribute significantly to the exchange bias properties.

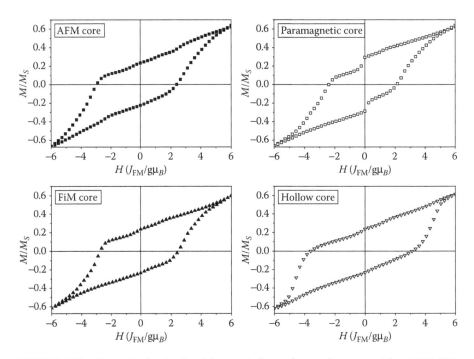

FIGURE 6.16 Monte Carlo simulated hysteresis loops for small nanoparticles: AFM–FiM (closed squares), paramagnetic–FiM (open squares), pure FiM (up triangles) and hollow-FiM (down triangles) nanoparticles of $R = 4.5$ lattice constants. (Modified from Estrader, M. et al., *Nanoscale*, 7, 3002–3015, 2015.)

This exchange bias is in a sense similar to the exchange bias observed in a hollow FiM NPs assuming that the inner shell is disordered in hollow structures.

The results clearly showed that different exchange bias behavior of AFM–FiM core–shell systems stems from a combination of factors such as size effects and structural characteristics.

We have extended our study of the inverted AFMsoft–FiMhard NPs to two main types of inverted structures, depending on the transition temperature of the materials: the inverted and doubly inverted. The "inverted" NPs are those in which the Curie temperature of the FiM shell, T_C, is larger than the Néel temperature of the AFM core, T_N, that is, $T_C > T_N$, for example, FeO/Fe$_3$O$_4$. On the other hand, if $T_N > T_C$, the systems are usually denoted by *doubly inverted*, for example, MnO/Mn$_3$O$_4$. Although this type of structure is rarely studied in thin film systems, the available results evidence rather interesting properties [124,125]. Similarly, doubly inverted core–shell NPs exhibit some novel properties such as a nonmonotonic dependence of the coercivity and the loop shift on the core size [35]. Here we present our MC simulations of the magnetic behavior of inverted and doubly inverted spherical AFM–FiM NPs. To take into account the difference in transition temperatures, for the first case we considered the exchange coupling constant of the core as $J_{core} = -0.5\,J_{FM}$ and that of the shell as $J_{shell} = -1.5\,J_{FM}$ and for the second case J_{shell} was considered the same, but J_{core} was increased to account for the larger T_N. Thus, the exchange coupling constants were set to $J_{core} = -3.0\,J_{FM}$ and $J_{shell} = -1.5\,J_{FM}$. The exchange coupling constant of the IF J_{IF} was taken equal to that of the shell J_{shell}. The anisotropy parameters were $K_{core} = 0.05\,J_{FM}$, $K_{IF} = K_{shell} = 0.5\,J_{FM}$, and $K_S = 1.4\,J_{FM}$ (of radial type) [126].

The hysteresis loops were calculated after a field cooling procedure starting at temperature $T = 7.0\,J_{FM}/k_B$ down to $T = 0.01\,J_{FM}/k_B$, at a constant rate under a static magnetic field $H_{cool} = 4.0\,J_{FM}/g\mu_B$ directed along the z-axis. The value of H_{cool} was selected as the optimum value to observe maximum H_E as Figure 6.17 shows.

In Figure 6.18 are plotted the MC results for the coercive and the exchange bias field as a function of the AFM core diameter D_{core} for shell thickness 0, 1, 4, 8 lattice spacings for the doubly inverted (a), (b) and inverted (c), (d), respectively. The shell thickness 0 corresponds to the pure AFM NPs with the same exchange coupling and anisotropy

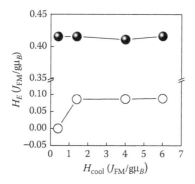

FIGURE 6.17 Cooling field, H_{cool} dependence of H_E for the doubly inverted AFM–FiM nanoparticles with core diameter $D_{core} = 8.2$ (closed symbols) and 26 (open symbols), and shell thickness 4 lattice spacings. (Modified from Vasilakaki, M. et al., *Sci. Rep.*, 5, 9609, 2015.)

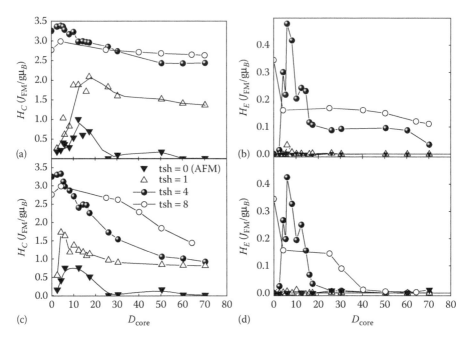

FIGURE 6.18 Dependence of the coercivity, H_C (a), (c) and the exchange bias field, H_E (b), (d) on the AFM core size (D_{core}) for the doubly inverted ($T_C < T_N - J_{core} > J_{shell}$) structures (upper panel) and inverted ($T_C > T_N - J_{core} < J_{shell}$) structures (lower panel) for FI shell thickness $t_{SH} = 0, 1, 4$, and 8 lattice constants.

constants with the doubly inverted ones. The exchange bias effect in this case is small, and it vanishes rapidly in the doubly inverted structures. Actually in the pure AFM NPs, the maxima in H_C and H_E appear for the sizes that have a big IF contribution but then H_E vanishes. Also Figure 6.18 for $t_{SH} = 1$ lattice spacing shows that the very thin shell thickness cannot contribute to an exchange bias phenomenon so H_E disappears.

For larger shell thickness of 4 and 8 lattice spacings, the results clearly showed that inverted structures can result in sizable loop shifts and coercivity enhancements, similar to conventional FM–AFM structures [2]. Nevertheless, contrary to conventional systems, a strong nonmonotonic behavior was observed. Significantly, both H_C and H_E exhibited maximum values for rather small core diameters (D_{core}). Moreover, the maximum H_C was obtained for very small D_{core} (e.g., $D_{core} = 4.2$ for $t_{SH} = 4$), whereas the maximum H_E was observed for slightly larger D_{core} (e.g., $D_{core} = 6$ for $t_{SH} = 4$). These results were in qualitative agreement with the experimental doubly inverse MnO/Mn_3O_4 NPs case, which also showed an analogous nonmonotonic dependence of H_C and H_E on D_{core} [35].

The simulations have revealed a number of notable results that, in some cases, are in clear contrast with the existing knowledge on the exchange bias behavior of thin film and FM–AFM core–shell NPs: (1) both H_C and H_E exhibited a strong nonmonotonic behavior with D_{core}; (2) the maximum H_C and H_E was obtained for rather small D_{core}, with sizes comparable to those of the shell; (3) the largest H_C was obtained for smaller D_{core} than for H_E; (4) although H_E and H_C were increased for thinner $t_{SH} = 4$ rather than $t_{SH} = 6$

(not shown here) or 8 [126] for small D_{core}, they showed an opposite behavior at large D_{core}. The decrease of H_E and H_C for large D_{core} was pushed to larger D_{core} as t_{SH} increased.

Significantly, the behavior of the exchange bias and the coercive field in doubly inverted AFM core–hard FiM shell NP systems has been shown to depend on the core size in a different way at various core size ranges. For very small core sizes, there was contribution on H_E from the surface uncompensated spins as our calculations showed in Reference [126]. For moderate core sizes, the uncompensated spins of the core and the shell IF also contributed to the exchange bias, resulting in a maximum H_E value. For even larger core diameters, the exchange in the core, J_{core} and the AFM character of the core determined H_E. For large D_{core}, the whole shell played the role of the shell IF; thus, the exchange bias effect for these core sizes has increased with shell thickness, in contrast to conventional systems. The study of the role of the shell thickness indicated that a sizable shell contribution was needed to ensure enhancement of the exchange bias properties.

Notably, this unusual behavior is to some extent different in single inverted AFM core–FiM shell NPs shown in Figure 6.18 (lower panel). Namely, although H_C exhibits a $H_C \propto t_{SH}$ dependence for large D_{cores} similar to the doubly inverted case, H_E shows this inverse behavior only in a very narrow range of D_{cores} as H_E vanishes at large D_{cores}. In this case, as J_{core} is weaker, there is no competition between the core and the shell. Thus, the shell drags the core spins with the consequent decrease of H_E and H_C. Consequently, the doubly inverted structures present improved properties compared with the single inverted ones, especially for large D_{cores}.

In order to clarify the role of the soft AFM core in the magnetic behavior of the inverted core–shell NPs, we have examined four different structures of NPs with a constant FiM shell thickness of $t_{SH} = 4$: (1) doubly inverted core–shell NP, (2) a hollow NP with a core IF (i.e., including core IF, shell IF, shell, and shell surface—removing the bulk part of the AFM core), (3) a hollow NP without a core IF (i.e., including shell IF, shell, and shell surface—removing the all the contributions from the AFM core), and (4) a NP with FiM core. We must notice that in the latter case, we have a larger core magnetic component due to the change in the spin magnitude of the one sublattice from 1 to 1.5. For all the studied structures, we have kept the same anisotropy and exchange coupling constants.

In Figure 6.19, we observe that, for the hollow NPs with the presence of the AFM IF, the absence of core (circles) affects only marginally the H_C and H_E behavior. This can be attributed to the fact that the core magnetization component is very small due to AFM coupling and cannot affect H_C and H_E. Significantly, the absence of the core and the core IF (completely hollow core) causes a large reduction in H_C and a moderate reduction in H_E as the core diameter increases.

This is due to the fact that the core IF contribution, which is absent in this case, is bigger for H_C due to its larger anisotropy. Also the reduction of H_E is smaller than we should expect in the case of a completely hollow core due to the small number of the core IF uncompensated spins. We must notice here that the fluctuations observed on the H_E curves (Figure 6.19) in the case of completely hollow NPs can be attributed to the change of the normalized number of uncompensated spins in each region as the total number of spins is decreased in this case.

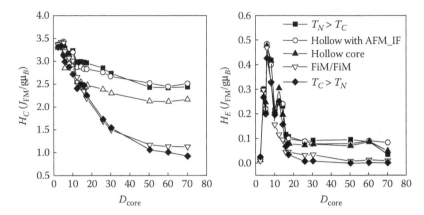

FIGURE 6.19 Dependence of the coercivity, H_C (left) and exchange bias field, H_E (right) on the AFM core size (D_{core}) for the doubly inverted ($T_C < T_N$) structures AFM core–FiM shell, for the hollow core with AFM_IF, for the hollow core with only the FiM–IF and doubly inverted FiM–FiM. In all cases, the parameters are the same with the doubly inverted AFM–FiM nanoparticles ones. The inverted ($T_C > T_N$) AFM core–FiM shell nanoparticle case is also included for comparison.

In Figure 6.19, we also show that the introduction of a FiM core (down triangles) instead of an AFM one influences the H_E and H_C a lot as the core size increases. From our calculations, we have seen that the remanent magnetization of the core and the core IF were larger in the case of FiM–FiM than in the AFM–FiM case for the same core size. This difference was getting larger as the core size increases causing also the larger decrease in the H_C and H_E values of FiM–FiM NP (Figure 6.19). The large FiM core magnetic component drags more core IF spins that are coupled stronger ferrimagnetically to the shell IF spins dragging them also more easily to follow the core spins. As a result, H_C falls off more rapidly from the doubly inverted AFM–FiM case, and H_E is reduced fast with the increase of the core diameter in the range of core sizes, in which H_E is stable in the case of doubly inverted AFM–FiM NP. This behavior is very similar to that of the inverted AFM–FiM NP in which again the core spins are loosely connected (J_{core} smaller than J_{shell}). The decrease of H_E with the increase of the soft core magnetization is similar to that of bilayer structures [2] in which the increase in the FM magnetization causes the decrease in the exchange bias field.

6.3 Mesoscopic Scale Modeling of the Exchange Bias Behavior of Assemblies of Core–Shell Nanoparticles

Interparticle interactions play an important role in the magnetic properties of assemblies of NPs [127]. Modeling assemblies of NPs with core–shell morphology becomes an excessively complicated issue. In this case, the model has to take into account together with the interparticle interactions the intraparticle characteristics. It has been demonstrated that the interplay of the intra- and interparticle characteristics affects the exchange bias behavior of the assembly [68,69,91].

Despite the importance of interparticle interactions, classical MC approaches are inadequate to simulate core–shell NP assemblies due to the prohibitive computational requirements. We have developed a simple mesoscopic model to simulate the magnetic properties of assemblies of bi-magnetic core–shell NPs [69,91,128]. Our mesoscopic method was based on the reduction of the amount of simulated spins to the minimum number necessary to describe the magnetic structure of the core–shell particles and on the introduction of the adequate exchange and anisotropy parameters between the different spin regions inside the NP. Our modeling is multiscale as the magnetic moments are evaluated using data from our atomistic simulations of the core–shell NPs. Then we integrate them properly into the mesoscopic model going in this way from the atomic scale to the mesoscopic modeling.

This mesoscopic model of core–shell NP assemblies has become a useful and effective tool to predict and explain their observed exchange bias properties.

Our model goes beyond the classical model of coherent rotation of a particle's magnetization of Stoner–Wohlfarth [73] in which each NP is described by a classical spin vector (S_i). Here, each core–shell NP in the assembly is described by a set of two, three, or six classical spin vectors depending on the exact morphology of the particle and the magnetic character (FM, AFM, FiM) of its constituents. The values of the different parameters in the simulation are set on the basis of their bulk values, if they exist, and their modifications are established considering the NPs' morphology (e.g., reduced symmetry and reduced size) using a mean field approach or even the data from atomistic simulations wherever it is possible.

In the general case of an assembly of core–shell NPs with FM core–AFM (FiM) shell morphology, each NP is described by six spins: one for the FM core, one for the FM IF, two for the AFM IF, and two for the AFM shell. Figure 6.20 describes nicely the multiscale approach going from the core–shell atomic model to the mesoscopic six-spin model of NPs in ordered and disordered NP arrays.

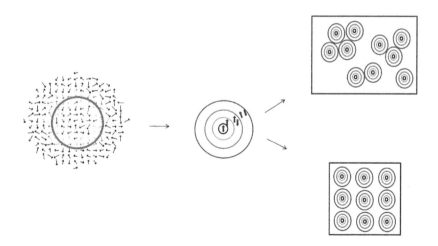

FIGURE 6.20 Multiscale approach going from the core–shell atomic model to the mesoscopic six-spin model of nanoparticles in ordered and random assemblies.

The total energy of the assembly of N core–shell particles is

$$
E = -\sum_{i=1}^{N}
\begin{bmatrix}
j_{\text{core}}\left(\hat{S}_{1i}\cdot\hat{S}_{2i}\right)+j_{\text{IF}}\left(\hat{S}_{2i}\cdot\hat{S}_{3i}\right)+j_{\text{IF}}\left(\hat{S}_{2i}\cdot\hat{S}_{4i}\right)+ \\
j_{\text{shell}}\left[\left(\hat{S}_{3i}\cdot\hat{S}_{4i}\right)+\left(\hat{S}_{3i}\cdot\hat{S}_{5i}\right)+\left(\hat{S}_{4i}\cdot\hat{S}_{6i}\right)+\left(\hat{S}_{5i}\cdot\hat{S}_{6i}\right)\right]
\end{bmatrix}
$$

$$
-k_{\text{core}}\sum_{i=1}^{N}\left(\hat{S}_{1i}\cdot\hat{e}_{1i}\right)^{2}-k_{\text{IF}}\sum_{i=1}^{N}\left[\left(\hat{S}_{2i}\cdot\hat{e}_{2i}\right)^{2}+\left(\hat{S}_{3i}\cdot\hat{e}_{3i}\right)^{2}+\left(\hat{S}_{4i}\cdot\hat{e}_{4i}\right)^{2}\right]
$$

$$
-k_{\text{shell}}\sum_{i=1}^{N}\left[\left(\hat{S}_{5i}\cdot\hat{e}_{5i}\right)^{2}+\left(\hat{S}_{6i}\cdot\hat{e}_{6i}\right)^{2}\right]
$$

$$
-H\sum_{i=1}^{N}\sum_{n=1}^{6}m_{n}\hat{S}_{ni}\cdot\hat{e}_{h}-g\sum_{\substack{i,j=1\\i\neq j}}^{N}\left(\sum_{n=1}^{6}m_{n}\hat{S}_{ni}\right)D_{ij}\left(\sum_{k=1}^{6}m_{k}\hat{S}_{kj}\right)
$$

$$
-j_{\text{out}}\sum_{\langle i,j\rangle}\sum_{n,k}\hat{S}_{ni}\cdot\hat{S}_{kj}
$$

Here S_{1i}, S_{2i}, S_{3i}, S_{4i}, S_{5i}, and S_{6i} are the mesoscopic classical six spins inside the ith particle, namely one for the FM core, one for the FM IF, two for the AFM IF, and two for the AFM shell. The $\hat{e}_{h},\hat{e}_{1i},\hat{e}_{2i},\hat{e}_{3i},\hat{e}_{4i},\hat{e}_{5i},\hat{e}_{6i}$, are the unit vectors in the direction of the magnetic field and the anisotropy easy axes of the 6 spins in the ith particle, respectively. The first seven energy terms correspond to the intraparticle Heisenberg exchange interactions of the spins inside each NP with exchange coupling constants for the core j_{core}, the shell j_{shell} and the IF j_{IF}. The next six energy terms are the anisotropy energy of the core the IF and the shell with anisotropy constants k_{core}, k_{IF}, k_{shell}, respectively. The following two energy terms are the Zeeman energy of the spins and the dipolar interactions among the ith and jth particle of the assembly, where D_{ij} is the dipolar interaction tensor [69] where $m = M_{s}V$ is the particle's magnetic moment, and g the dipolar energy strength. The last energy term corresponds to the short-range exchange interactions that decay within a few lattice constants of the parent magnetic material; therefore, it is restricted in our model to the particles in contact where $-i, j-$ denotes summation over nearest neighbors only to the ith particle, and n, k are the exchange coupled internal spins of the ith and jth particles. Consequently, this last term does not exist in the model that describes ordered arrays, in which the particles are not in contact. Notably, the number of the mesoscopic spins that describe each NP can be reduced depending on the characteristics of the NPs in the system.

The dipolar interactions were calculated with the Ewald summation technique [69,129]. We assume periodic boundaries in all directions for the 3D model and mixed periodic boundaries (xy plane) and open boundaries (z axis) in a 2D model. The spin configuration is obtained by a Metropolis MC algorithm. At a given temperature and applied field, the system is allowed to relax toward equilibrium for the first 10^{3} MCSS, and thermal averages are calculated over the subsequent 10^{4} steps. The results are averaged over 10–30 samples with different realizations of the easy-axes distribution and in the case of random assemblies with different spatial configurations for NPs.

The hysteresis loop and ZFC–FC magnetization curves were calculated and repeated for a large number of different random numbers in order to produce uncorrelated data and thus independent configurations to perform an ensemble average. A constant step rate has been kept of the magnetic field in the calculation of the hysteresis loops and of the temperature in the calculation of the ZFC–FC magnetization curves [130].

6.3.1 Results and Discussion on the Exchange Bias Behavior of FM Core–AFM Shell Nanoparticle Assemblies: Interplay of Intraparticle Characteristics and Interparticle Interactions

6.3.1.1 Diluted FM–AFM Nanoparticle Assemblies

We review our study on the magnetic NP assemblies by examining first the effect of the interplay between single-particle characteristics and interparticle interactions on the exchange bias effect of a dilute assembly (particle volume density 5%) of NPs with FM core–AFM shell morphology [91,131] that were randomly distributed on a lattice corresponding to Co/Mn granular assemblies. In this dilute system, experimental measurements indicated that there is a strong IF coupling between the FM Co core and the surrounding AFM Co_xMn_x shell, responsible for transmitting long-range interparticle interactions [82,91].

In our six spins model for the description of each NP in the assembly intraparticle, nearest neighbors exchange interactions with exchange coupling constants $j_{core} = 1$, $j_{IF} = 0.5$, $j_{shell} = -0.5$, and interparticle dipolar interactions ($g = 0.1$) have been considered. Interparticle exchange interactions have been neglected because the system has low particle concentration. The anisotropy strengths have been set $k_c = 0.1$, $k_{IF} = 0.5$, and $k_{shell} = 1.0$ for the core, the IF and the shell, respectively. The particles assembly has been assumed monodispersed in accordance with experimental evidence [82].

Our MC simulations have demonstrated exchange bias behavior of the Co/Mn. We calculated shifted hysteresis loops. Our results were compared with the ones of a diluted assembly of Co/Ag NPs, and they showed higher coercive fields. In the case of the Co/Ag system, the IF exchange coupling was absent [82], and no shifting in the loops was observed. Also we found that the exchange bias field and the coercive field of Co/Mn system decay exponentially with temperature whereas in the single-spin NP assembly, the decay of H_C was monotonic in agreement with the experimental results [128]. To investigate the effect of the dipolar interactions, in Figure 6.21, we plotted the hysteresis loop of the FM–AFM NP assembly (circles) together with the hysteresis loop of the system without taking into account the interparticle dipolar interactions (squares). We observe that even in the absence of dipolar interactions, an exchange bias field is produced, confirming the frustration along the core–shell. Importantly, in this system, we have shown in Reference [132] that the increase in dipolar strength results in the decrease in the exchange bias field, whereas the coercive field tends to increase. We have attributed this behavior to the competition between dipolar and anisotropy energy [132].

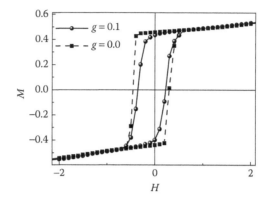

FIGURE 6.21 Simulated hysteresis loops of the magnetization for the FM–AFM nanoparticle assembly (circles) and for the system without dipolar interactions (squares) at a low temperature ($T = 0.01$).

Finally from our MC simulations on the ZFC–FC magnetization curves, we have observed that the blocking temperature of the system (the maximum of ZFC) of the core–shell NP assembly is higher than that of the single-spin system. This is attributed to the additional IF exchange anisotropy in agreement with experimental findings [132].

6.3.1.2 Dense FM–AFM Nanoparticle Assemblies

Next we studied the exchange bias behavior of dense random assemblies of FM core–AFM shell NPs. In this case, we took into account the interparticle dipolar and exchange interactions. Considering a dense assembly of Co/CoO NPs with very thin CoO layer, we developed a three-spin model for the Co/CoO NP to simulate dense 2D and 3D assemblies [69]. In each NP, the FM core was described by one spin, and two spins described the thin AFM shell with the appropriate anisotropy ($k_{core} = 0.1$, $k_{shell} = 8.0$) and exchange parameters ($j_{core-shell1} = 0.32$, $j_{core-shell2} = 0.3$, $j_{shell} = -6$). The anisotropy axis of each NP was assumed randomly oriented. The interparticle dipolar strength was taken as $g = 0.1$ and the interparticle exchange interactions with strengths ($j_{core-shell1} = 2.0$, $j_{core-shell2} = 0.5$, and $j_{shell1-shell1} = j_{shell2-shell2} = 2.5$) have been also introduced. The shell thickness of the NP in our model is very small, so if the NPs are in direct contact not only the shell spins of the one NP interact with those of the other but also the core spin of the one NP interact with the shell spins of the other NP. Thus, the interparticle exchange interactions in these systems are particularly strong.

We have simulated the hysteresis loops at a low temperature, and we show in Figure 6.22, the results for the exchange bias and the coercive field as a function of the particle density for 2D and 3D random assemblies of Co/CoO core–shell NPs. The most prominent result of the simulation was the significant increase of both H_C and H_E where H_E exceeds H_C as the concentration increases. We have attributed this increase in H_C and H_E as the number of particles increases to the increase of the *effective thickness* of the AFM layer for the particles in direct contact. Notably, the core of each NP instead of feeling one shell feels two shells and thus the effective anisotropy increases due to the strong interparticle exchange interactions. The effect is larger in the case of 3D assemblies in

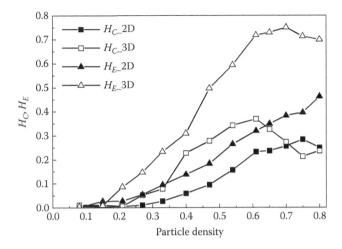

FIGURE 6.22 MC calculations of the coercive H_C and exchange bias field H_E for 2D and 3D dense FM core–thin AFM shell nanoparticle assemblies as a function of the particle density.

which the number of neighboring particles is larger. This increase in the AFM effective thickness results in the enhancement of H_E and H_C similar to what is observed for thin-film systems when the thickness of the AFM layer increases. Moreover, in the 3D assemblies a peak at H_C and H_E appears well above the percolation threshold and then they both decrease whereas in the 2D assemblies, where the number of neighboring particles is smaller, the increase of H_C and H_E insists for higher values of particle concentrations. Consequently, in random dense assemblies of core–shell NPs, the most important role in the observed magnetic behavior is played by the interparticle exchange interactions. Our MC simulation results on the effect of inter- and intraparticle exchange interactions on the exchange bias behavior of NPs were in very good agreement with the experimental findings of the study of random assemblies of Co/CoO core–shell NPs [68,69].

Notably, by switching off the exchange interactions and keeping all the other parameters of the NP the same in the 2D [69] and 3D [132] dense FM–AFM NP assemblies, our simulations showed that the behavior of H_C and H_E is totally deferent in this case. The H_C and the H_E decrease with the increase in the particle density due to the competition between the anisotropy and the dipolar energy confirming the importance of the interparticle exchange coupling.

Similarly, the MC simulations of the ZFC–FC magnetization curves either for 2D [69] or 3D [132] FM–AFM dense random assemblies have shown that at higher densities, the blocking temperature T_B increases dramatically, in agreement with experiments [68].

Finally, in the case of ordered arrays of core–shell, we have developed a two-spin model to describe each NP, one spin for the FM core and one for the AFM fully uncompensated IF assuming that the net magnetic moment of the shell is negligible [128,133].

In the study of the effect of intraparticle characteristics and its interplay with the interparticle interactions, we examined the role of the IF exchange coupling strength in the exchange bias field of an ordered array of FM–AFM NPs in the presence of dipolar interactions. We have calculated the H_C and H_E as a function of the dipolar strength (g) for

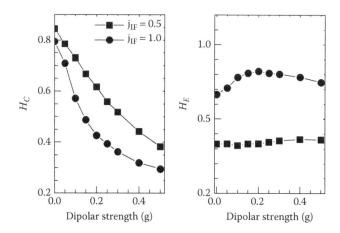

FIGURE 6.23 Dependence of the coercivity (H_C) and the exchange bias field (H_{ex}) on the dipolar strength (g) at low temperature ($T = 0.01$) for two different interface exchange coupling constants.

two different 2D triangular arrays of composite NPs with a low $j_{IF} = 0.5$ and an intermediate value $j_{IF} = 1.0$ of the IF exchange coupling constant comparing to the core anisotropy constant [133]. We observed that for a given IF exchange coupling the coercive field was reduced as the dipolar strength (g) increased, due to the collective response of the magnetic moments, which leads to a reduction of the energy barrier for magnetization reversal. Our calculations indicated that this reduction of the coercive field with dipolar interactions is sensitive to the strength of the IF exchange [132]. Notably, a weaker reduction of H_C was observed in systems with stronger IF exchange coupling ($j_{IF} = 1.5$) (Figure 6.23).

The behavior of the exchange bias field with increasing dipolar coupling strength appeared more complex [133] in our MC results. In the case of a weak IF coupling ($j_{IF} = 1.0$), the H_E values go through a maximum at intermediate dipolar strength ($g \approx 0.2$) before decreasing to a constant value in the strong dipolar limit ($g \gg 1$). The observed enhancement of H_E is attributed to the competition between the dipolar energy and the anisotropy energy that leads to an increased loop shift. Indeed, in the case of stronger j_{IF} our calculations have shown that the H_E drops linearly with increasing dipolar strength due to the dominance of the dipolar energy over the anisotropy energy term [132]. Finally, we noticed that for noninteracting arrays, as in the case of random arrays increase of the IF exchange causes reduction of coercivity and increase of exchange bias field values. This behavior was in agreement with the behavior of single NPs, as described by the atomic-scale models [25]. Consequently, in an ordered array of magnetic NPs with core–shell morphology, interparticle dipolar interactions cause suppression of the coercive field, whereas the behavior of the exchange field depends strongly on the IF exchange coupling.

6.4 Conclusions

In this chapter, we review our MC simulations results on the exchange bias properties of NPs with a core–shell morphology and in their assemblies. We start with a review on our study of the intraparticle characteristics that influence the exchange bias behavior of noninteracting FM–AFM and FM–FiM, exchange spring nanomagnets of FiM–FiM, inverted and doubly inverted AFM–FiM core–shell NPs. These NP characteristics include the IF and surface structure, the exchange coupling sign, and strength along the IF and its competition with the anisotropy strengths, the size, the shape, and the shell thickness. In addition, the influence of the applied cooling field on the exchange bias, the vertical shift of the hysteresis loop as well as the training and aging effect which are associated with a disordered shell are reviewed.

Significantly, the agreement of our MC results with the experimental findings indicates that MC simulations can shed light into the microscopic origin of the magnetic behavior of core–shell NPs and dictate the conditions for optimized exchange bias properties. In particular, the doubly inverted AFM–FiM core–shell NPs and the exchange spring nanomagnets of FiM–FiM core–shell NPs have shown promising results for their exploitation in novel exchange bias applications.

Taking into account our results of single core–shell particles, we proceeded with the study of assemblies of core–shell NPs including both interparticle and intraparticle characteristics. We have developed a novel mesoscopic model to study the influence of these characteristics and their interplay in the magnetic behavior of structurally ordered and disordered NP assemblies. Our simulations showed that in the dilute random assemblies, in which only dipolar interactions are present, the exchange bias behavior depends on the competition between anisotropy and dipolar energy. The additional IF exchange anisotropy increases the blocking temperature. The exchange interparticle interactions can play the dominant role in the exchange bias properties of dense arrays of core–shell NPs causing a remarkable increase in the exchange and coercive field with $H_E > H_C$ for $p > 0.2$ and in their blocking temperature in agreement with experimental results.

In the case of the ordered arrays, the interparticle dipolar interactions decrease the coercive field, whereas they produce a more complex behavior of the exchange field that depends mainly on the IF exchange coupling strength. Our results showed that our mesoscopic model reproduces properly the trends that are observed experimentally, but the exact behavior is very much system dependent.

Importantly though our mesoscopic model was developed of FM–AFM NP assemblies, it can be easily extrapolated to the study of the magnetic and magnetotransport properties of other bi-magnetic systems by the proper choice of spins characterizing the systems and their interactions as it was demonstrated in the Co/Mn and Co/CoO systems.

References

1. A. López-Ortega, M. Estrader, G. Salazar-Alvarez, A. G. Roca, J. Nogués, *Phys. Rep.*, (2015), 553: 1–32.
2. J. Nogués, J. Sort, V. Langlais, V. Skumryev, S. Suriñach, J. S. Muñoz et al., *Phys. Rep.*, (2005), 422: 65–117.
3. D. K. Mishra, inventor; University of Louisiana at Lafayette, assignee. FeRh-FePt core shell nanostructure for ultra-high density storage media. United States patent 7, 964, 013 (2011) June 21.

4. A. Rida, inventor; S. A. Spinomix, assignee. Tailored magnetic particles comprising a non-magnetic component and a magnetic core-shell component, and method to produce same. United States patent 8, 142, 892 2012 March 27.

5. C. Binns et al., *J. Phys. D: Appl. Phys.*, (2005), 38: R357–R379.

6. A. Boatwright, C. Feng, D. Spence, E. Latimer, C. Binns, A. M. Ellis, S. Yang, *Farad. Discuss.*, (2013), 162: 113–124.

7. H. Zeng, J. Li, J. Liu, Z. Wang, S. Sun, *Nature*, (2002), 420: 395–398.

8. V. Skumryev, S. Stoyanov, Y. Zhang, *Nature*, (2003), 423: 19–22.

9. J. Kools, *Magn. IEEE Trans.*, (1996), 32: 3165–3184.

10. Z. Zhou, Y. Sun, J. Shen, J. Wei, C. Yu, B. Kong, W. Liu, H. Yang, S. Yang, W. Wang, *Biomaterials*, (2014), 35: 7470–7478.

11. C. Binns, P. Prieto, S. Baker, P. Howes, R. Dondi, G. Burley, L. Lari, R. Kroger, A. Pratt, S. Aktas, J. K. Mellon, *J. Nanopart. Res.*, (2012), 14: 1136.

12. J. H. Lee et al., *Nat. Nanotechnol.*, (2011), 6: 418–422.

13. D. Fiorani, L. Del Bianco, A. M. Testa, K. N. Trohidou, *Phys. Rev. B*, (2006), 73: 092403.

14. S. Giri, M. Patra, S. Majumdar, *J. Phys.: Condens. Matter.*, (2011), 23: 073201.

15. R. Skomski, J. M. D. Coey, *Phys. Rev. B*, (1993), 48: 15812.

16. H. Zeng, S. Sun, *Adv. Funct. Mater.*, (2008), 18: 391.

17. L. Carbone, P. D. Cozzoli, *Nano Today.*, (2010), 5: 449.

18. W. H. Meiklejohn, C. P. Bean, *Phys. Rev.*, (1957), 105: 904.

19. J. Nogués, I. K. Schuller, *J. Magn. Magn. Mater.*, (1999), 192: 203.

20. A. E. Berkowitz, K. Takano, *J. Magn. Magn. Mater.*, (1999), 200: 552–570.

21. K. O'Grady, L. E. Fernandez-Outon, G. Vallejo-Fernandez, *J. Magn. Magn. Mater.*, (2010), 322: 883–899.

22. A. P. Malozemoff, *Phys. Rev. B*, (1987), 35: 3679.

23. T. C. Schulthess, W. H. Butler, *Phys. Rev. Lett.*, (1998), 81: 4516.

24. U. Nowak, K. D. Usadel, J. Keller, P. Miltenyi, B. Beschoten, G. Guntherodt, *Phys. Rev. B*, (2002), 66: 014430.

25. E. Eftaxias, K. N. Trohidou, *Phys. Rev. B*, (2005), 71: 134406.

26. O. Iglesias, X. Batlle, A. Labarta, *Phys. Rev. B*, (2005), 72: 212401.

27. M. Estrader et al., *Nat. Commun.*, (2013), 4: 2960.

28. M. Vasilakaki, K. N. Trohidou, *Phys. Rev. B*, (2009), 79: 144402.

29. A. López-Ortega et al., *Nanoscale*, (2012), 4: 5138–5147.

30. M. Vasilakaki, E. Eftaxias, K. N. Trohidou, *Phys. Stat. Solid. A*, (2008), 205: 1865–1871.

31. M. H. Wu, Q. C. Li, J. M. Liu, *J. Phys.: Condens. Matter*, (2007), 19: 186202.

32. X. Zianni, K. N. Trohidou, *J. Phys.: Condens. Matter*, (1998), 10: 7475.

33. E. Eftaxias, M. Vasilakaki, K. N. Trohidou, *Mod. Phys. Lett. B*, (2007), 21(21): 1169–1177.

34. E. C. Passamani, C. Larica, C. Marques, J. R. Provetti, A. Y. Takeuchi, F. H. Sanchez, *J. Magn. Magn. Mater.*, (2006), 299: 11.

35. G. Salazar-Alvarez, J. Sort, S. Suriñách, M. D. Baró, J. Nogués, *J. Am. Chem. Soc.*, (2007), 129: 9102–9108.

36. G. Salazar-Alvarez et al., *J. Am. Chem. Soc.*, (2011), 133: 16738–16741.

37. K. L. Krycka et al., *J. Appl. Phys.*, (2013), 113: 17B531.
38. I. V. Golosovsky et al., *Phys. Rev. Lett.*, (2009), 102: 247201.
39. A. López-Ortega et al., *J. Am. Chem. Soc.*, (2010), 132: 9398–9407.
40. A. E. Berkowitz et al., *Phys. Rev. B*, (2008), 77: 024403.
41. A. E. Berkowitz et al., *J. Phys. D*, (2008), 41: 134007.
42. D. W. Kavich, J. H. Dickerson, S. V. Mahajan, S. A. Hasan, J. H. Park, *Phys. Rev. B*, (2008), 78: 174414.
43. R. Chalasani, S. J. Vasudevan, *Phys. Chem. C*, (2011), 115: 18088–18093.
44. E. Wetterskog, C. W. Tai, J. Grins, L. Bergström, G. Salazar-Alvarez, *ACS Nano*, (2013), 7: 7132–7144.
45. H. Khurshid et al., *Nanoscale*, (2013), 5: 7942–7952.
46. H. Khurshid et al., *J. Appl. Phys.*, (2013), 113: 17B508.
47. X. Sun, N. F. Huls, A. Sigdel, S. Sun, *Nano Lett.*, (2012), 12: 246–251.
48. A. Lak et al., *Nanoscale*, (2013), 5: 12286–12295.
49. S. K. Sharma et al., *J. Alloy. Compd.*, (2011), 509: 6414–6417.
50. M. I. Bodnarchuk et al., *Small*, (2009), 5: 2247–2252.
51. M. Sytnyk et al., *Nano Lett.*, (2013), 13: 586–593.
52. P. Torruella et al., *Nano Lett.*, (2016), 16: 5068.
53. S. K. Sharma, J. M. Vargas, K. R. Pirota, Shalendra Kumar, C. G. Lee, M. Knobel, *J. Alloy Compd.*, (2011), 509: 6414–6417.
54. C. J. Chen, R. K. Chiang, H. Y. Lai, C. R. Lin, *J. Phys. Chem. C*, (2010), 114: 4258–4263.
55. E. Lima Jr. et al., *Chem. Mater.*, (2012), 24: 512–516.
56. E. L. Winkler et al., *Appl. Phys. Lett.*, (2012), 101: 252405.
57. G. C. Lavorato et al., *Nanotechnol.*, (2014), 25: 355704.
58. W. Baaziz et al., *Chem. Mater.*, (2014), 26: 5063–5073.
59. N. Fontaiña Troitiño, B. Rivas-Murias, B. Rodriguez-González, V. Salgueiriño, *Chem. Mater.*, (2014), 26: 5566–5575.
60. C. Liu, J. Cui, X. He, H. Shi, *J. Nanopart. Res.*, (2014), 16: 2320.
61. G. C. Lavorato, E. Lima Jr., H. E. Troiani, R. D. Zysler, E. L. Winkler, *J. Alloy Compd.*, (2015), 633: 333–337.
62. B. K. Yun, Y. S. Koo, J. H. Jung, *J. Magn.*, (2009), 14: 147–149.
63. P. Z. Si et al., *Thin Solid Films*, (2011), 519: 8423–8425.
64. C. H. Jin et al., *Mater. Lett.*, (2013), 92: 213–215.
65. L. Basit et al., *Appl. Phys. A*, (2009), 94: 619–625.
66. S. M. Yusuf et al., *J. Appl. Phys.*, (2013), 113: 173906.
67. K. D. Sung, Y. A. Park, K. Y. Kim, N. Hur, J. H. Jung, *J. Appl. Phys.*, (2013), 114: 103902.
68. J. Nogués, V. Skumryev, J. Sort, D. Givord, *Phys. Rev. Lett.*, (2006), 97: 157203.
69. G. Margaris, K. N. Trohidou, J. Nogués, *Adv. Mater.*, (2012), 24: 4331–4336.
70. Q. Song, Z. J. Zhang, *J. Am. Chem. Soc.*, (2012), 134: 10182–10190.
71. R. Cabreira-Gomes, F. G. Silva, R. Aquino, P. Bonville, F. A. Tourinho, R. Perzynski, J. Depeyrot, *J. Magn. Magn. Mater.*, (2014), 368: 409–414.
72. J. L. Dormann, D. Fiorani, E. Tronc, *Adv. Chem. Phys.*, (1997), 98: 283.
73. E. Stoner, E. Wohlfarth, *Philos. Trans. R. Soc. London, Ser. A*, (1948), 240: 599.

74. O. Michele, J. Hesse, H. Bremers, *J. Phys.: Condens. Matter*, (2006), 18: 4921–4934.
75. D. Kechrakos, K. N. Trohidou, *Appl. Surf. Sci.*, (2004), 226: 261–264.
76. D. Kechrakos, K. N. Trohidou, *J. Magn. Magn. Mater.*, (2003), 262: 107–110.
77. G. Margaris, K. Trohidou, H. Kachkachi, *Phys. Rev. B*, (2012), 85: 024419.
78. V. Iannotti et al., *Phys. Rev. B*, (2011), 214422.
79. E. Lima, Jr., J. M. Vargas, H. R. Rechenberg, R. D. Zysler, *J. Nanosci. Nanotechnol.*, (2008), 8: 5913–5920.
80. G. Margaris, K. N. Trohidou, V. Iannotti, G. Ausanio, L. Lanotte, D. Fiorani, *Phys. Rev. B*, (2012), 86: 214425.
81. S. K. Sharma et al., *J. Appl. Phys.*, (2010), 107: 09D725.
82. C. Binns et al., *J. Phys.: Condens. Matter*, (2010), 22: 436005.
83. M. T. Qureshi, S. H. Baker, C. Binns, M. Roy, S. Laureti, D. Fiorani, D. Peddis, *J. Magn. Magn. Mater.*, (2015), 378: 345–352.
84. F. G. Silva, R. Aquino, F. A. Tourinho, V. I. Stepanov, Yu L. Raikher, R. Perzynski, J. Depeyrot, *J. Phys. D: Appl. Phys.*, (2013), 46: 285003 (9pp).
85. L. Del Bianco, D. Fiorani, A. M. Testa, E. Bonetti, L. Signorini, *Phys. Rev. B*, (2004), 70: 052401.
86. N. Domingo, D. Fiorani, A. M. Testa, C. Binns, S. Baker, J. Tehada, *J. Phys. D: Appl. Phys.*, (2008), 41: 134009.
87. H. Bi, S. Li, X. Jiang, Y. Du, C. Yang, *Phys. Lett. A*, (2003), 307: 69.
88. E. Girgis, R. D. Portugal, H. Loosvelt, M. J. V. Bael, I. Gordon, M. Malfait, K. Temst, C. V. Haesendonck, L. H. A. Leunissen, R. Jonckheere, *Phys. Rev. Lett.*, (2003), 91: 187202.
89. N. Metropolis, A. Rosenbluth, M. Rosenbluth, A. Teller, E. Teller, Equation of state calculations by fast computing machines. *J. Chem. Phy.*, (1953), 21(6): 1087–1092, ISSN 0021-9606.
90. K. Binder, A. P. Young, *Rev. Mod. Phys.*, (1986), 58: 801–976.
91. M. Vasilakaki, K. N. Trohidou, D. Peddis, D. Fiorani, R. Mathieu, M. Hudl, P. Nordblad, C. Binns, S. Baker, *Phys. Rev. B*, (2013), 88: 140402.
92. K. N. Trohidou, M. Vasilakaki, L. Del Bianco, D. Fiorani, A. M. Testa, *J. Magn. Magn. Mater.*, (2007), 316: e82–e85
93. J.-O. Andersson, C. Djurberg, T. Jonsson, P. Svedlindh, P. Nordblad, *Phys. Rev. B*, (1997), 56: 13983.
94. D. P. Landau, K. Binder, *A Guide to Monte Carlo Simulations in Statistical Physics*, (2000), Cambridge, MA: Cambridge University Press.
95. K. Binder. *Applications of the Monte-Carlo Method in Statistical Physics*, (1987), New York: Springer-Verlag.
96. J. Garcia-Otero, M. Porto, J. Rivas, A. Bunde, *J. Appl. Phys.*, (1999), 85: 2287–2292.
97. D. A. Dimitrov, G. M. Wysin, *Phys. Rev. B*, (1996), 54: 9237–9241.
98. R W. Chantrell, N. Walmsley, J. Gore, M. Maylin, *Phys. Rev. B*, (2001), 63: 024410–024414.
99. D. Hinzke, U. Nowak, *Comp. Phys. Comm.*, (1999), 121–122: 334–337.
100. H. F. Du, A. Du, *J. Appl. Phys.*, (2006), 99: 104306.
101. T. Kaneyoshi, *J Phys.: Condens. Matter*, (1991), 3: 4497.
102. M. Respaud et al., *Phys. Rev. B*, (1998), 57: 2925.

103. D. W. Wood, N. W. Dalton, *Phys. Rev.*, (1967), 159: 384.
104. S. T. Chui, T. De-Cheng, *J. Appl. Phys.*, (1995), 78: 3965.
105. M. Feygenson, Y. Yiu, A. Kou, K. Kim, M. C. Aronson, *Phys. Rev. B*, (2010), 81: 195445.
106. O. Iglesias, A. Labarta, *Phys. Rev. B*, (2001), 63: 184416.
107. R. Morel, A. Brenac, C. Portemont, *J. Appl. Phys.*, (2004), 95: 3757.
108. S. K. Baker, Hasanain, S. Ismat Shah, *J. Appl. Phys.*, (2004), 96: 6657.
109. E. Tronc et al., *J. Magn. Magn. Mater.*, (2004), 272–276: 1474.
110. T. Ambrose, K. Leifer, K. J. Hemker, C. L. Chien, *J. Appl. Phys.*, (1997), 81: 5007.
111. S. Gangopadhyay, G. C. Hadjipanayis, C. M. Sorensen, K. J. Klabunde, *Nanostruct. Mater.*, (1992), 1: 449.
112. C. Leighton, J. Nogués, B. J. Jonsson-Akerman, I. K. Schuller, *Phys. Rev. Lett.*, (2000), 84: 3466.
113. B. Kagerer, Ch. Binek, W. Kleemann, *J. Magn. Magn. Mater.*, (2000), 217: 139; A. Paul, C. M. Schneider, J. Stahn, *Phys. Rev. B*, (2007), 76: 184424.
114. D. Fiorani, L. Del Bianco, A. M. Testa, *J. Magn. Magn. Mater.*, (2006), 300: 179.
115. S. M. Zhou, D. Imhoff, K. Yu-Zhang, Y. Leprince-Wang, *Appl. Phys. A: Mater. Sci. Process.*, (2005), 81: 115.
116. Yan-kun Tang, Young Sun, Zhao-hua Cheng, *J. Appl. Phys.*, (2006), 100: 023914; *Phys. Rev. B*, (2006), 73: 174419.
117. D. L. Peng, K. Sumiyama, T. Hihara, S. Yamamuro, T. J. Konno, *Phys. Rev. B*, (2000), 61: 3103.
118. D. L. Peng, T. Hihara, K. Sumiyama, H. Morikawa, *J. Appl. Phys.*, (2002), 92: 3075.
119. C. Binns, P. Prieto, S. Baker, P. Howes, R. Dondi, G. Burley, L. Lari, R. Kröger, A. Pratt, S. Aktas, J. K. Mellon, *J. Nanopart. Res.*, (2012), 14: 1136.
120. P. Umek, A. Zorko, D. Arčon, Magnetic properties of transition-metal oxides: From bulk to nano, in *Ceramics Science and Technology. Volume 2: Materials and Properties* (Eds. R. Riedel and I.-W. Chen), (2010), Weinheim, Germany: Wiley-VCH Verlag GmbH & Co. KGaA.
121. M. Estrader et al., *Nanoscale*, (2015), 7: 3002–3015.
122. S. Chikazumi, in AIP Conf. Proc., AIP 29 (1976) pp. 382–387.
123. G. Salazar-Alvarez et al., *J. Am. Chem. Soc.*, (2008), 130: 13234–13239.
124. J. W. Cai, K. Liu, C. L. Chien, *Phys. Rev. B*, (1999), 60: 72–75.
125. K. D. Sossmeier, L. G. Pereira, J. E. Schmidt, J. Geshev, *Appl. Phys.*, (2011), 109: 083938.
126. M. Vasilakaki, K. Trohidou, J. Nogués, *Sci. Rep.*, (2015), 5: 9609.
127. K. Trohidou, *Magnetic Nanoparticle Assemblies*, (2014), Singapore: Pan Stanford Publishing, 306 pp.
128. K. N. Trohidou, M. Vasilakaki, *Acta Phys. Pol. A*, (2010), 117: 374.
129. A. Grzybowski, E. Gwoźdź, A. Brodka, (2000). *Phys. Rev. B*, (2014), 61: 6706–6712.
130. M. Bahiana, J. P. Pereira Nunes, D. Altbir, P. Vargas, M. Knobel, *J. Magn. Magn. Mater.*, (2004), 281: 372–377.
131. D. Peddis, M. Vasilakaki, K. Trohidou, D. Fiorani, *IEEE Trans. Magn.*, (2014), 50: 6971627.

132. M. Vasilakaki, G. Margaris, K. Trohidou, Magnetic behavior of composite nanoparticle assemblies, in *Magnetic Nanoparticle Assemblies* (Ed. K. N. Trohidou), (2014), Singapore: Pan Stanford Publishing, Ch. 8, pp. 251–284.
133. M. Vasilakaki, G. Margaris, K. Trohidou, Monte Carlo simulations on the magnetic behaviour of nanoparticle assemblies: Interparticle interactions effects, in *Nanoparticles Featuring Electromagnetic Properties: From Science to Engineering*, (Ed. Alessandro Chiolerio and Paolo Allia), (2012), Kerala, India: Signpost, pp. 105–132.

7

All-Electric Spintronics through Surface/ Interface Effects

Wen-Yi Tong and
Chun-Gang Duan

7.1 Introduction

Spintronics is an emerging area of nanoscale electronics involving the spin injection, transport, modulation and detection, and so on (Wolf et al., 2001; Žutić et al., 2004; Bader and Parkin, 2010; Ohno, 2010). It aims at developing devices based on the control of spin degree of freedom and is of great importance in revolutionizing conventional electronics. Spintronics devices are considered a major candidate for next-generation electronic devices with ultra-high speed and ultra-low power consumption, as spin responds quickly and generates less thermal energy. Different systems, including semiconductors, metals, semiconductor/metal interface, transition metal oxides, and carbon nanostructures, can be good functional materials for spintronics.

Very recently, a new research branch, that is, all-electric spintronics, which requires spin manipulation via electric means, has received much focus. Compared with the control of magnetization by the traditional magnetic field or the more advanced spin-current method, the advantage of using electric approach is apparent. As pointed out by Ohno (2010), electrical reversal, if realized, could reduce the magnetization switching energy down to 1 fj/m³, which is two orders of magnitude smaller than that required by spin-transfer torque method. Due to the lower power consumption, it attracts large amounts of attention.

In this section, we will briefly summarize recent progress relating to the all-electric spintronics, including physical mechanism for magnetoelectric coupling, as well as various forms of spin states manipulation through electric means proposed theoretically and/or realized experimentally. In consideration of greater design flexibility, multifunctionality, and generally stronger magnetoelectric coupling, special focus is given to surface/interface systems here.

7.1.1 Mechanism for Magnetoelectric Coupling at Surface/Interface

To eventually realize the practical application of tuning spin states electrically, understanding the mechanisms responsible for the magnetoelectric coupling is both fundamental and crucial. Various origins have been proposed.

7.1.1.1 Elastic Interaction Induced Magnetoelectric Effect

Historically, the concept of magnetoelectric composites was proposed by van Suchtelen (1972) and experimentally realized in $BaTiO_3/CoFe_2O_4$ ceramic composites (Van Den Boomgaard et al., 1974). The magnetoelectric effect here is a direct result of the product from the magnetostrictive effect in the magnetic phase and the piezoelectric effect in the ferroelectric constituent, which can be expressed as (Nan, 1994)

$$ME_E \text{ effect} = \frac{\text{Electrical}}{\text{Mechanical}} \times \frac{\text{Mechanical}}{\text{Magnetic}} \tag{7.1}$$

or

$$ME_H \text{ effect} = \frac{\text{Magnetic}}{\text{Mechanical}} \times \frac{\text{Mechanical}}{\text{Electrical}} \tag{7.2}$$

This kind of extrinsic magnetoelectric coupling is mediated via elastic interaction at the interface. When an electric field is applied to a composite, the strain induced in the electrical component is mechanically transferred to the magnetostrictive or piezomagnetic materials, in which it changes the magnetization (and vice versa).

As the one main route, the strain-mediated magnetoelectric effect in composites has three common connectivity schemes (Nan et al., 2008). For 0–3 particulate type (Figure 7.1a), magnetic nanoparticles are embedded into ferroelectric films, in which strain exists in grain boundaries. In the 1–3 fiber type (Figure 7.1b), ferromagnetic nanopillars are inserted in ferroelectric films. Strain transfers in vertical ferromagnetic/ferroelectric interfaces. The 2–2 laminate type (Figure 7.1c) with alternating ferroelectric and ferromagnetic layers exhibits strain transfer through horizontal interfaces.

Although the mechanism is quite simple in physics, its magnetoelectric response appears to be quite large. As shown in Figure 7.2, electrically induced giant, sharp,

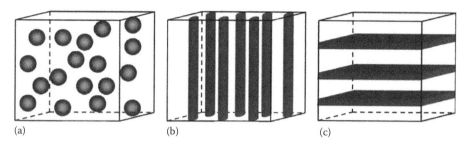

(a)   (b)   (c)

FIGURE 7.1 Schematic illustration of three bulk composites with the three common connectivity schemes: (a) 0–3 particulate composite, (b) 1–3 fiber composite, and (c) 2–2 laminate composite. (From Nan, C.-W. et al., *J. Appl. Phys.*, 103, 031101, 2008.)

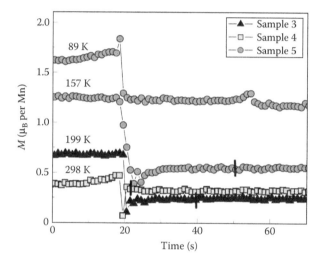

FIGURE 7.2 Large sharp magnetic switching due to an applied electric field in multiferroic $La_{0.67}Sr_{0.33}MnO_3/BaTiO_3$ heterostructure. On ramping the voltage, there is initially no significant change in magnetization of ferromagnetic $La_{0.67}Sr_{0.33}MnO_3$ film. At some threshold voltage, a large, sharp, and persistent magnetic transition is observed. (From Eerenstein, W. et al., *Nat. Mater.*, 6, 348–351, 2007.)

and persistent magnetic changes are reported at an epitaxial ferromagnetic 40-nm $La_{0.67}Sr_{0.33}MnO_3$ films on ferroelectric $BaTiO_3$ substrates over a wide range of temperatures including room temperature (Eerenstein et al., 2007). X-ray diffraction confirms the existence of strain coupling at interface, indicating the exciting possibilities for technological applications in electric-field control of magnetism utilizing the magnetoelectric effect induced by elastic interaction.

7.1.1.2 Interface Bonding Induced Magnetoelectric Effect

The coupling between elastic components of the magnetic constituents and the piezoelectric ones through the strain effect is undoubtedly important. It is not, however, the only source of the magnetoelectric effect in composite multiferroics. There is

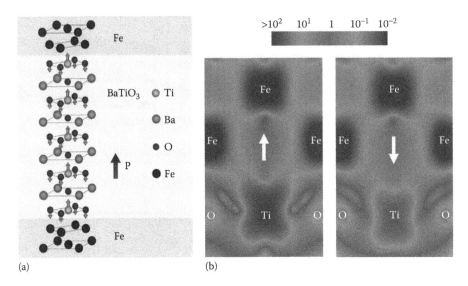

(a)   (b)

FIGURE 7.3   (a) Atomic structure of $Fe/BaTiO_3$ multilayer, (b) majority-spin charge density at the $Fe/BaTiO_3$ interface for two opposite polarizations in $BaTiO_3$. (From Duan, C.-G. et al., *Phys. Rev. Lett.*, 97, 047201, 2006.)

quantum mechanical origin at the interface to control magnetic properties of thin-film layered structures by electric fields.

In 2006, by performing first-principles calculations, Duan et al. found a strong coupling between ferroelectricity and magnetism in $Fe/BaTiO_3(0\ 0\ 1)$ heterostructure as displayed in Figure 7.3a. It derives from atomic bonding at the ferromagnet/ferroelectric interface. Due to the structural inversion symmetry broken by the upward polarized $BaTiO_3$, the hybridization between Fe-$3d$ and Ti-$3d$ orbitals at the top surface is enhanced (Figure 7.3b), which increases the induced magnetic moment on top Ti atoms but reduces the magnetic moment of top Fe ones. Note that the sizeable induced magnetic moments in Ti atoms are resulted from the $Fe–TiO_2$ bonding. They are, thus, sensitive to the interfacial bonding strength and naturally can be controlled by the direction of ferroelectric polarization in $BaTiO_3$. The magnetoelectric coefficient driven by interface bonding is predicted up to 0.01 G cm/V, as large as that induced by elastic interaction measured in epitaxial $BiFeO_3/CoFe_2O_4$ columnar nanostructures (Zheng et al., 2004). Later, similar effects were investigated *ab initio* at the interface between half-metal Co_2MnSi and ferroelectric $BaTiO_3$ (Yamauchi et al., 2007), as well as ferromagnetic/ferroelectric $Fe_3O_4/BaTiO_3$ interfaces (Niranjan et al., 2008; Park et al., 2009).

Further study (Duan et al., 2008a) revealed that besides the change of the interface magnetization, the orbital magnetic moments, and in turn, the magnetocrystalline anisotropy energy (MAE) of Fe atoms at the interface can also be affected by the electric polarization reversal in such multilayers. As shown in Figure 7.4, the interface bonding induced magnetoelectric effect alters the MAE of the Fe monolayer by as much as 50%. With the magnetocrystalline anisotropy and the additional thickness-dependent shape

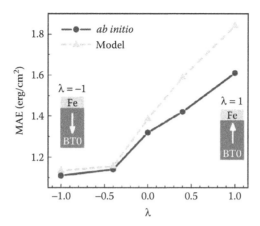

FIGURE 7.4 MAE as a function of a polarization scaling factor λ. Here, $\lambda = 1$ and $\lambda = -1$ correspond to the spontaneous ferroelectric polarization up and down, respectively. (From Duan, C.-G. et al., *Appl. Phys. Lett.*, 92, 122905–122903, 2008a.)

anisotropy, the discovery offers new mechanism in the writing on high coercivity perpendicular media utilizing electric field.

Motivated by the theoretical achievements, Sahoo et al. (2007) deposited a 10-nm thick Fe film on a single-crystal $BaTiO_3(100)$ substrate by molecular beam epitaxy. Up to 20% coercivity change is achieved via electrical control at room temperature. Nevertheless, the polycrystalline nature of the top Fe film reveals that the primary mechanism of the magnetoelectric effect is the interface strain coupling. In 2010, Garcia et al. fabricated $Fe/BaTiO_3/La_{0.67}Sr_{0.33}MnO_3$ multiferroic tunnel junction. By controlling the ferroelectric polarization of $BaTiO_3$, reversible changes of the tunnel resistance by about 30% are observed, providing direct evidence on the existence of the new magnetoelectric effect.

Compared with the elastic-mediated magnetoelectric effect, that is, merely sensitive to the intensity of ferroelectric polarization, the effect driven by the ferromagnetic/ferroelectric interface bonding is closely related to the polarization direction of ferroelectric constituents (Meyerheim et al., 2011). In such case, the all-electric magnetic data storage without external magnetic field appears very promising based on the latter mechanism.

7.1.1.3 Spin-Dependent Screening Induced Magnetoelectric Effect

The interface bonding induced magnetoelectric effect mentioned above mainly comes from ferroelectric displacements at the interface between ferroelectric and ferromagnetic constituents under the electric polarization. Even without displacement atoms, there exists a totally different carrier-mediated magnetoelectricity at interface/surface, which is driven by the screening effect.

7.1.1.3.1 Surface Magnetoelectric Effect

For a ferromagnetic metal, due to the well-known spin-dependent screening effect (Zhang, 1999), that is, the spin-up and spin-down electrons will have quite different responses to electric field penetrating into the ferromagnet. The spin accumulation

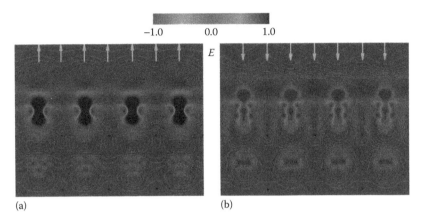

FIGURE 7.5 Surface magnetoelectric effect of Fe film. The arrows indicate directions of the electric field: (a) increase of the magnetization when the electric field is pointed away from the surface and (b) decrease of the magnetization when the electric field is pointed toward the surface. (From Duan, C.G., *Prog. Phys.* [in Chinese], 29, 215, 2009.)

of screening charges at surface of the ferromagnetic metal directly affects the surface magnetization. As this type of magnetoelectric effect is essentially limited to the metal surface, it is named the surface magnetoelectric effect (Figure 7.5). Using density-functional calculations, Duan et al. (2008b) revealed the new mechanism in ferromagnetic Fe(001), Ni(001), and Co(001) films, exploring the direct influence of an external electric field on magnetic properties of ferromagnetic metals besides the interface magnetoelectric effect in heterostructures.

Although the screening effect of the external electric field in ferromagnetic metals is too complex to be directly described, the phenomena resulted from the surface magnetoelectric effect can be quantitatively understood within a simple semiclassical model. By assuming the localization of the screening charge on the metal surface and a rigid shift of the chemical potential on the surface in response to the applied electric field, the surface magnetoelectric coefficient can be written as follows (Duan, 2012):

$$\alpha_s = -P_s \frac{\varepsilon \mu_B}{ec^2} \tag{7.3}$$

The expression is referenced to the definition of the one in bulk, where $P_s = n^\uparrow - n^\downarrow / n^\uparrow + n^\downarrow$ is the spin polarization rate of the conduction electron at the surface. ε, μ_B, e, and c are dielectric constant, Bohr magneton, electron charge, and speed of light, respectively. Note that dominant spin electrons at the surface are not necessarily the same as that in the bulk material. The sign of the magnetoelectric coefficient reflects the relative spin orientation of the dominant conduction electron at the surface. The modification of d

electron states at surface of ferromagnetic metals due to applied electric field was also confirmed by later theoretical studies (Nakamura et al., 2009; Tsujikawa and Oda, 2009; Zhang et al., 2009).

The notable changes in the surface magnetization are certainly not the only consequence of the effect originating from spin-dependent screening of the external electric field. The orbital moment anisotropy and the surface magnetocrystalline anisotropy are also influenced, which is of considerable interest in adjusting MAE electrically. Experimentally, Weisheit et al. (2007) observed that the magnetocrystalline anisotropy of ordered intermetallic compounds FePt (FePd) can be reversibly modified by an applied electric field when immersed in an electrolyte. Upon the voltage change of −0.6 V, the coercivity of 2-nm-thick FePt and FePd films is altered by −4.5% and +1%, respectively.

Above introductions mainly focused on the surface magnetoelectric effect in the normal ferromagnetic films, in which both the majority- and minority-spin states are located near the Fermi level at the surface. When it comes to the case of a half-metal with 100% spin polarization rate, what would happen? As illustrated in Figure 7.6, for half-metals, conducting electrons are present only in one spin channel, whereas the other is insulating. In the vacuum, the surface magnetoelectric coefficient evolves to

$$\alpha_s = \pm \frac{\mu_B}{ec^2} = \pm \frac{\hbar}{2mc^2} \approx \pm 6.44 \times 10^{-14} \frac{G \cdot cm^2}{V} \tag{7.4}$$

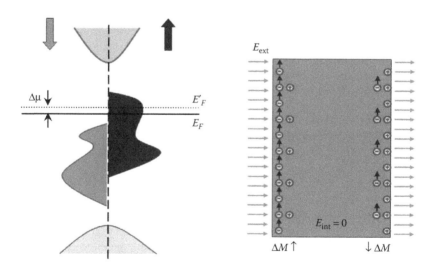

FIGURE 7.6 Schematic of the surface magnetoelectric effect in half-metals. Due to the 100% spin polarization rate, the accumulation of charge carriers Δμ at the half-metal surface induced by an electric field directly corresponds to the change of the surface magnetization ΔM. (From Duan, C.G., *Prog. Phys.* [in Chinese], 29, 215, 2009.)

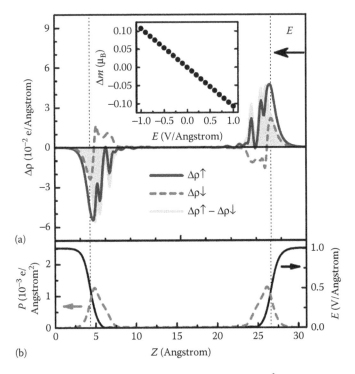

FIGURE 7.7   Effects of electric field on electronic properties of a 22-Å-thick $CrO_2(001)$ film along the z-direction normal to the film surface: (a) induced spin-dependent charge densities $\Delta\rho = \rho(E)-\rho(0)$ for majority-(↑) and minority-(↓) spin electrons and spin density $\Delta\rho\uparrow-\Delta\rho\downarrow$ averaged over the film plane. Inset: induced magnetic moment Δm per unit cell as a function of the applied electric field and (b) variation in the electric field (solid line) and the induced minority-spin polarization (dashed line) across the film. (From Duan, C.G., *Prog. Phys.* [in Chinese], 29, 215, 2009.)

Here, the positive (negative) sign corresponds to the conducting minority- (majority-) spin states. It suggests a surprising conclusion that the coefficient is a universal constant in a half-metal and is independent of its specific components, electronic, crystal, and/ or surface structures. A theoretical research (Duan et al., 2009) on the CrO_2 thin film confirmed it.

As clearly seen from Figure 7.7, with presence of external electric field, the spin-dependent screening causes the opposite sign of the charge densities localized in the vicinity of the surfaces. The excess spin densities imply electrically induced surface magnetization. By calculating the slope of the induced magnetic moment as a function of the applied electric field, the surface magnetoelectric coefficient for the $CrO_2(001)$ film is estimated as -6.41×10^{-14} G cm²/V, in good agreement with the value predicted by Equation 7.4. As the universal surface magnetoelectric coefficient of half-metals is distinguishable from that of ordinary ferromagnetic metals, it offers the unique feature to detect half-metallicity in ferromagnetic metals.

It is obvious that the predicted coefficient of the surface magnetoelectric effect in half-metals is too small to be applied in practice. Moreover, in an ordinary ferromagnetic metal, as a fraction of the universal constant mentioned in Equation 7.4, the surface magnetoelectric coefficient would be even less. It seems that the surface magnetoelectric effects are significant only when the applied electric field is very large, for example, of a few hundred mV/Å. In order to make the surface effect useful in application, a possible way to increase it is to improve the dielectric constant ε. In 2010, by studying the influence of an external electric field on magnetic properties of the Fe/MgO(001) interface, Niranjan et al. (2010) found that the magnetoelectric effect on the interface magnetization and magnetocrystalline anisotropy can be substantially enhanced if the electric field is applied across a dielectric material with a large dielectric constant.

In particular, they proposed that the surface magnetoelectric coefficient is larger than that for the Fe(001) surface by a factor of 3.8, which is approximately equal to high frequency dielectric constant of MgO. In addition, they pointed out that the change in the relative occupancy of the $3d$-orbitals of Fe atoms at the Fe/MgO surface significantly increases the electric field effect on the surface MAE. The enhancement of the interface magnetocrystalline anisotropy energy has been experimentally proved in a thin body-centered cubic Fe(001)/MgO(001) junction made by Maruyama et al. (2009). These experimental and theoretical results show the possibility to develop the electrically written magnetic information technology based on surface magnetoelectric effect.

7.1.1.3.2 *Screening Effect in Ferromagnetic/Ferroelectric Interfaces*

Similar spin-dependent screening effect of ferromagnetic metals induces magnetoelectric interaction in interfaces between ferromagnetic and dielectric (ferroelectric) materials as well. In 2008, Rondinelli et al. explored the influence of electric field on magnetic properties in $SrTiO_3/SrRuO_3/SrTiO_3$ heterostructures by first-principles calculations. As shown in Figure 7.8a, when an external voltage is applied, accumulation of spin is localized at the interfaces with an exactly equal magnitude and opposite sign for the two electrodes. The spin response of the interface is assessed by the parameter η, defined as the ratio of the surface spin polarization to the charge density. It is about 0.37 for the system. The high-frequency-limit result (Figure 7.8b) determines the origin of the calculated linear magnetoelectric effect. It is only ascribed to the accumulation of spin carriers at the interface and doesn't originate from the spin–lattice interaction. Also by replacing $SrTiO_3$ to $BaTiO_3$, the ratio η is unchanged; demonstrating that the larger amount of charge to screen the ferroelectric polarization can induce a larger change of the interface magnetic moment. Note that the phenomena that spin-polarized charge is stored at the interface asymmetrically are very similar to the behavior of the charge in traditional capacitor. This kind of carrier-mediated accumulation of spin-polarized carriers is a general characteristic of all ferromagnetic electrodes in contact with ferroelectric insulators, which provides the possibility to fabricate a new spin capacitor for the analogous spintronic devices in the future.

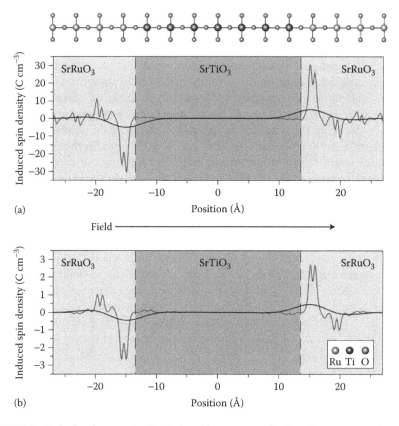

(a)

Field ⟶

(b)

FIGURE 7.8 Calculated magnetization induced by an external voltage in a nanocapacitor consisting of seven layers of $SrTiO_3$ alternating with seven layers of ferromagnetic, metallic $SrRuO_3$ electrodes with parallel magnetic alignment. The static response including ionic and electronic contributions (a) and the high-frequency (electronic only) response (b) are shown. The gray and black lines show the planar averaged and macroscopically averaged induced magnetizations, respectively. (From Rondinelli, J.M. et al., *Nat. Nanotech.*, 3, 46–50, 2008.)

7.1.2 Electric Field Control of Spin States Based on Magnetoelectric Coupling

With the deep understanding of mechanisms for coupling between magnetic and electric orders at surface/interface, many efforts have been made toward all-electric spintronics, including electric field control of magnetic anisotropy, exchange bias, spin transport, and the generalized electrically controlled magnetism.

7.1.2.1 Electric Field Control of Magnetic Anisotropy

As one of the most important properties of magnetic materials, magnetic anisotropy introduces preferential directions for the magnetization of a crystal. The magnetic moment of magnetically anisotropic systems will tend to align with an easy axis, which is an energetically favorable direction of spontaneous magnetization. Conversely, there exists a hard axis, corresponding to the hard direction of the magnetization. Generally speaking, ferromagnetic materials have intrinsic easy and hard axis. For low-dimensional magnetic systems, including surfaces, interfaces and thin-films, their magnetic anisotropy is usually larger and quite different with bulk phase. More excitingly, it appears to be modulated by electric method, which crucially reduces the required magnetic field to switch the magnetization. Thus, the development of electric-field assisted magnetic recording technology is expected.

The impact of electric field on the magnetic anisotropy has been widely reported in ferromagnetic metals combined with ferroelectric materials, such as $BaTiO_3$ (Sahoo et al., 2007; Meyerheim et al., 2011; Shu et al., 2012), $Pb(Zr, Ti)O_3$ (Weiler et al., 2009), $(PbMg_{1/3}Nb_{2/3}O_3)_{1-x}-(PbTiO_3)_x$ (Hu et al., 2011; Wu et al., 2011), and organic polyvinylidene fluoride with trifluoroethylene (Mardana et al., 2011). Strain-mediated and interface bonding origin of magnetoelectric effect are dominant in such ferroelectric/ferromagnetic hybrids.

In parallel with these reports, efforts have been also made on manipulation through an electric field effect of the magnetic anisotropy in ferromagnetic films adjacent to dielectrics (Maruyama et al., 2009; Yoichi et al., 2009; Endo et al., 2010; Bonell et al., 2011). Figure 7.9 shows a good example for a bcc Fe(001)/MgO(001) junction. Maruyama et al. (2009) found a relatively small electric field (less than 100 mV/nm) can greatly affect (~40%) the magnetic anisotropy here. The effect is attributed to the change of the relative electron occupancy in the $3d_{xy}$ and d_{x2-y2} states of Fe atoms adjacent to the MgO barrier. Although the nonferroelectric nature of dielectrics (typically MgO) makes the magnetoelectric effect in such systems volatile, these materials are readily compatible with existing technologies for magnetic random access memories and spin logic applications.

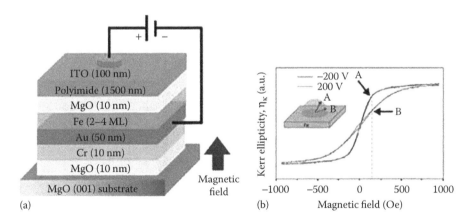

FIGURE 7.9 (a) Schematic of the sample used for a voltage-induced magnetic anisotropy change. A positive voltage is defined as a positive voltage on the top electrode with respect to the bottom electrode. A perpendicular magnetic anisotropy was induced by a negative voltage. The magnetic field was applied perpendicular to the film plane for Kerr ellipticity measurements and (b) representative magneto-optical Kerr ellipticity, η_K, curves measured under the bias voltage application of $U = \pm200$ V. (From Maruyama, T. et al., *Nat. Nanotech.*, 4, 158–161, 2009.)

In 2012, Wang et al. demonstrated in CoFeB/MgO/CoFeB magnetic tunnel junction (MTJ) that the electric field, both the magnitude and its direction, has a direct effect on the perpendicular magnetic anisotropy of CoFeB layers. The coercivity, the magnetic configuration, as well as the tunneling magnetoresistance (TMR) can be switched by voltage pulses associated with much smaller current densities. Figure 7.10a shows the schematic drawing of the p-MTJ and the effect of electric field through a small voltage supplied by a battery, and Figure 7.10b displays that the switching characteristics of the MTJ depend explicitly on the value and sign of the bias voltage V_{bias}. These results represent a crucial step toward ultralow energy switching in MTJs, and even voltage-controlled spintronic devices.

It is known that an electric field does not break time-reversal symmetry. Thus, external magnetic field seems to be necessary to realize the magnetization switching. Intriguingly, it is possible to switch in-plane magnetization utilizing the macro spin procession motion and voltage pulse, which is called coherent precessional magnetization switching. Shiota et al. (2012) realized the effect in a FeCo(001)/MgO(001)/Fe(001) MTJ (Figure 7.11a). For the multilayers, perpendicular magnetic anisotropy of the bottom ultrathin FeCo layer can be modulated by an applied voltage. Then, a bistable toggle switching using the coherent precession can be observed. As shown in Figure 7.11b and c, assuming the tunneling resistance measured after 50 successive negative and positive pulse voltage applications, the magnetization switching occurs only for the negative voltage pulse. When the positive voltage pulse is applied, no switching event is observed. This is because the voltage effect in the positive bias direction cannot excite coherent switching, owing to the suppression of the perpendicular magnetic anisotropy.

It is worth mentioning that the above structure could also be used to induce ferromagnetic resonances directly with an electric field, exerting coherent control over electron spin dynamics. As an example, Nozaki et al. (2012) revealed electric-field-induced ferromagnetic resonance excitation by means of voltage control over the magnetic anisotropy

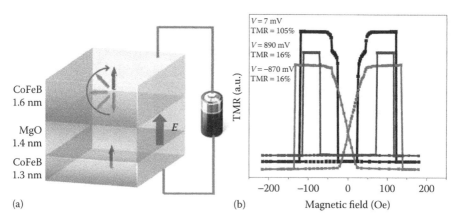

(a)   (b)

FIGURE 7.10 Electric-field-assisted switching in a CoFeB/MgO/CoFeB MTJ with interfacial perpendicular magnetic anisotropy: (a) schematic drawing of the p-MTJ and the effect of electric field through a small voltage supplied by a battery and (b) TMR curves under different bias voltages. (From Wang, W.-G. et al., *Nat. Mater.*, 11, 64–68, 2012.)

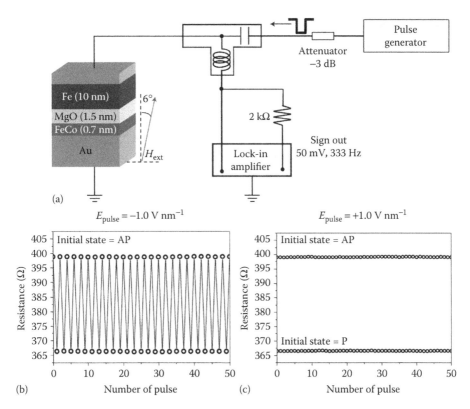

FIGURE 7.11 Pulse-voltage-induced coherent magnetization switching. (a) Schematic of the measurement set-up and sample structure of the MTJ device. Voltage pulses are applied to the MTJ using a pulse generator, and the ac resistance of the sample is monitored using a standard lock-in technique; (b) and (c) are examples of magnetization switching events induced by 50 successive pulse voltage applications. (From Shiota, Y. et al., *Nat. Mater.*, 11, 39–43, 2012.)

in a few monolayers of FeCo at room temperature, providing another low-power, highly localized, and coherent means to manipulate electron spin dynamics. Similar resonance response was also investigated in CoFeB/MgO/CoFeB MTJ (Zhu et al., 2012).

In theoretical aspects, possible effects on the magnetocrystalline anisotropy in MgO/FePt-based tunnel junctions and their origins have been systematically researched (Zhu et al., 2011, 2013). Based on them, Zhu et al. (2014) explored the dependence of MAE on the external electric field in the MgO/FePt/Pt(001) films. Using first-principles calculations, the surface magnetoelectric coefficient is estimated to be 10 times larger than that of the Fe(001) surface (Duan et al., 2008b), indicating that the MAE can be linearly manipulated in a broad range by applying electric field. Considering the linear tuning effect, the Laudau–Lifshitz–Gilbert macrospin simulation is used to investigate the coherent magnetization switching triggered by short pulses of electric-field. The final state of the magnetization switching is proved to depend on the pulse time width τ. When the pulse amplitude is enhanced, the minimal critical pulse width decreases, whereas the maximal one increases. In addition, by increasing the initial processional

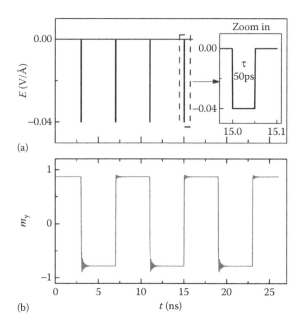

(a)

(b)

FIGURE 7.12   The in-plane magnetization switching by multiple pulses: (a) the applied electric field pulses during the switching and (b) the magnetization switching between parallel and anti-parallel states. The beginning state is the parallel state. (From Zhu, W. et al., *Sci. Rep.*, 4, 4117, 2014.)

angle of the magnetization, the irregular switching would be restricted under the situation of short pulse width. A successive magnetization switching can be achieved by a series of electric field pulses, as displayed in Figure 7.12.

7.1.2.2 Electric Field Control of Exchange Bias

In addition to the composites mentioned above, magnetoelectric effect has been demonstrated in ferromagnets coupled to multiferroics (magnetoelectrics). As most multiferroics (magnetoelectrics) are antiferromagnetic, an exchange bias effect may be obtained on the adjacent ferromagnetic layers. In detail, through magnetoelectric coupling, the application of voltage across the structure is expected to change the magnetic exchange coupling across the interface and to thus modify the magnetic response of the ferromagnets.

Borisov et al. (2005) first reported electric field control of exchange bias using magnetoelectric Cr_2O_3 single crystals as pinning layers. The switching of perpendicular exchange bias field H_{EB} in the heterostructure $Cr_2O_3(111)/(Co/Pt)_3$ is explained by magnetoelectrically induced antiferromagnetic single domains that extend to the interface, in which the direct of their end spins controls the sign of H_{EB}. Although the effect is reproducible, it requires cooling the system below the Néel temperature of Cr_2O_3.

Strong static exchange interaction is observed in heterostructures based on multiferroic material $YMnO_3$ (Laukhin et al., 2006; Marti et al., 2006a, b) as well. For room-temperature operations, most efforts have focused on $BiFeO_3$ (Bea et al., 2006, 2008; Dho et al., 2006), an excellent multiferroic with ferroelectric and antiferromagnetic ordering temperatures well above 300 K. In 2008, Martin et al. demonstrated a direct correlation

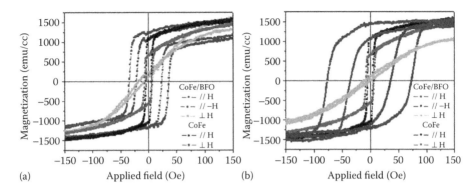

FIGURE 7.13 Room temperature magnetic properties for CoFe/BiFeO$_3$ heterostructures exhibiting: (a) exchange enhancement in stripe-like BiFeO$_3$ films and (b) exchange bias properties in mosaic-like BiFeO$_3$ films. (From Martin, L.W. et al., *Nano. Lett.*, 8, 2050–2055, 2008.)

between the domain structure of multiferroic antiferromagnet BiFeO$_3$ thin films and exchange bias of Co$_{0.9}$Fe$_{0.1}$/BiFeO$_3$ heterostructures. For the case of stripe-like ferroelectric domain structures corresponding to arrays of 71° domain walls (Figure 7.13a), a significant enhancement of the coercive field (exchange enhancement) is observed as compared to that of CoFe grown directly on SrTiO$_3$(001). When the BiFeO$_3$ film is of mosaic-like architecture (Figure 7.13b), comprising a mixture of all possible domain wall types, that is, large fractions of 109° domain walls and smaller fractions of 71° and 180° walls, there exists an enhancement of the coercive field combined with large shifts of the hysteresis loop, confirming the behavior to be an exchange bias interaction. The distinctly different types of room temperature magnetic responses of CoFe, which depend directly on the type and crystallography of the nanoscale (~2 nm) domain walls in the BiFeO$_3$ film, prove the existence of magnetoelectric effect here. It is later be used to realize electric control of magnetization rotation (Chu et al., 2008).

7.1.2.3 Electric Field Control of Spin Transport

As a well-known phenomenon in quantum mechanics, tunnel effect not only has fundamental scientific interests, that is, it demonstrates the wave-particle dualism, but also is of great importance to modern technology. The successful application of TMR effect (Julliere, 1975) or giant magnetoresistance effect (Baibich et al., 1988) in hard-disk and magnetoresistive random-access memory (Åkerman, 2005) drives the research of other types of tunnel effects. Considering the coexistence of electric and magnetic orders in multiferroics, which implies the feasibility of multistate memory, multiferroic tunnel junctions serve as candidates in the field of all-electric spintronics.

In 2007, Gajek et al. fabricated such tunnel junctions with multiferroic material La$_{0.1}$Bi$_{0.9}$MnO$_3$ as a barrier. The epitaxial La$_{0.1}$Bi$_{0.9}$MnO$_3$ films are both ferromagnetic and ferroelectric and retain these ferroic properties down to a thickness of 2 nm. As schematically diagrammed in Figure 7.14, this allows the tunneling electrons to be efficiently filtered according to their spin, as well as the ferroelectric polarization states. The combination of the tunnel magnetoresistance (Figure 7.14a) and the electroresistance effect

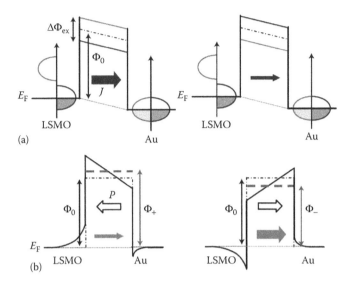

FIGURE 7.14 Schematic representation of the tunnel barrier potential profiles and tunnel currents for a ferromagnetic and a ferroelectric barrier: (a) spin-dependent tunneling for parallel (left) or antiparallel (right) configurations of the nonferroelectric $La_{0.1}Bi_{0.9}MnO_3$ and the half-metallic $La_{2/3}Sr_{1/3}MnO_3$ magnetizations and (b) assuming a nonmagnetic $La_{0.1}Bi_{0.9}MnO_3$ barrier, the tunneling electrons for the two electric polarization directions of the barrier. Φ_0 is the barrier in the absence of magnetic or electric polarization. $\Delta\Phi_{ex}$ is the exchange splitting. Φ_+ and Φ_- are the average barrier heights seen by the carriers when ferroelectric polarization P points towards LSMO and Au, respectively. (From Gajek, M. et al., *Nat. Mater.*, 6, 296–302, 2007.)

(Figure 7.14b) produces a four-resistance-state system. The four-state resistive phenomenon is, excitingly, observed in the junctions, constituting an important step toward the integration of nanometric multiferroic elements in spintronics devices.

Because single-phase multiferroics are very rare in nature and just a few of them retain multiferroic properties at room temperature, it is advantageous to make multiferroic tunnel junctions from the combination of ferroelectric and ferromagnetic materials. On the basis of first-principles calculations, Velev et al. (2009) demonstrated four well-defined resistance states in $SrRuO_3/BaTiO_3/SrRuO_3$ multiferroic tunnel junctions. The two asymmetrical interfaces in the tunnel junction provide the crucial structure basis for the coexistence of tunneling electroresistance effect and TMR effect. As displayed in Figure 7.15, due to the sensitivity of conductance to the magnetization alignment of the electrodes and the polarization orientation in the ferroelectric barrier, the resistance of such a multiferroic tunnel junction is significantly changed when the electric polarization of the barrier is reversed and/or when the magnetizations of the electrodes are switched from parallel to antiparallel. Both the calculated TMR and tunneling electroresistance ratios are very large and dependent on the other ferroic order, revealing the strong effect of ferroelectricity on spin transport properties. The higher working temperature and relatively significant difference among conductance offer their potential application in multilevel nonvolatile memories, tunable electric and magnetic field sensors, multifunctional resistive switches, and some other multifunctional spintronic devices.

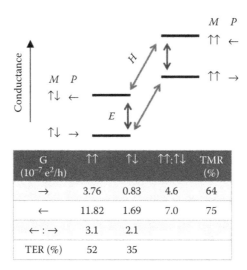

G $(10^{-7}\,e^2/h)$	↑↑	↑↓	↑↑:↑↓	TMR (%)
→	3.76	0.83	4.6	64
←	11.82	1.69	7.0	75
← : →	3.1	2.1		
TER (%)	52	35		

FIGURE 7.15 Conductance of the SrRuO₃/BaTiO₃/SrRuO₃ multiferroic tunnel junctions. The four conductance states, which can be controlled via electric E and magnetic H fields, are distinguished by polarization P in the barrier pointing to the left (←) or right (→) and magnetization M of the electrodes being parallel (↑↑) or antiparallel (↓↓). (From Velev, J.P. et al., *Nano. Lett.*, 9, 427–432, 2009.)

Numerous theoretical and experimental efforts have been input to study the electric field manipulation of spin transport in multiferroic tunnel junctions, including ferromagnetic metal/BaTiO₃/LSMO (Garcia et al., 2010; Burton and Tsymbal, 2011; Valencia et al., 2011), Co/PbZr₀.₂Ti₀.₈O₃/LSMO (Pantel et al., 2012), La₀.₁Bi₀.₉MnO₃/BiFeO₃/LSMO (Hambe et al., 2010), La₀.₅Ca₀.₅MnO₃/BaTiO₃/LSMO (Yin et al., 2013), and Co/PbTiO₃/LSMO (Quindeau et al., 2015). A more comprehensive account for the recent progress in multiferroic tunnel junctions can be found in the review by Fang et al. (2013).

7.1.2.4 Electric Field Control of Magnetization Rotation

Compared with great advances in electric-field control of macroscopic magnetic properties demonstrated above, to realize the electrical manipulation of micromagnetic elements, especially the direct magnetization rotation, is the most fundamental and important task in all-electric spintronics. It represents the critical enhancement toward the realization of practical devices in information processing industry.

Chu et al. (2008) reported a more complex approach (Figure 7.16) for electrical control of magnetization switching in ferromagnet/multiferroic heterostructure. The first ingredient is the internal magnetoelectric coupling between ferroelectricity and antiferromagnetism in the multiferroic BiFeO₃ thin film. The second one is the interfacial exchange coupling between the antiferromagnetic BiFeO₃ and ferromagnetic Co₀.₉Fe₀.₁, which leads to a significant enhancement in the coercive field, as well as an exchange-biased hysteresis loop for the ferromagnetic pad deposited on top of BiFeO₃. The intensity distribution in the photoemission electron microscopy image shows that the direction of the magnetization in the ferromagnet can be rotated by 90° with an electric field,

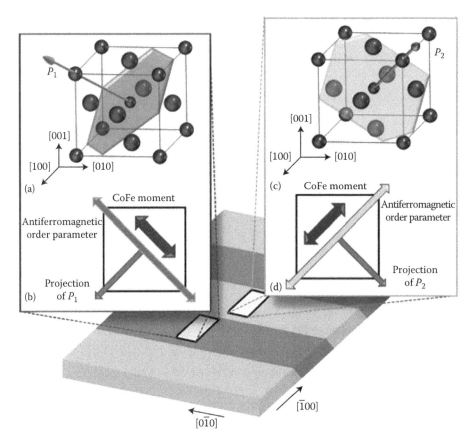

FIGURE 7.16 Mechanism of coupling in $Co_{0.9}Fe_{0.1}/BiFeO_3$ heterostructures. (a,c) Schematic diagrams of two adjacent domains in the [001]-oriented $BiFeO_3$ crystal, in which the [111] polarization directions as well as the antiferromagnetic plane (that is perpendicular to this *P* direction) are identified, and (b,d) the corresponding projections of the polarization direction, the antiferromagnetic plane onto the [001], and the corresponding magnetization directions in the CoFe layer. (From Chu, Y.-H. et al., *Nat. Mater.*, 7, 478–482, 2008.)

illustrating the possibility of marrying intrinsic coupling and interfacial coupling to realize electrically controlled magnetization switching.

As a further step, an 180°reversal of magnetization is realized in the $Co_{0.9}Fe_{0.1}$ layer in contact with multiferroic $BiFeO_3$ by Heron et al. (2011). As shown in Figure 7.17, using a striped two-variant $BiFeO_3$ film, each of the polarization variants is rotated by 71° corresponding to a 90° in-plane rotation. One rotates clockwise, whereas the other is counter-clockwise. The combination of the variants' rotation leads to a unique path for a reversal of net polarization. Due to the one-to-one magnetic interface coupling in the $Co_{0.9}Fe_{0.1}/$ $BiFeO_3$ heterostructures, the net polarization reversal naturally results in a reversal of the in-plane projection of $BiFeO_3$'s magnetic order, and consequently, the magnetization of the exchange coupled $Co_{0.9}Fe_{0.1}$ layer. It may be the first pure electric-field control of room-temperature magnetization switching realized experimentally. The work, thus, stands out as

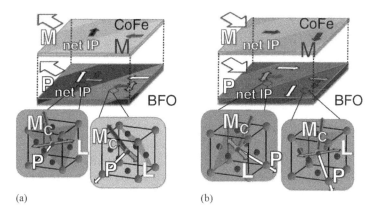

FIGURE 7.17 Representations of the one-to-one magnetic interface coupling in the $Co_{0.9}Fe_{0.1}$/$BiFeO_3$ heterostructure in the (a) as-grown state and (b) after the polarization reversal. (From Heron, J.T. et al., *Phys. Rev. Lett.*, 107, 217202, 2011.)

one of the landmarks in all-electric spintronics. However, in-plane electric fields were applied, which intrinsically prohibits the high storage capacity for memories (Hu et al., 2011).

Theoretically, Fechner et al. (2012) proposed a way to switch magnetization by 180° with external electric field perpendicular to multilayers. They designed a hybrid structure consisting of a ferroelectric thin film in contact with a ferromagnetic/nonmagnetic/ferromagnetic trilayer, as displayed in Figure 7.18. The electric polarization $P\uparrow$ favors parallel alignment of the two Fe layers spaced by Au film of certain thickness, whereas $P\downarrow$ corresponds to antiparallel alignment of layer magnetizations. Combining the

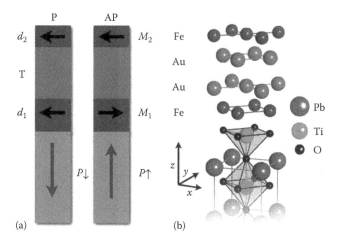

FIGURE 7.18 (a) Hybrid design to allow for switching the interlayer exchange coupling by polarization reversal. If the magnetoelectric coupling at the interface is strong enough, it allows controlling the alignment of magnetizations M_1 and M_2, (b) hybrid structure on the atomic scale, consisting of a TiO_2 terminated $PbTiO_3$ layer covered by a Fe/Au/Fe trilayer. (From Fechner, M. et al., *Phys. Rev. Lett.*, 108, 197206, 2012.)

effects of magnetoelectric coupling at the multiferroic interface and magnetic interlayer exchange coupling in the trilayer, polarization switching of ferroelectric film can induce a 180° magnetization switching of the magnetic Fe free layer.

In general, the elastic coupling mechanism is promising to realize controllable magnetization rotation with vertical electric fields. It is a pity that strains only result in at best 90° magnetization switching in an individual magnetic domain, which is also nondeterministic (Hu and Nan, 2009). Using phase-field simulations, Wang et al. (2014) received reproducible and controllable full 180° magnetization reversal in strain-driven magnetoelectric heterostructures based on both the in-plane piezostrains and magnetic shape anisotropy. They examined an artificial multiferroic heterostructure (Figure 7.19a), in which a patterned single-domain nanomagnet with four-fold magnetic axis is placed on a ferroelectric layer with electric-field-induced uniaxial strains. Under a negative electric field, the tensile strain on the Ni nanodot along the y-axis arises from the piezostrain of the PMN-PT layer. Then, the magnetization m is clockwise switched away from its initial state ① lying in the first quadrant with $\alpha = 23°$ to the energy minima in the fourth quadrant (see ② in Figure 7.19c). A continuous positive electric field, generating a compressive strain, makes m continuously switch clockwise to finish an almost 180° magnetization reversal (see ③ in Figure 7.19c). After removal of

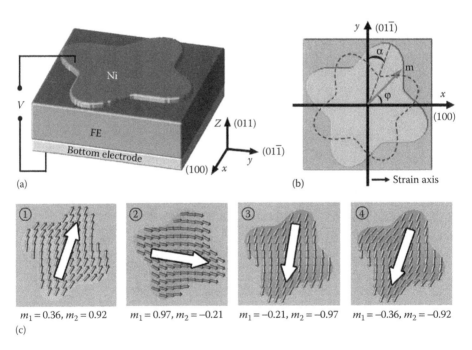

$m_1 = 0.36, m_2 = 0.92$ $m_1 = 0.97, m_2 = -0.21$ $m_1 = -0.21, m_2 = -0.97$ $m_1 = -0.36, m_2 = -0.92$

(c)

FIGURE 7.19 (a) Schematic of the morphologically engineered artificial multiferroic heterostructure, using Ni/PMN-PT as an example, (b) top view. The long axis of Ni nanomagnet is tilted at an angle α away from the main direction of in-plane anisotropic piezostrain, that is, y-axis. The dashed line represents the shape anisotropy energy contour, and (c) magnetization vector diagrams. (From Wang, J.J. et al., *Sci. Rep.*, 4, 7507, 2014.)

the electric field, m switches a small angle from state ③ to state ④, that is, to the energy minimum in the third quadrant on the reverse direction of the initial m. This 180° magnetization full reversal from state ① to state ④ driven by a pair of negative and/or positive electric field pulses is repeatable. Similar strategy is experimentally adopted in Co/Pb(Mg$_{1/3}$Nb$_{2/3}$)$_{0.7}$O$_3$Ti$_{0.3}$O$_3$ multiferroic heterostructure (Yang et al., 2014a). The in-plane magnetic anisotropy of ferromagnetic film is, however, fulfilled by the interface effect rather than the shape ones. The nonvolatile magnetization reversal is, thus, still driven by the undesirable in-plane electric field. In addition, the reversal cannot be repeatable before applying an external static magnetic field.

Very recently, repeatable local magnetization reversal manipulated by purely perpendicular electric field is experimentally realized in a NiFe thin film grown directly on a rhombohedral Pb(Mg$_{1/3}$Nb$_{2/3}$)$_{0.7}$Ti$_{0.3}$O$_3$ ferroelectric crystal (Gao et al., 2016). The magnetization switching path in the simple magnetic/ferroelectric bilayer is strain and exchange-bias comediated. On the contrary, the observed magnetization switching is of 135°, rather than full 180°.

Various theoretical and experimental attempts have been made in this regard (Liu et al., 2011; Hu et al., 2012, 2015; Ghidini et al., 2013; Lei et al., 2013; Yang et al., 2014b), greatly promoting the realization of magnetoelectric switching and technologically pertinent functionality for nanometer-scale, low-energy-consumption, and nonvolatile magnetoelectronics.

7.1.2.5 Generalized Electrically Controlled Magnetism

In a more general sense, the electric field control of magnetism not only includes the electrically modulated magnetic anisotropy, spin transport, exchange bias, and magnetization mentioned above, but also magnetic ordering (Ohno et al., 2000; Burton and Tsymbal, 2009; Ding and Duan, 2012), domain structures (Gerhard et al., 2010; Lahtinen et al., 2012), transition temperature (Ohno et al., 2000; Thiele et al., 2007; Stolichnov et al., 2008; Molegraaf et al., 2009), and so on, which provide various interesting magnetoelectric effects at interfaces of multiferroic multilayers.

Electrical control of the magnetic order, as well as the Curie temperature was demonstrated as early as 2000 in diluted magnetic semiconductors with carrier-mediated ferromagnetism. Using an insulating-gate (In, Mn)As field-effect transistor structure (see Figure 7.20), Ohno et al. (2000) demonstrated tuning of the magnetic state by an applied electric field. At 22 K, without any electric field, the (In, Mn)As channel exhibits weak ferromagnetism due to the vicinity of its Curie temperature. At the same temperature, when the positive exchange bias V_G is applied, holes are partially depleted from the channel, resulting in decrease of the ferromagnetic interaction among magnetic Mn ions. A clear paramagnetic response is observed at $V_G = +125$ V. The negative V_G has an opposite effect. In such case, the ferromagnetism is reinforced with the enhancement of hole concentration. This corresponds to an isothermally and reversibly shift of the transition temperature of hole-induced ferromagnetism by 1 K upon application of electric fields.

Practical interfacial magnetoelectric devices rely on the feasibility of enduring reversible manipulation of domains. In BaTiO$_3$/CoFe heterostructures, Lahtinen et al. (2012) provided spectacular proof for writing and erasing ferromagnetic domains in an electric

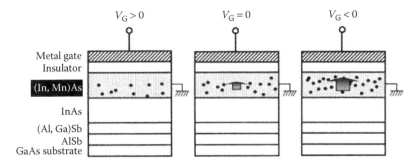

FIGURE 7.20 Field-effect control of the hole-induced ferromagnetism in magnetic semiconductor (In, Mn)As field-effect transistors. Shown are the cross-sections of a metal–insulator–semiconductor structure under gate biases V_G. The applied electric field controls the hole concentration in the magnetic semiconductor channel (filled circles). (From Ohno, H. et al., *Nature*, 408, 944–946, 2000.)

field and accurately controlling the motion of magnetic domain walls. As shown in Figure 7.21a, the homogeneous ferroelectric polarization of the $BaTiO_3$ substrate is saturated by an out-of-plane bias voltage of 120 V, which produces a regular ferromagnetic stripe pattern in the CoFe film. Reverting back to zero bias cause the ferroelectric microstructure to a mixed-domain pattern, resulting in the erasure of magnetic stripes on top of the ferroelectric a_1 domains with in-plain polarization (Figure 7.21b). Correlations

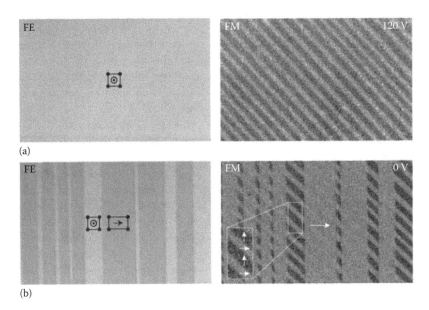

FIGURE 7.21 Electric-field writing and erasure of ferromagnetic domain patterns. Polarization microscopy images of the ferroelectric switching in $BaTiO_3$ domains (left) coupled with the magnetic domains in $Co_{0.9}Fe_{0.1}$ while applying an out-of-plane bias of (a) 120 V or (b) 0 V. (From Lahtinen, T.H.E. et al., *Sci. Rep.*, 2, 258, 2012.)

between the ferromagnetic microstructure and the underlying ferroelectric lattice indicate that it is indeed possible to control lateral domain growth and shrinking by applying an out-of-plane electric field followed by relaxation. The approach is based on recurrent strain transfer from ferroelastic domains in ferroelectric media to continuous magnetostrictive films with negligible magnetocrystalline anisotropy.

Besides the external electric field, charge injection is another effective electric approach to control magnetism. Using relativistic density-functional theory, Gong et al. (2012) investigated the MAE of freestanding Fe monolayer in Fe/graphene complex system. Because of the strong hybridization effect between Fe's $3d$ states and graphenes' p_z states, the MAE of Fe atoms is drastically suppressed from meV/atom in freestanding Fe monolayer to µeV/atom in Fe/graphene system. More significantly, the suppressed MAE of Fe atoms in Fe/graphene can be restored back through injecting charge, which provides an additional electric approach to control MAE. Similarly, Ruiz-Diaz et al. (2013) used the method of charge injection to tune MAE of Fe-Pt multilayers and found that MAE and the direction of magnetization in metallic magnetic multilayers can be tailored by surface charging.

7.1.3 Electric Field Control of Spin States Based on Spin–Orbit Coupling Effect

Spin–orbit coupling (SOC) (Winkler, 2003), which acts as a bridge between the orbital motion and spin degree of freedom even in nonmagnetic materials, has attracted considerable research attention in the field of spintronics. For the electron, SOC can be regarded as an effective magnetic field, making spin of the electron respond to its orbital environment. Its advantage of manipulating the spin degree of freedom in absence with external magnetic fields and/or magnetic materials offers the potential application in all-electric spintronics.

7.1.3.1 Rashba Spin–Orbit Coupling

SOC interaction is a kind of the relativistic effect, which can be expressed as follows in single-electron approximation:

$$H_{SOC} = \frac{\hbar}{4m_0^2 c^2} \sigma \cdot V_0(r) \times p \qquad (7.5)$$

It describes an effective magnetic field experienced by an electron with momentum p and spin σ moving in an electric field $V_0(r)/e$.

Among various kinds of SOCs, Rashba-type SOC (Yu and Rashba, 1984) has attracted the most research focus due to its tunability through external electric filed, which has been experimentally implemented in semiconductor heterostructures (Engels et al., 1997; Nitta et al., 1997). Rashba SOC arises from the structural inversion asymmetry and thus generally exists at the interface/surface. Its Hamiltonian can be written as

$$H_R = \frac{\alpha}{\hbar}(\sigma \times p) \cdot \hat{z} \qquad (7.6)$$

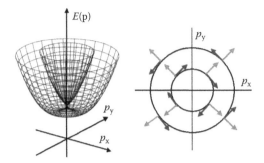

FIGURE 7.22 Schematic of the energy dispersion for a two-dimensional electron system with spin-momentum locking induced by Rashba SOC. For each momentum (light gray arrows), the two eigen spinors point in the azimuthal direction (dark gray arrows). (From Sinova, J. et al., *Phys. Rev. Lett.*, 92, 126603, 2004.)

where $\boldsymbol{\sigma}$ is the Pauli matrices, α as the Rashba coupling parameter, and $\hat{z}$ is the unit vector perpendicular to the interface/surface. Rashba SOC effect locks spin to the linear momentum and split the spin subbands in energy (Figure 7.22).

7.1.3.2 Electric Field Control of Rashba Spin Splitting

7.1.3.2.1 Rashba Spin Splitting in Semiconductors

The use of Rashba SOC in the research field of spintronics can be traced back at least to the theoretical proposal of the famous spin field-effect-transistor (FET) by Datta and Das (1990). As shown in Figure 7.23, the InAlAs/InGaAs heterostructure provides a two-dimensional channel for electron transport between two ferromagnetic electrodes. One electrode acts as a spin emitter, and the other the spin collector. Here, the electron spins precess in the effect magnetic field due to the Rashba SOC effect. By tuning the strength of Rashba SOC via gate voltage, one can thus control the flow of electron spins between spin polarized source and drain contacts and then determines the transport

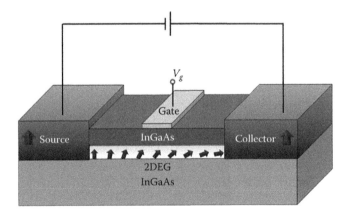

FIGURE 7.23 Schematic diagram of the Datta–Das type spin FET.

current. In detail, an electron passes through the collector if its spin is parallel to the magnetization of the collector and does not under antiparallel configuration.

Although the Datta–Das spin FET was proposed as early as in 1990, it remains a theoretical model. Various difficulties, including the low-spin injection efficiency from ferromagnetic source to the semiconductor channel as well as the serious spin decoherence effect induced by strong strength of Rashba SOC, restrict its realization in experiments. Nevertheless, the gate-controlled device using Rashba SOC effect lies at the heart of potential applications for spintronics.

As an unusual magnetoelectric effect, the well-known spin Hall effect (SHE) offers strong evidence of electrical spin manipulation based on Rashba SOC. Analogous to the conventional Hall effect, the SHE refers to spin accumulation at the edges of the sample when applying a pure charge current j_x (Figure 7.24a) (Dyakonov and Perel, 1971). This occurs through two types of mechanism. The extrinsic one relates to spin-dependent Mott scattering (Hirsch, 1999). Electrons with different spin projections diffuse toward opposite directions after scattering against SOC impurities. The intrinsic one concerns the spin-dependent distortion of electron trajectories in the presence of a SOC band structure (Murakami et al., 2003; Sinova et al., 2004).

Without any external magnetic field, an out-of-plane spin polarization with opposite sign on opposite edges was experimentally detected in n-doped GaAs (Figure 7.24b)

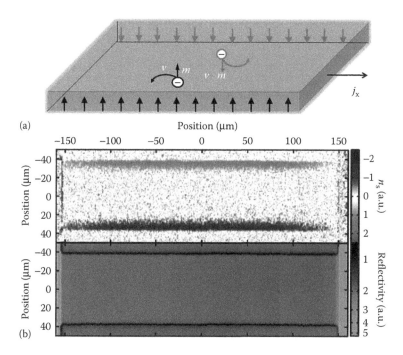

FIGURE 7.24 (a) Schematic of the spin Hall effect. (From Dyakonov, M.I., Perel, V.I., *Phys. Lett. A*, 35, 459–460, 1971.) (b) Two-dimensional images of spin density n_s and reflectivity of the unstrained GaAs sample, respectively. The images are gained using Kerr rotation microscopy. (From Kato, Y.K. et al., *Science*, 306, 1910–1913, 2004.)

(Kato et al., 2004). The weak spin polarization attributes the SHE observed here to the extrinsic mechanism. In light-emitting diodes based on a GaAs two-dimensional hole system, the SHE was also observed (Wunderlich et al., 2005). The magnitude of the spin polarization in this experiment is in good agreement with that predicted by the intrinsic SHE.

Combining spin transistor and the SHE, an all-semiconductor SHE transistor was realized by Wunderlich et al. (2010). In the active semiconductor channel in the device, spin precession induced by Rashba SOC is controlled by external gate electrodes, and detection is provided by transverse SHE voltage. The study realizes the spin transistor with electrical detection directly along the gated semiconductor channel and, more importantly, provides an experimental tool for exploring spin Hall and spin procession phenomena in an electrically tunable nonmagnetic semiconductor layer. More utilities of the SHE in microelectronic device geometries are available in the excellent reviews (Jungwirth et al., 2012; Sinova et al., 2015).

Even without any external magnetic fields or using ferromagnetic contacts, almost 100% spin polarization can be realized in a Stern–Gerlach spin separation (Ohe et al., 2005). The spin separation roots in a spatial gradient of the effective magnetic field due to the Rashba SOC effect. The experimental manifestation of such an inhomogeneous SOC-induced electronic spin device was demonstrated by Kohda et al. (2012) in InGaAsP/InGaAs-based field FETs. As schematically shown in Figure 7.25, a spin-dependent force is generated by the lateral potential confinement, which separates the spin-up and spin-down electrons, and the selective filtering of the spin components at

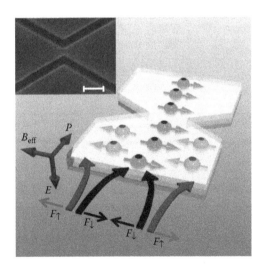

FIGURE 7.25 Stern–Gerlach type spin separation in a quantum point contact structure. The device comprises an InGaAs/InGaAsP two-dimensional electron gas with strong Rashba SOC. The arrows labelled $F_\uparrow$ and $F_\downarrow$ are the directions of the spin-dependent force for up and down spins. The directions of the electron momentum P, the internal electric field E and the effective magnetic field B_{eff} are shown in thick arrows. The inset shows an image of a quantum point contact channel region taken by a scanning electron microscope. Channel conductance is controlled by the trench-type side gates. Scale bar = 1 μm. (From Kohda, M. et al., *Nat. Commun.*, 3, 1082, 2012.)

the quantum point contact potential results in a spin–split single channel. Note that the spin-polarized orientation is in-plane here, in contrast with the SHE.

In addition to the Rashba SOC dependent semiconductor devices mentioned above, many others have been proposed in recent years, including spin polarizer (Debray et al., 2009), spin filter (Nitta and Koga, 2003; Gong and Yang, 2007), spin interferometer (Koga et al., 2004). Especially, the spin–orbit qubits have been established in GaAs/AlGaAs gate-defined quantum dots (Nowack et al., 2007), InAs nanowires (Nadj-Perge et al., 2010), InSb nanowires (van den Berg et al., 2013), and so on, paving a way for a scalable quantum computing architecture using electric field rather than static or oscillating magnetic ones.

7.1.3.2.2 Rashba Spin Splitting in Metal Surface

In the above discussion, we have highlighted semiconductor spintronics related to the Rashba SOC effect. On the contrary, the strength of Rashba effect in most common semiconductors, such as silicon and gallium arsenide, is too small to effectively manipulate electron spin. Even if strong enough SOC is available, the momentum-changing scattering of an electron moving through a semiconductor causes sudden changes in the effective SOC magnetic field. The spin randomization (D'Yakonov and Perel, 1971) appears to be a major challenge in practical applications. Under the circumstances, research interests of the Rashba effect extended from semiconductors to a new material family, that is, surfaces of metals (LaShell et al., 1996; Nicolay et al., 2001; Krupin et al., 2005; Bihlmayer et al., 2006; Sakamoto et al., 2009), whose strength of SOC effect is much stronger and can be directly detected through angle-resolved photoelectron spectroscopy (ARPES). The first observation for the Rashba spin splitting of metal surfaces succeeded in 1996 on the Au(111) surface (LaShell et al., 1996). A considerable spin splitting in Figure 7.26,

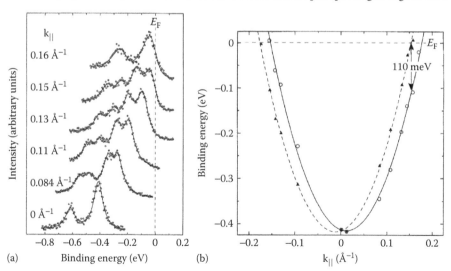

FIGURE 7.26 Rashba spin splitting of an Au(111) surface: (a) spectra taken along the Γ–M at indicated values of parallel momenta and (b) energy dispersion. Open circles and filled triangles are experimental data gained from (a). The solid and dashed lines are parabolic fits. (From LaShell, S. et al., *Phys. Rev. Lett.*, 77, 3419–3422, 1996.)

about 110 meV at the Fermi level, was obtained in the *sp*-derived surface states through ARPES. The authors already pointed out the importance of both the surface gradient and the steep nuclear Coulomb potential in heavy elements that made the observation possible. The role of both contributions has been theoretically clarified by tight-binding models (Petersen and Hedegård, 2000) and density-functional theory calculations (Nicolay et al., 2001). An experimental confirmation was also obtained by comparing Li-covered (110) surfaces of W and Mo (Rotenberg et al., 1999). Besides the Au(111), similar Rashba-type spin splitting was investigated in other metal surfaces as well, such as Bi(111), Bi(110), and Bi(100) (Koroteev et al., 2004), Sb(111) (Sugawara et al., 2006), Ir(111) (Varykhalov et al., 2012), and Gd(0001) (Krupin et al., 2005). The largest Rashba SOC effect so far has been found for a surface alloy of Bi/Ag(111) (Ast et al., 2007), in which a Rashba coupling parameter α ≈ 3.05 eV/Å was obtained. There were studies on various surface alloys, including Ag/Au(111) (Nuber et al., 2011), Ag/Pt(111) (Bendounan et al., 2011), and Bi(Pb)/Cu(110) (Alberto et al., 2013), aiming to tune the magnitude of the Rashba splitting strength by surface modification.

The influence of atomic and interfacial potential gradients on the Rashba splitting has been revealed by several work (Krupin et al., 2005; Ast et al., 2007). However, the electric field, as the most promising and popular approach to manipulate SOC effect in semiconductor heterostructures, is seldom exploited in the metal surfaces. In 2006, Bihlmayer et al. (2006) first tried to explore the electric field control of Rashba splitting at Lu(0001) surface, yet they just provided very simple discussion. Park et al. (2011) calculated surface bands of high-Z materials. A single metal layer (a single Bi layer as an example) itself shows no band splitting. When an external electric field ($E = 3$ V/Å) perpendicular to the film is applied, the Rashba-type surface band splitting appears. The orbital angular momentum in combining with the electron momentum causes asymmetric charge distribution or electric dipole moment in such high-Z materials due to the strong atomic SOC. The surface-normal electric field then aligns the electric dipole and results in the relevant Rashba-type splitting. In general, the thinner the film is, the larger an applied electric filed should be.

In 2010, Mirhosseini et al. (2010) conceptually demonstrated the possibility of manipulating the Rashba spin splitting by the electric polarization in Bi/BaTiO$_3$(001) multilayers. Switching of the intrinsic electric polarization in the ferroelectric by external electric field, to some extent, affects the spin–electric coupling and then changes the strength of the Rashba splitting in the adsorbed heavy *p* metal. Although the calculated spin–electric effect is moderate, with a relative change in the splitting of about 5%, it is conceivable to increase it by larger atomic displacements in particular at the ferroelectric/adlayer interface.

The direct influence of the electric field on Rashba spin splitting was investigated by Gong et al. (2013a) for the first time in Au(111) surface states. Combining density-functional theory calculations and theoretical analyses, they found that the applied electric field can tune the surface electrostatic potential and its gradient. Significantly, the change of the former can induce an upward or downward shift of the surface Rashba splitting bands. Meanwhile, the change of the latter increases or decreases the Rashba splitting energy, as plotted in Figure 7.27. For the electric field ranging from −0.4 to +0.4 V/Å, a good linear relationship between the Rashba splitting energy and the external field is revealed, attributing to the linear response of the first-order Rashba parameter to the

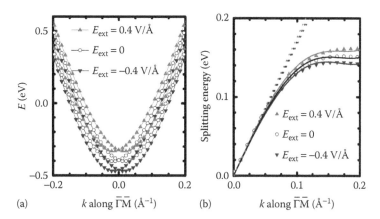

FIGURE 7.27 (a) The surface Rashba splitting bands and (b) the Rashba splitting energies under various electric fields of 22-layer Au(111) slab. (From Gong, S.-J. et al., *Phys. Rev. B*, 87, 035403, 2013a.)

electric field. The higher order Rashba components, however, are less influenced by the electric field, which causes the nonlinear relationship between the splitting energy and wave vectors.

7.1.3.2.3 Rashba Spin Splitting in Other Systems

Over the past 30 years, Rashba SOC effect has inspired a great number of predictions, discoveries and innovative concepts in semiconductors and metals. More recently, the progress in low-dimensional Dirac systems, ranging from graphene to layered graphene-like materials, offers attracting opportunities for research on Rashba physics.

Graphene, a two-dimensional carbon hexagonal crystal with remarkable electron mobility, has been considered as a promising candidate for spintronic applications since its successful exfoliation by Novoselov et al. (2004). An obvious SOC is desirable in graphene to open up a topological nontrivial band gap and then induce the quantum SHE, a quantized response of a transverse spin current to an electric field (Kane and Mele, 2005a,b). Due to its quite small atomic number, the intrinsic SOC in pristine graphene, however, corresponds to an unrealistic spin–orbit gap of ~0.00086 meV (Min et al., 2006; Yao et al., 2007). It is believed to be too weak for effectively manipulating electron spin states under present experimental conditions, which makes it essential to explore large extrinsic SOC. The extrinsic one, known as the Rashba SOC effect, arises from breaking the mirror symmetry of the graphene plane. In addition to directly apply a strong perpendicular electric field, many approaches, such as introducing impurities (Castro Neto and Guinea, 2009), curving the graphene plane (Huertas-Hernando et al., 2006), and building interfaces by depositing graphene on substrates (Dedkov et al., 2008; Rader et al., 2009; Gong et al., 2010, 2011). As shown in Figure 7.28, a 225 meV of Rashba spin splitting has been observed by means of ARPES for graphene grown on Ni(111) (Dedkov et al., 2008), indicating the possibility to realize the quantum SHE, and even the more intriguing topological phenomenon, namely the quantum anomalous Hall effect (Qiao et al., 2010) in the fascinating single-layer material.

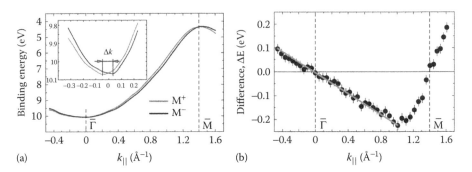

(a) $k_{||}$ (Å^{-1})   (b) $k_{||}$ (Å^{-1})

FIGURE 7.28 Manifestation of the Rashba effect in the graphene layer on Ni(111) (a) energy dispersions extracted from the photoemission spectra measured for two opposite directions of magnetization and (b) Rashba splitting obtained from the data shown in (a) as function of the wave vector k along the Γ–M direction of the surface Brillouin zone. (From Dedkov, Y.S. et al., *Phys. Rev. Lett.*, 100, 107602, 2008.)

In addition to the above discussions, the Rashba effect plays a crucial role in more exotic fields of physics and materials science such as topological classes of materials, Majorana fermions at semiconductor/superconductor interfaces, and ultracold atomic Bose and Fermi gases. Figure 7.29 illustrates the various subfields in which magnetization

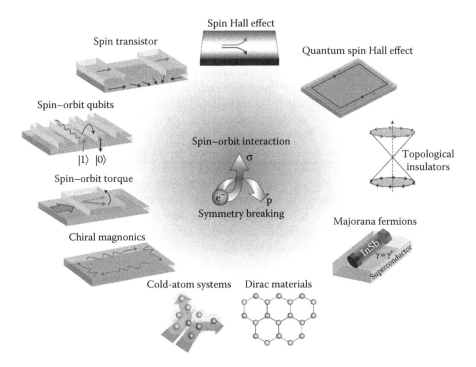

FIGURE 7.29 Tremendous active fields of all-electric spintronics based on Rashba SOC effect. (From Manchon, A. et al., *Nat. Mater.*, 14, 871–882, 2015.)

and spin directions can be manipulated electrically, and in which novel states of matter have been revealed. More details and recent developments on this topic are summarized in the excellent reviews by Bihlmayer et al. (2015) and Manchon et al. (2015).

7.1.4 Other Forms for Spin Manipulation via Electric Means

There are also some interesting systems in presence with spin manipulation via electrical approaches. Their mechanisms, yet, cannot be simply ascribed to anyone mentioned in the above sections. They provide us comprehensive understanding on the possible influence of the electric-field on the magnetic properties of the matter.

A type of magnetoelectric coupling, such as the half-metallicity in graphene ribbons, is closely related to edge states of low-dimensional systems. Using first-principles calculations, in 2006, Son et al. (2007) found that magnetic properties of the zigzag-shaped edges in nanometer-scale graphene ribbon can be controlled by an external electric field. The ground state of the system is antiferromagnetic spin configuration. As displayed in Figure 7.30a, the occupied and unoccupied localized edge states on left side are α-spin and β-spin states, respectively, and vice versa on right side with the same energy gap for both sides. With a transverse electric field, the valance and conduction edge-state bands associated with β spin orientation close their gap, whereas those with the α-spin widen theirs (Figure 7.30b). The nanoribbon is forced into a half-metallic state, resulting in insulating behavior for one spin and metallic behavior for the other. Due to the change of the magnetic moment in edge states, as well as the obtained spin polarized current with presence of electric field, the phenomenon can be regarded as a magnetoelectric effect, which originates from the electrostatic potential difference between the two edge states.

Electric field control of the magnetic state in a material enables new memory devices. Bauer et al. (2012) proposed a magnetoelectric charge trap memory in 2012

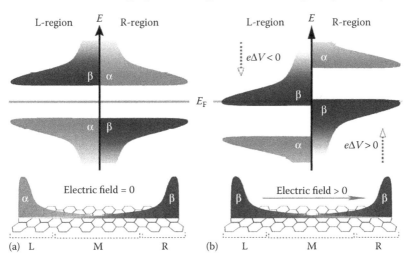

FIGURE 7.30 Schematic density of states diagram of the electronic states of a zigzag graphene nanoribbon (a) in the absence of an applied electric field and (b) with an in-plane homogeneous electric field across the zigzag-shaped edges. (From Son, Y.-W. et al., *Nature*, 446, 342–342, 2007.)

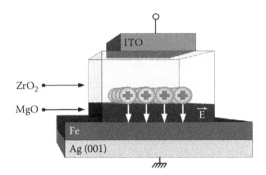

FIGURE 7.31 Schematic of the magnetoelectric charge trap memory. The bottom electrode is ferromagnetic and consists of a high-quality epitaxial Fe film grown in wedge geometry on an Ag(001) single crystalline substrate. The bottom electrode is covered by a double-layer dielectric consisting of MgO and ZrO$_2$. Transparent conducting gate electrodes are made of indium tin oxide. (From Bauer, U. et al., *Nano. Lett.*, 12, 1437–1442, 2012.)

(Figure 7.31). They designed a simple capacitor with a charge-trapping double-layer placed in proximity to a ferromagnetic metal, enabling efficient electrical and optical control of the metal's magnetic properties. When the device is under a positive gate bias, holes from gate electrode can be optically pumped into ZrO$_2$ charge trapping layer with high charge trap density. Trapped holes remain in ZrO$_2$ trapping layer after external bias is removed and generate internal electric field E across MgO barrier. Fe/MgO interface exhibits a strong interfacial magnetic anisotropy. Thus, the internal electric field of the same polarity as the external one can greatly modify the magnetic properties of the Fe film. Note that the trapped charges rather than the external bias play the key role in the magnetoelectric effect, which makes it totally different from those we mentioned above. Retention of charge trapped inside the charge trapping layer provides nonvolatility to the magnetoelectric effect. Based on this point, the engineered charge-trapping layer is of great importance to promote today's pervasive charge trap flash memory technology.

Magnetoelectric coupling was also observed in low-dimensional transition metal-based dichalcogenides that are at the heart of the growing and flourishing field of valleytronics (Xiao et al., 2007, 2012), a cousin concept of spintronics. Very recently, electrical tuning of valley magnetic moment (Wu et al., 2013) and the valley Hall effect (Lee et al., 2016) were realized in bilayer MoS$_2$ transistors. When an electric field is applied perpendicular to the plane of the system, the inversion symmetry present in bilayer MoS$_2$ is broken. The valley polarization can be induced by the longitudinal electrical current. Significantly, due to the strong SOC effect in MoS$_2$, valley and spin are tightly coupled. As displayed in Figure 7.32, Kerr rotation microscopy images that the spin polarization (or equivalently valley polarization) exists near the edges of the device channel with opposite sign for the two edges, and strongly dependent on the gate voltage. The mechanism of tuning the spin splitting at the two edges roots in controlling the potential difference between two single layers by vertical electric field. The work opens a new path for studies of all-electric spintronics, all-electric valleytronics and even crossing areas.

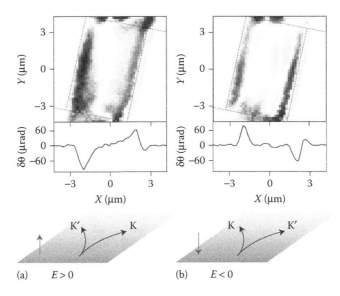

FIGURE 7.32 Gate dependence of the valley Hall effect in bilayer MoS_2. Spatial two-dimensional map and selected horizontal linecut of the Kerr rotation angle $\delta\theta$, as well as the illustration of the field-controlled valley Hall effect under (a) a positive and (b) a negative gate voltage. Vertical arrows show the direction of the electric field. (From Lee, J. et al., *Nat. Nanotech.*, 11, 421–425, 2016.)

7.2 Conclusion

Here, we briefly summarize the recent progress in the booming and fascinating research branch, that is, all-electric spintronics through surface/interface effect. Both the physical mechanisms and improvements including experimental and theoretical aspects are discussed. More details about the achievements in this and other related field based on magnetoelectric composites can be found in these references (Nan et al., 2008; Vaz et al., 2010; Bibes et al., 2011; Ma et al., 2011; Gong et al., 2013b; Fang et al., 2015; Tong et al., 2016). So far, various amazing breakthrough has been made in both the computer simulations and the experiments. Particularly, the concepts of multistate memory (Israel et al., 2008; Velev et al., 2009), multiferroic random access memory (Scott, 2007; Bibes and Barthelemy, 2008; Chu et al., 2008), and electrically assisted magnetic recording (Zavaliche et al., 2007) have been put forward and some of them have already been experimentally realized. Nevertheless, to make magnetoelectric materials useful in practical application, especially in data storage technologies, it is still far from enough. More efforts should be paid in the research of all-electric spintronics. Even if much attempts are required, we can already feel the rapid pulse of this exciting frontier. With the cooperation among theoretical research, materials designing and devices fabrication, we foresee that the application of all-electric spintronic devices in data storage with unique function and low-energy consumption will be realized in the near future.

References

Åkerman J (2005) Toward a universal memory. *Science* 308:508–510.

Alberto C, Gustav B, Klaus K, Marco G (2013) Combined large spin splitting and one-dimensional confinement in surface alloys. *New J Phys* 15:105013.

Ast CR, Henk J, Ernst A, Moreschini L, Falub MC, Pacilé D, Bruno P, Kern K, Grioni M (2007) Giant spin splitting through surface alloying. *Phys Rev Lett* 98:186807.

Bader SD, Parkin SSP (2010) Spintronics. *Annu Rev Condens Matter Phys* 1:71–88.

Baibich MN, Broto JM, Fert A, Van Dau FN, Petroff F, Etienne P, Creuzet G, Friederich A, Chazelas J (1988) Giant magnetoresistance of (0 0 1)Fe/(0 0 1)Cr magnetic superlattices. *Phys Rev Lett* 61:2472–2475.

Bauer U, Przybylski M, Kirschner J, Beach GSD (2012) Magnetoelectric charge trap memory. *Nano Lett* 12:1437–1442.

Bea H, Bibes M, Cherifi S, Nolting F, Warot-Fonrose B, Fusil S, Herranz G, Deranlot C, Jacquet E, Bouzehouane K, Barthelemy A (2006) Tunnel magnetoresistance and robust room temperature exchange bias with multiferroic $BiFeO_3$ epitaxial thin films. *Appl Phys Lett* 89:2402204.

Bea H, Bibes M, Ott F, Dupe B, Zhu XH, Petit S, Fusil S, Deranlot C, Bouzehouane K, Barthelemy A (2008) Mechanisms of exchange bias with multiferroic $BiFeO_3$ epitaxial thin films. *Phys Rev Lett* 100(1):017204.

Bendounan A, Aït-Mansour K, Braun J, Minár J, Bornemann S, Fasel R, Gröning O, Sirotti F, Ebert H (2011) Evolution of the Rashba spin–orbit–split Shockley state on Ag/Pt(1 1 1). *Phys Rev B* 83:195427.

Bibes M, Barthelemy A (2008) Multiferroics: Towards a magnetoelectric memory. *Nat Mater* 7:425–426.

Bibes M, Villegas JE, Barthélémy A (2011) Ultrathin oxide films and interfaces for electronics and spintronics. *Adv Phys* 60:5–84.

Bihlmayer G, Koroteev YM, Echenique PM, Chulkov EV, Blügel S (2006) The Rashba-effect at metallic surfaces. *Surf Sci* 600:3888–3891.

Bihlmayer G, Rader O, Winkler R (2015) Focus on the Rashba effect. *New J Phys* 17:050202.

Bonell F, Murakami S, Shiota Y, Nozaki T, Shinjo T, Suzuki Y (2011) Large change in perpendicular magnetic anisotropy induced by an electric field in FePd ultrathin films. *Appl Phys Lett* 98:232510.

Borisov P, Hochstrat A, Chen X, Kleemann W, Binek C (2005) Magnetoelectric switching of exchange bias. *Phys Rev Lett* 94:117203.

Burton JD, Tsymbal EY (2009) Prediction of electrically induced magnetic reconstruction at the manganite/ferroelectric interface. *Phys Rev B* 80:174406.

Burton JD, Tsymbal EY (2011) Giant tunneling electroresistance effect driven by an electrically controlled spin valve at a complex oxide interface. *Phys Rev Lett* 106:157203.

Castro Neto AH, Guinea F (2009) Impurity-induced spin-orbit coupling in graphene. *Phys Rev Lett* 103:026804.

Chu Y-H, Martin LW, Holcomb MB, Gajek M, Han S-J, He Q, Balke N, Yang C-H, Lee D, Hu W, Zhan Q, Yang P-L, Fraile-Rodriguez A, Scholl A, Wang SX, Ramesh R (2008) Electric-field control of local ferromagnetism using a magnetoelectric multiferroic. *Nat Mater* 7:478–482.

D'Yakonov MI, Perel VI (1971) Spin relaxation of conduction electrons in noncentro-symmetric semiconductors. *Fizika Tverdogo Tela* 13:3581–3585.

Datta S, Das B (1990) Electronic analog of the electro-optic modulator. *Appl Phys Lett* 56:665–667.

Debray P, Rahman SMS, Wan J, Newrock RS, Cahay M, Ngo AT, Ulloa SE, Herbert ST, Muhammad M, Johnson M (2009) All-electric quantum point contact spin-polarizer. *Nat Nano* 4:759–764.

Dedkov YS, Fonin M, Rüdiger U, Laubschat C (2008) Rashba effect in the graphene/Ni(1 1 1) system. *Phys Rev Lett* 100:107602.

Dho JH, Qi XD, Kim H, MacManus-Driscoll JL, Blamire MG (2006) Large electric polarization and exchange bias in multiferroic BiFeO₃. *Adv Mater* 18:1445.

Ding H-C, Duan C-G (2012) Electric-field control of magnetic ordering in the tetragonal-like BiFeO₃. *Europhys Lett* 97:57007.

Duan CG (2009) Progress in the study of magnetoelectric effect. *Prog Phys* (in Chinese) 29:215.

Duan C-G (2012) Interface/surface magnetoelectric effects: New routes to the electric field control of magnetism. *Front Phys* 7:375–379.

Duan C-G, Jaswal SS, Tsymbal EY (2006) Predicted magnetoelectric effect in Fe/BaTiO₃ multilayers: Ferroelectric control of magnetism. *Phys Rev Lett* 97:047201.

Duan C-G, Nan C-W, Jaswal SS, Tsymbal EY (2009) Universality of the surface magnetoelectric effect in half-metals. *Phys Rev B* 79:140403.

Duan C-G, Velev JP, Sabirianov RF, Mei WN, Jaswal SS, Tsymbal EY (2008a) Tailoring magnetic anisotropy at the ferromagnetic/ferroelectric interface. *Appl Phys Lett* 92:122903–122905.

Duan C-G, Velev JP, Sabirianov RF, Zhu Z, Chu J, Jaswal SS, Tsymbal EY (2008b) Surface magnetoelectric effect in ferromagnetic metal films. *Phys Rev Lett* 101:137201.

Dyakonov MI, Perel VI (1971) Current-induced spin orientation of electrons in semiconductors. *Phys Lett A* 35:459–460.

Eerenstein W, Wiora M, Prieto JL, Scott JF, Mathur ND (2007) Giant sharp and persistent converse magnetoelectric effects in multiferroic epitaxial heterostructures. *Nat Mater* 6:348–351.

Endo M, Kanai S, Ikeda S, Matsukura F, Ohno H (2010) Electric-field effects on thickness dependent magnetic anisotropy of sputtered MgO/Co(40)Fe(40)B(20)/Ta structures. *Appl Phys Lett* 96:212503.

Engels G, Lange J, Schäpers T, Lüth H (1997) Experimental and theoretical approach to spin splitting in modulation-doped. $In_xGa_{1-x}As/InP$ quantum wells for B→0. *Phys Rev B* 55:R1958–R1961.

Fang Y-W, Ding H-C, Tong W-Y, Zhu W-J, Shen X, Gong S-J, Wan X-G, Duan C-G (2015) First-principles studies of multiferroic and magnetoelectric materials. *Sci Bull* DOI 10.1007/s11434-014-1628-4.

Fang Y-W, Gao Y-C, Gong S-J, Duan C-G (2013) Advances in ferroelectric and multiferroic tunnel junctions. *Prog Phys* (Chinese) 33:382–413.

Fechner M, Zahn P, Ostanin S, Bibes M, Mertig I (2012) Switching magnetization by 180° with an electric field. *Phys Rev Lett* 108:197206.

Gajek M, Bibes M, Fusil S, Bouzehouane K, Fontcuberta J, Barthelemy A, Fert A (2007) Tunnel junctions with multiferroic barriers. *Nat Mater* 6:296–302.

Gao Y, Hu J-M, Nelson CT, Yang TN, Shen Y, Chen LQ, Ramesh R, Nan CW (2016) Dynamic in situ observation of voltage-driven repeatable magnetization reversal at room temperature. *Sci Rep* 6:23696.

Garcia V, Bibes M, Bocher L, Valencia S, Kronast F, Crassous A, Moya X, Enouz-Vedrenne S, Gloter A, Imhoff D, Deranlot C, Mathur ND, Fusil S, Bouzehouane K, Barthelemy A (2010) Ferroelectric control of spin polarization. *Science* 327:1106–1110.

Gerhard L, Yamada TK, Balashov T, Takacs AF, Wesselink RJH, Dane M, Fechner M, Ostanin S, Ernst A, Mertig I, Wulfhekel W (2010) Magnetoelectric coupling at metal surfaces. *Nat Nanotech* 5:792–797.

Ghidini M, Pellicelli R, Prieto JL, Moya X, Soussi J, Briscoe J, Dunn S, Mathur ND (2013) Non-volatile electrically-driven repeatable magnetization reversal with no applied magnetic field. *Nat Commun* 4:1453.

Gong C, Lee G, Shan B, Vogel EM, Wallace RM, Cho K (2010) First-principles study of metal—Graphene interfaces. *J Appl Phys* 108:123711–123718.

Gong S, Ding H, Zhu W, Duan C, Zhu Z, Chu J (2013b) A new pathway towards all-electric spintronics: Electric-field control of spin states through surface/interface effects. *Sci China Phys Mech Astron* 56:232–244.

Gong S-J, Duan C-G, Zhu Y, Zhu Z-Q, Chu J-H (2013a) Controlling Rashba spin splitting in Au(1 1 1) surface states through electric field. *Phys Rev B* 87:035403.

Gong SJ, Duan C-G, Zhu Z-Q, Chu J-H (2012) Manipulation of magnetic anisotropy of Fe/graphene by charge injection. *Appl Phys Lett* 100:122410.

Gong SJ, Li ZY, Yang ZQ, Gong C, Duan CG, Chu JH (2011) Spintronic properties of graphene films grown on Ni(1 1 1) substrate. *J Appl Phys* 110:043704.

Gong SJ, Yang ZQ (2007) Spin filtering implemented through Rashba spin–orbit coupling and weak magnetic modulations. *J Appl Phys* 102:033704–033706.

Hambe M, Petraru A, Pertsev NA, Munroe P, Nagarajan V, Kohlstedt H (2010) Crossing an interface: Ferroelectric control of tunnel currents in magnetic complex oxide heterostructures. *Adv Funct Mater* 20:2436–2441.

Heron JT, Trassin M, Ashraf K, Gajek M, He Q, Yang SY, Nikonov DE, Chu YH, Salahuddin S, Ramesh R (2011) Electric-field-induced magnetization reversal in a ferromagnet-multiferroic heterostructure. *Phys Rev Lett* 107:217202.

Hirsch JE (1999) Spin hall effect. *Phys Rev Lett* 83:1834–1837.

Hu J-M, Li Z, Chen L-Q, Nan C-W (2011) High-density magnetoresistive random access memory operating at ultralow voltage at room temperature. *Nat Commun* 2:553.

Hu J-M, Li Z, Chen L-Q, Nan C-W (2012) Design of a voltage-controlled magnetic random access memory based on anisotropic magnetoresistance in a single magnetic layer. *Adv Mater* 24:2869–2873.

Hu J-M, Nan CW (2009) Electric-field-induced magnetic easy-axis reorientation in ferromagnetic/ferroelectric layered heterostructures. *Phys Rev B* 80:224416.

Hu J-M, Yang T, Wang J, Huang H, Zhang J, Chen L-Q, Nan C-W (2015) Purely electric-field-driven perpendicular magnetization reversal. *Nano Lett* 15:616–622.

Huertas-Hernando D, Guinea F, Brataas A (2006) Spin–orbit coupling in curved graphene, fullerenes, nanotubes, and nanotube caps. *Phys Rev B* 74:155426.

Israel C, Mathur ND, Scott JF (2008) A one-cent room-temperature magnetoelectric sensor. *Nat Mater* 7:93–94.

Julliere M (1975) Tunneling between ferromagnetic films. *Phys Lett A* 54:225–226.

Jungwirth T, Wunderlich J, Olejnik K (2012) Spin Hall effect devices. *Nat Mater* 11:382–390.

Kane CL, Mele EJ (2005a) Z_2 topological order and the quantum spin Hall effect. *Phys Rev Lett* 95:146802.

Kane CL, Mele EJ (2005b) Quantum spin Hall effect in graphene. *Phys Rev Lett* 95:146802.

Kato YK, Myers RC, Gossard AC, Awschalom DD (2004) Observation of the spin Hall effect in semiconductors. *Science* 306:1910–1913.

Koga T, Nitta J, van Veenhuizen M (2004) Ballistic spin interferometer using the Rashba effect. *Phys Rev B* 70:161302.

Kohda M, Nakamura S, Nishihara Y, Kobayashi K, Ono T, Ohe J-I, Tokura Y, Mineno T, Nitta J (2012) Spin–orbit induced electronic spin separation in semiconductor nanostructures. *Nat Commun* 3:1082.

Koroteev YM, Bihlmayer G, Gayone JE, Chulkov EV, Blügel S, Echenique PM, Hofmann P (2004) Strong spin-orbit splitting on Bi surfaces. *Phys Rev Lett* 93:046403.

Krupin O, Bihlmayer G, Starke K, Gorovikov S, Prieto JE, Döbrich K, Blügel S, Kaindl G (2005) Rashba effect at magnetic metal surfaces. *Phys Rev B* 71:201403.

Lahtinen THE, Franke KJA, van Dijken S (2012) Electric-field control of magnetic domain wall motion and local magnetization reversal. *Sci Rep* 2:258.

LaShell S, McDougall BA, Jensen E (1996) Spin splitting of an Au(1 1 1) surface state band observed with angle resolved photoelectron spectroscopy. *Phys Rev Lett* 77:3419–3422.

Laukhin V, Skumryev V, Martí X, Hrabovsky D, Sánchez F, García-Cuenca MV, Ferrater C, Varela M, Lüders U, Bobo JF, Fontcuberta J (2006) Electric-field control of exchange bias in multiferroic epitaxial heterostructures. *Phys Rev Lett* 97:227201.

Lee J, Mak KF, Shan J (2016) Electrical control of the valley Hall effect in bilayer MoS_2 transistors. *Nat Nanotech* 11:421–425.

Lei N, Devolder T, Agnus G, Aubert P, Daniel L, Kim J-V, Zhao W, Trypiniotis T, Cowburn RP, Chappert C, Ravelosona D, Lecoeur P (2013) Strain-controlled magnetic domain wall propagation in hybrid piezoelectric/ferromagnetic structures. *Nat Commun* 4:1378.

Liu M, Lou J, Li S, Sun NX (2011) E-Field control of exchange bias and deterministic magnetization switching in AFM/FM/FE multiferroic heterostructures. *Adv Funct Mater* 21:2593–2598.

Ma J, Hu J, Li Z, Nan C-W (2011) Recent progress in multiferroic magnetoelectric composites: From bulk to thin films. *Adv Mater* 23:1062–1087.

Manchon A, Koo HC, Nitta J, Frolov SM, Duine RA (2015) New perspectives for Rashba spin–orbit coupling. *Nat Mater* 14:871–882.

Mardana A, Ducharme S, Adenwalla S (2011) Ferroelectric control of magnetic anisotropy. *Nano Lett* 11:3862–3867.

Marti X, Sanchez F, Fontcuberta J, Garcia-Cuenca MV, Ferrater C, Varela M (2006a) Exchange bias between magnetoelectric $YMnO_3$ and ferromagnetic $SrRuO_3$ epitaxial films. *J Appl Phys* 99(80):08P302–08P302-3.

Marti X, Sanchez F, Hrabovsky D, Fabrega L, Ruyter A, Fontcuberta J, Laukhin V, Skumryev V, Garcia-Cuenca MV, Ferrater C, Varela M, Vila A, Luders U, Bobo JF (2006b) Exchange biasing and electric polarization with $YMnO_3$. *Appl Phys Lett* 89:032510.

Martin LW, Chu Y-H, Holcomb MB, Huijben M, Yu P, Han S-J, Lee D, Wang SX, Ramesh R (2008) Nanoscale control of exchange bias with $BiFeO_3$ thin films. *Nano Lett* 8:2050–2055.

Maruyama T, Shiota Y, Nozaki T, Ohta K, Toda N, Mizuguchi M, Tulapurkar AA, Shinjo T, Shiraishi M, Mizukami S, Ando Y, Suzuki Y (2009) Large voltage-induced magnetic anisotropy change in a few atomic layers of iron. *Nat Nanotech* 4:158–161.

Meyerheim HL, Klimenta F, Ernst A, Mohseni K, Ostanin S, Fechner M, Parihar S, Maznichenko IV, Mertig I, Kirschner J (2011) Structural secrets of multiferroic interfaces. *Phys Rev Lett* 106:087203.

Min H, Hill JE, Sinitsyn NA, Sahu BR, Kleinman L, MacDonald AH (2006) Intrinsic and Rashba spin–orbit interactions in graphene sheets. *Phys Rev B* 74:165310.

Mirhosseini H, Maznichenko IV, Abdelouahed S, Ostanin S, Ernst A, Mertig I, Henk J (2010) Toward a ferroelectric control of Rashba spin–orbit coupling: Bi on $BaTiO_3(001)$ from first principles. *Phys Rev B* 81:073406.

Molegraaf HJA, Hoffman J, Vaz CAF, Gariglio S, van der Marel D, Ahn CH, Triscone J-M (2009) Magnetoelectric effects in complex oxides with competing ground states. *Adv Mater* 21:3470–3474.

Murakami S, Nagaosa N, Zhang S-C (2003) Dissipationless quantum spin current at room temperature. *Science* 301:1348–1351.

Nadj-Perge S, Frolov SM, Bakkers EPAM, Kouwenhoven LP (2010) Spin–orbit qubit in a semiconductor nanowire. *Nature* 468:1084–1087.

Nakamura K, Shimabukuro R, Fujiwara Y, Akiyama T, Ito T, Freeman AJ (2009) Giant modification of the magnetocrystalline anisotropy in transition-metal monolayers by an external electric field. *Phys Rev Lett* 102:187201.

Nan C-W (1994) Magnetoelectric effect in composites of piezoelectric and piezomagnetic phases. *Phys Rev B* 50:6082–6088.

Nan C-W, Bichurin MI, Dong S, Viehland D, Srinivasan G (2008) Multiferroic magnetoelectric composites: Historical perspective, status, and future directions. *J Appl Phys* 103:031101.

Nicolay G, Reinert F, Hüfner S, Blaha P (2001) Spin–orbit splitting of the L-gap surface state on Au(1 1 1) and Ag(1 1 1). *Phys Rev B* 65:033407.

Niranjan MK, Duan CG, Jaswal SS, Tsymbal EY (2010) Electric field effect on magnetization at the Fe/MgO(0 0 1) interface. *Appl Phys Lett* 96:222504.

Niranjan MK, Velev JP, Duan CG, Jaswal SS, Tsymbal EY (2008) Magnetoelectric effect at the $Fe_3O_4/BaTiO_3$ (0 0 1) interface: A first-principles study. *Phys Rev B* 78:104405.

Nitta J, Akazaki T, Takayanagi H, Enoki T (1997) Gate control of spin–orbit interaction in an inverted $In_{0.53}Ga_{0.47}As/In_{0.52}Al_{0.48}As$ heterostructure. *Phys Rev Lett* 78:1335–1338.

Nitta J, Koga T (2003) Rashba spin–orbit interaction and its applications to spin-interference effect and spin-filter device. *J Supercond* 16:689–696.

Novoselov KS, Geim AK, Morozov SV, Jiang D, Zhang Y, Dubonos SV, Grigorieva IV, Firsov AA (2004) Electric field effect in atomically thin carbon films. *Science* 306:666–669.

Nowack KC, Koppens FHL, Nazarov YV, Vandersypen LMK (2007) Coherent control of a single electron spin with electric fields. *Science* 318:1430–1433.

Nozaki T, Shiota Y, Miwa S, Murakami S, Bonell F, Ishibashi S, Kubota H, Yakushiji K, Saruya T, Fukushima A, Yuasa S, Shinjo T, Suzuki Y (2012) Electric-field-induced ferromagnetic resonance excitation in an ultrathin ferromagnetic metal layer. *Nat Phys* 8:492–497.

Nuber A, Braun J, Forster F, Minár J, Reinert F, Ebert H (2011) Surface versus bulk contributions to the Rashba splitting in surface systems. *Phys Rev B* 83:165401.

Ohe J, Yamamoto M, Ohtsuki T, Nitta J (2005) Mesoscopic Stern–Gerlach spin filter by nonuniform spin–orbit interaction. *Phys Rev B* 72:041308.

Ohno H (2010) A window on the future of spintronics. *Nat Mater* 9:952–954.

Ohno H, Chiba D, Matsukura F, Omiya T, Abe E, Dietl T, Ohno Y, Ohtani K (2000) Electric-field control of ferromagnetism. *Nature* 408:944–946.

Pantel D, Goetze S, Hesse D, Alexe M (2012) Reversible electrical switching of spin polarization in multiferroic tunnel junctions. *Nat Mater* 11:289–293.

Park MS, Song J-H, Freeman AJ (2009) Charge imbalance and magnetic properties at the $Fe_3O_4/BaTiO_3$ interface. *Phys Rev B* 79:024420.

Park SR, Kim CH, Yu J, Han JH, Kim C (2011) Orbital-angular-momentum based origin of Rashba-type surface band splitting. *Phys Rev Lett* 107:156803.

Petersen L, Hedegård P (2000) A simple tight-binding model of spin–orbit splitting of sp-derived surface states. *Surf Sci* 459:49–56.

Qiao Z, Yang SA, Feng W, Tse W-K, Ding J, Yao Y, Wang J, Niu Q (2010) Quantum anomalous Hall effect in graphene from Rashba and exchange effects. *Phys Rev B* 82:161414.

Quindeau A, Fina I, Marti X, Apachitei G, Ferrer P, Nicklin C, Pippel E, Hesse D, Alexe M (2015) Four-state ferroelectric spin-valve. *Sci Rep* 5:9749.

Rader O, Varykhalov A, Sánchez-Barriga J, Marchenko D, Rybkin A, Shikin AM (2009) Is there a Rashba effect in graphene on 3d ferromagnets? *Phys Rev Lett* 102:057602.

Rondinelli JM, Stengel M, Spaldin NA (2008) Carrier-mediated magnetoelectricity in complex oxide heterostructures. *Nat Nanotech* 3:46–50.

Rotenberg E, Chung JW, Kevan SD (1999) Spin-orbit coupling induced surface band splitting in Li/W(1 1 0) and Li/Mo(1 1 0). *Phys Rev Lett* 82:4066–4069.

Ruiz-Díaz P, Dasa TR, Stepanyuk VS (2013) Tuning magnetic anisotropy in metallic multilayers by surface charging: An *ab initio* study. *Phys Rev Lett* 110:267203.

Sahoo S, Polisetty S, Duan CG, Jaswal SS, Tsymbal EY, Binek C (2007) Ferroelectric control of magnetism in $BaTiO_3/Fe$ heterostructures via interface strain coupling. *Phys Rev B* 76:092108.

Sakamoto K, Oda T, Kimura A, Miyamoto K, Tsujikawa M, Imai A, Ueno N, Namatame H, Taniguchi M, Eriksson PEJ, Uhrberg RIG (2009) Abrupt rotation of the Rashba spin to the direction perpendicular to the surface. *Phys Rev Lett* 102(9):096805.

Scott JF (2007) Data storage—Multiferroic memories. *Nat Mater* 6:256–257.

Shiota Y, Nozaki T, Bonell F, Murakami S, Shinjo T, Suzuki Y (2012) Induction of coherent magnetization switching in a few atomic layers of FeCo using voltage pulses. *Nat Mater* 11:39–43.

Shu L, Li Z, Ma J, Gao Y, Gu L, Shen Y, Lin Y, Nan CW (2012) Thickness-dependent voltage-modulated magnetism in multiferroic heterostructures. *Appl Phys Lett* 100:022405.

Sinova J, Culcer D, Niu Q, Sinitsyn NA, Jungwirth T, MacDonald AH (2004) Universal intrinsic spin Hall effect. *Phys Rev Lett* 92:126603.

Sinova J, Valenzuela SO, Wunderlich J, Back CH, Jungwirth T (2015) Spin Hall effects. *Rev Mod Phys* 87:1213–1260.

Son Y-W, Cohen ML, Louie SG (2007) Half-metallic graphene nanoribbons. *Nature* 446:342.

Stolichnov I, Riester SWE, Trodahl HJ, Setter N, Rushforth AW, Edmonds KW, Campion RP, Foxon CT, Gallagher BL, Jungwirth T (2008) Non-volatile ferroelectric control of ferromagnetism in (Ga, Mn)As. *Nat Mater* 7:464–467.

Sugawara K, Sato T, Souma S, Takahashi T, Arai M, Sasaki T (2006) Fermi surface and anisotropic spin-orbit coupling of Sb(1 1 1) studied by angle-resolved photoemission spectroscopy. *Phys Rev Lett* 96:046411.

Thiele C, Dorr K, Bilani O, Rodel J, Schultz L (2007) Influence of strain on the magnetization and magnetoelectric effect in $La_{0.7}A_{0.3}MnO_3$/PMN-PT(0 0 1) (A=Sr, Ca). *Phys Rev B* 75:054408.

Tong W-Y, Fang Y-W, Cai J, Gong S-J, Duan C-G (2016) Theoretical studies of all-electric spintronics utilizing multiferroic and magnetoelectric materials. *Comput Mater Sci* 112:467–477.

Tsujikawa M, Oda T (2009) Finite electric field effects in the large perpendicular magnetic anisotropy surface Pt/Fe/Pt(0 0 1): A first-principles study. *Phys Rev Lett* 102:247203.

Valencia S, Crassous A, Bocher L, Garcia V, Moya X, Cherifi RO, Deranlot C, Bouzehouane K, Fusil S, Zobelli A, Gloter A, Mathur ND, Gaupp A, Abrudan R, Radu F, Barthélémy A, Bibes M (2011) Interface-induced room-temperature multiferroicity in $BaTiO_3$. *Nat Mater* 10:753–758.

van den Berg JWG, Nadj-Perge S, Pribiag VS, Plissard SR, Bakkers EPAM, Frolov SM, Kouwenhoven LP (2013) Fast spin–orbit qubit in an indium antimonide nanowire. *Phys Rev Lett* 110:066806.

Van Den Boomgaard J, Terrell DR, Born RAJ, Giller HFJI (1974) An in situ grown eutectic magnetoelectric composite material. I. Composition and unidirectional solidification. *J Mater Sci* 9:1705–1709.

van Suchtelen J (1972) Product properties: A new application of composite materials. *Philips Res Rep* 27:28–37.

Varykhalov A, Marchenko D, Scholz MR, Rienks EDL, Kim TK, Bihlmayer G, Sánchez-Barriga J, Rader O (2012) Ir(1 1 1) Surface state with giant Rashba splitting persists under graphene in air. *Phys Rev Lett* 108:066804.

Vaz CAF, Hoffman J, Anh CH, Ramesh R (2010) Magnetoelectric coupling effects in multiferroic complex oxide composite structures. *Adv Mater* 22:2900–2918.

Velev JP, Duan C-G, Burton JD, Smogunov A, Niranjan MK, Tosatti E, Jaswal SS, Tsymbal EY (2009) Magnetic tunnel junctions with ferroelectric barriers: Prediction of four resistance states from first principles. *Nano Lett* 9:427–432.

Wang JJ, Hu JM, Ma J, Zhang JX, Chen LQ, Nan CW (2014) Full 180° magnetization reversal with electric fields. *Sci Rep* 4:7507.

Wang W-G, Li M, Hageman S, Chien CL (2012) Electric-field-assisted switching in magnetic tunnel junctions. *Nat Mater* 11:64–68.

Weiler M, Brandlmaier A, Geprags S, Althammer M, Opel M, Bihler C, Huebl H, Brandt MS, Gross R, Goennenwein STB (2009) Voltage controlled inversion of magnetic anisotropy in a ferromagnetic thin film at room temperature. *New J Phys* 11:013021.

Weisheit M, Fahler S, Marty A, Souche Y, Poinsignon C, Givord D (2007) Electric field-induced modification of magnetism in thin-film ferromagnets. *Science* 315:349–351.

Winkler R (2003) *Spin–Orbit Coupling Effects in Two-dimensional Electron and Hole Systems*. Berlin, NY: Springer.

Wolf SA, Awschalom DD, Buhrman RA, Daughton JM, von Molnar S, Roukes ML, Chtchelkanova AY, Treger DM (2001) Spintronics: A spin-based electronics vision for the future. *Science* 294:1488–1495.

Wu S, Ross JS, Liu G-B, Aivazian G, Jones A, Fei Z, Zhu W, Xiao D, Yao W, Cobden D, Xu X (2013) Electrical tuning of valley magnetic moment through symmetry control in bilayer MoS_2. *Nat Phys* 9:149–153.

Wu T, Bur A, Zhao P, Mohanchandra KP, Wong K, Wang KL, Lynch CS, Carman GP (2011) Giant electric-field-induced reversible and permanent magnetization reorientation on magnetoelectric Ni/(0 1 1) [Pb(Mg(1/3)Nb(2/3))O(3)]((1-x))-[PbTiO(3)] (x) heterostructure. *Appl Phys Lett* 98:012504.

Wunderlich J, Kaestner B, Sinova J, Jungwirth T (2005) Experimental observation of the spin-hall effect in a two-dimensional spin–orbit coupled semiconductor system. *Phys Rev Lett* 94:047204.

Wunderlich J, Park B-G, Irvine AC, Zârbo LP, Rozkotová E, Nemec P, Novák V, Sinova J, Jungwirth T (2010) Spin Hall effect transistor. *Science* 330:1801–1804.

Xiao D, Liu G-B, Feng W, Xu X, Yao W (2012) Coupled spin and valley physics in monolayers of MoS_2 and other Group-VI dichalcogenides. *Phys Rev Lett* 108:196802.

Xiao D, Yao W, Niu Q (2007) Valley-contrasting physics in graphene: Magnetic moment and topological transport. *Phys Rev Lett* 99:236809.

Yamauchi K, Sanyal B, Picozzi S (2007) Interface effects at a half-metal/ferroelectric junction. *Appl Phys Lett* 91:062506.

Yang S-W, Peng R-C, Jiang T, Liu Y-K, Feng L, Wang J-J, Chen L-Q, Li X-G, Nan C-W (2014a) Non-volatile 180° magnetization reversal by an electric field in multiferroic heterostructures. *Adv Mater* 26:7091–7095.

Yang TN, Hu J-M, Nan CW, Chen LQ (2014b) Predicting effective magnetoelectric response in magnetic-ferroelectric composites via phase-field modeling. *Appl Phys Lett* 104:052904.

Yao Y, Ye F, Qi X-L, Zhang S-C, Fang Z (2007) Spin–orbit gap of graphene: First-principles calculations. *Phys Rev B* 75:041401.

Yin YW, Burton JD, Kim YM, Borisevich AY, Pennycook SJ, Yang SM, Noh TW, Gruverman A, Li XG, Tsymbal EY, Li Q (2013) Enhanced tunnelling electroresistance effect due to a ferroelectrically induced phase transition at a magnetic complex oxide interface. *Nat Mater* 12:397–402.

Yoichi S, Takuto M, Takayuki N, Teruya S, Masashi S, Yoshishige S (2009) Voltage-assisted magnetization switching in ultrathin $Fe_{80}Co_{20}$ Alloy layers. *Appl Phys Exp* 2:063001.

Yu AB, Rashba EI (1984) Oscillatory effects and the magnetic susceptibility of carriers in inversion layers. *J Phys C: Solid State Phys* 17:6039–6045.

Zavaliche F, Zhao T, Zheng H, Straub F, Cruz MP, Yang PL, Hao D, Ramesh R (2007) Electrically assisted magnetic recording in multiferroic nanostructures. *Nano Lett* 7:1586–1590.

Zhang HB, Richter M, Koepernik K, Opahle I, Tasnadi F, Eschrig H (2009) Electric-field control of surface magnetic anisotropy: A density functional approach. *New J Phys* 11:043007.

Zhang S (1999) Spin-dependent surface screening in ferromagnets and magnetic tunnel junctions. *Phys Rev Lett* 83:640–643.

Zheng H, Wang J, Lofland SE, Ma Z, Mohaddes-Ardabili L, Zhao T, Salamanca-Riba L, Shinde SR, Ogale SB, Bai F, Viehland D, Jia Y, Schlom DG, Wuttig M, Roytburd A, Ramesh R (2004) Multiferroic $BaTiO_3$–$CoFe_2O_4$ nanostructures. *Science* 303:661–663.

Zhu J, Katine JA, Rowlands GE, Chen Y-J, Duan Z, Alzate JG, Upadhyaya P, Langer J, Amiri PK, Wang KL, Krivorotov IN (2012) Voltage-induced ferromagnetic resonance in magnetic tunnel junctions. *Phys Rev Lett* 108:197203.

Zhu W, Ding H-C, Gong S-J, Liu Y, Duan C-G (2013) First-principles studies of the magnetic anisotropy of the Cu/FePt/MgO system. *J Phys: Condens Matter* 25:396001.

Zhu W, Liu Y, Duan C-G (2011) Modeling of the spin-transfer torque switching in FePt/MgO-based perpendicular magnetic tunnel junctions: A combined ab initio and micromagnetic simulation study. *Appl Phys Lett* 99:032508.

Zhu W, Xiao D, Liu Y, Gong SJ, Duan C-G (2014) Picosecond electric field pulse induced coherent magnetic switching in MgO/FePt/Pt(001)-based tunnel junctions: A multiscale study. *Sci Rep* 4:4117.

Žutić I, Fabian J, Das Sarma S (2004) Spintronics: Fundamentals and applications. *Rev Mod Phys* 76:323–410.

8

Understanding of Exchange Bias in Ferromagnetic/ Antiferromagnetic Bilayers

M. Pankratova,
A. Kovalev, and
M. Žukovič

8.1 Introduction

In this chapter, we present the theoretical analysis and numerical simulation of the exchange bias (EB) phenomenon in the ferromagnetic/antiferromagnetic bilayers (FM/AFM) [1–3]. The EB phenomenon was discovered experimentally by Meiklejohn and Bean in 1956 [1]. The phenomenon has numerous applications in nanoelectronics and spintronics. A comprehensive explanation of the EB does not exist yet [4–17]. The effect manifests itself in the shift of the magnetization dependence on the external magnetic field $M(H)$ along the field axis. The shift of the hysteresis loop is proportional to the exchange interaction through the FM/AFM interface J_0, whereas the width of the hysteresis loop is proportional to the magnetic anisotropy in the FM layer. The phenomenon appears in the systems with contacting FM/AFM subsystems: layered systems, nanoparticles [2], multiferroics [18]. In the simplest model of EB, the AFM is assumed to be a hard magnetic material [19,20], the AFM interface is uncompensated, and the FM film has a uniform magnetization. The easy-plane anisotropy and an additional weak anisotropy in the easy plane are taken into account. The model of the phenomenon gives an expression for the EB $H_{bias} = J_0 S_{AFM} S_{FM}/L_F M_F$, where S_{AFM} and S_{FM} are the AFM and FM magnetic moments, respectively, M_{FM} is the magnetization of the FM that is constant along the FM/AFM interface, and J_0 is the interlayer exchange interaction. Such an expression for H_{bias}, however, gives a significantly larger value of the hysteresis loop shift than that observed in experiments. It is important to note that the same value of the shift is observed in the case of compensated FM/AFM interface [21–23]. Since EB was discovered, a lot of theoretical models have been proposed [4]. Among them, the possibility of the domain walls appears both in the FM and the AFM subsystems [7,8,24]; the impact of the interface defects on the EB effect [5], the influence of the domain and polycrystalline structure of materials [6]. In References [25,26], the authors presented an explanation of the nature of EB phenomenon in systems with a compensated FM/AFM interface. The recent works also study the geometrical frustration in the FM/AFM-layered system and its impact on the shift of the magnetization curves [27–29]. Recently, new features of the EB phenomenon were observed experimentally [30–33]. The shifted hysteresis loop becomes asymmetric $M(2H_{bias}-H) \neq -M(H)$ and additional horizontal plateaus appear [31,33]. The slope of the magnetization curve is different on different parts of the curve. These features can be explained by different spin-flip kinetics in different magnetic configurations. Here, we propose an explanation to the magnetization curves features, such as horizontal plateaus, asymmetry, split of the hysteresis loop. We use the assumption that these features can be associated with an inhomogeneous magnetization distribution in the FM film. We consider the cases of the perfect and rough FM/AFM interface. The horizontal plateaus, asymmetry, and other features of the magnetic hysteresis loops are obtained in the framework of such models.

The outline of the chapter is as follows. In Section 8.2, we consider a layered system made of FM and AFM monolayers with a perfect and rough FM/AFM interface.

Section 8.3 is devoted to the study of the magnetization curves for FM/AFM bilayers. The paper is completed by the concluding remarks.

8.2 The Models of Exchange Bias in Ferromagnetic/ Antiferromagnetic Bilayers

To explain features of the EB phenomenon and shapes of the hysteresis loops, we propose several simple theoretical models with a corresponding energy functions [9–17]. The energy functions will be used in Section 8.3 deriving expression for the magnetization. From now on, we assume that bulk inhomogeneous states of the magnetization in the FM film appear. These structures are responsible for the features of the EB phenomenon such as horizontal plateaus, curve asymmetry, and inclined parts of the magnetization curve with different slopes [30,31]. We consider the classical Heisenberg models for discrete and continuous media with easy-plane anisotropies. We take into account an additional weak anisotropy in the easy plane of the FM film and a surface anisotropy β_1. We assume the isotropic exchange interaction in the FM and AFM, collinear direction of the anisotropy axis, and external magnetic field axis. The AFM is a hard magnetic material and the magnetic state in the AFM is assumed to be fixed in all the cases except for the geometrically frustrated case.

Even if the magnetization in the AFM, anisotropy in the FM and the external magnetic field H has the same direction; there are four parameters characterizing this model: the exchange interaction J (in the FM), the exchange interaction J_0 through the interface, the weak magnetic anisotropy β in the easy plane of the AFM, and the external magnetic field H. The magnetic structure of the FM is characterized by the rotational angles φ of the magnetization vectors $\vec{M}$. These angles φ are measured from the anisotropy axis in the easy plane of the FM. To describe the interface roughness, we use the simplest types of the interface inhomogeneity, such as periodically distributed along the FM/AFM interface atomic steps and point magnetic contacts (PMC).

We use local coordinates x, y, z, where y, z are the coordinates in the plane, parallel to the interface, and x is perpendicular to the interface. The inhomogeneity of the states perpendicular to the interface may be associated with a nonuniform interface. The nonuniform interface can appear because of the domain structure of the AFM. With the interface roughness, the additional length parameter/L_s is associated. This parameter describes the period of the interface structure, for example, the period of the atomic steps on the interface. Unfortunately, the values of some parameters (J_0, β) are complicated to obtain from experiments. Therefore, it is difficult to explain theoretically the experimental data. We use another approach. We specify the model and the parameters ratio for which the corresponding behavior of the magnetic systems takes place. The ratio between the parameters is essential. In different models, for different values of the system parameters, it can lead to the same types of the magnetization curves. In addition, this enables us to estimate the values of the parameters that are not known from the experiments.

8.2.1 The Discrete Models of the Ferromagnetic/ Antiferromagnetic Bilayers Layered Systems: Perfect Interface

In Figure 8.1, the scheme of the simplest EB model with uncompensated AFM interface is presented. Here, the FM film consists only of one monolayer. The energy of this system has the form

$$E = \frac{\beta}{2}\cos^2 \varphi - J_0 \cos\varphi - H\cos\varphi \qquad (8.1)$$

The exchange interaction through the interface is ferromagnetic, that is, $J_0 > 0$. This model can be used for a multilayered FM film made of N layers assuming that the exchange interaction in the FM subsystem is strong enough and the magnetization is varying homogeneously. For a multilayered film in the energy Equation 8.1, we replace $\beta \to N\beta$ and $H \to NH$ (Figure 8.2).

To generalize this model, one can take into account an inhomogeneous magnetization states (nonuniform magnetization) in the direction orthogonal to the FM/AFM interface (Figure 8.1) [9,11]. We assume that the FM layer consists only of two atomic layers (or two thin FM films). The rotational angles φ_1 and φ_2 can take different values in the interface plane. Nontrivial results appear even in the absence of the magnetic anisotropy in the easy plane. The appearance of a horizontal plateau is associated with

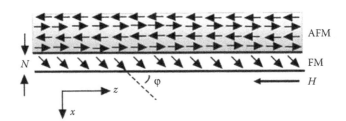

FIGURE 8.1 The simplest exchange bias model of the FM/AFM bilayer with uncompensated interface.

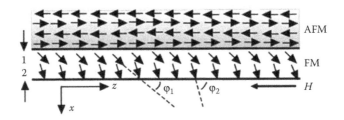

FIGURE 8.2 The model of FM/AFM bilayer with FM consisting only of two atomic layers.

the collinear structure $\varphi_1 = 0$, $\varphi_2 = \pi\,(\uparrow\downarrow)$, whereas the inclined parts with the noncollinear structures $\varphi_{1,2} \neq 0, \pi$ in the FM film. In the presence of the magnetic anisotropy, Equation 8.1 becomes

$$E = -J_0 \cos\varphi_1 - J\cos(\varphi_1 - \varphi_2) - H(\cos\varphi_1 + \cos\varphi_2)$$
$$- \frac{\beta_1}{2}\cos^2\varphi_1 - \frac{\beta_2}{2}\cos^2\varphi_2 \tag{8.2}$$

Here, β_1 and β_2 stand for anisotropy parameters for the first (closest to the interface FM/AFM) and the second FM layer. In general, these parameters do not coincide $\beta_1 \neq \beta_2$. The parameter β_2 corresponds to the anisotropy in the FM film, and the parameter β_1 is responsible for the surface anisotropy. In order to study this model, we start with the case of the absence of the magnetic anisotropy. In this case, the energy is given by

$$E = -J_0 \cos\varphi_1 - J\cos(\varphi_1 - \varphi_2) - H(\cos\varphi_1 + \cos\varphi_2) \tag{8.3}$$

Then we introduce the anisotropy and assume $\beta_1 = \beta_2 = \beta$

$$E = -J_0 \cos\varphi_1 - J\cos(\varphi_1 - \varphi_2) - H(\cos\varphi_1 + \cos\varphi_2)$$
$$- \frac{\beta}{2}\left(\cos^2\varphi_1 + \cos^2\varphi_2\right) \tag{8.4}$$

Finally, we turn to the case of $\beta_1 \neq \beta_2$ described in Equation 8.2. The proposed model also describes the trilayer system AFM/FM1/FM2 in the cases when the exchange interaction in the two FM films is essentially stronger than the exchange interaction through the interfaces. We assume here that the magnetization vectors rotate as whole and each layer is characterized only by the angle φ_i.

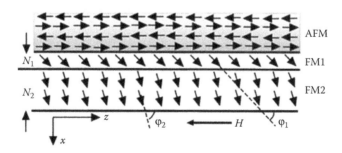

FIGURE 8.3 The model of FM/AFM trilayer system with various FM layer thicknesses.

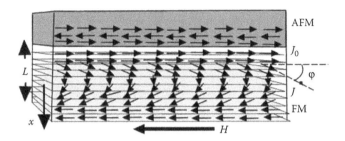

FIGURE 8.4   The continuum model of the FM film on an AFM substrate.

Furthermore, we can take into account the thickness of the FM layer (Figure 8.3). We assume that the number of the atomic layers in the FM1 is N_1, and in the FM2 is N_2 (Figure 8.3). Then the energy reads

$$E = -J_0 \cos\varphi_1 - J\cos(\varphi_1 - \varphi_2) - H(N_1\cos\varphi_1 + N_2\cos\varphi_2)$$

$$-\frac{\beta_1}{2}N_1\cos^2\varphi_1 - \frac{\beta_2}{2}N_2\cos^2\varphi_2 \qquad (8.5)$$

Given that the normal FM film thickness in the experiments is about 20–50 atomic layers, our next step is to study the FM film of a finite thickness $L = aN$, where a is the interatomic distance and N is the number of the FM atomic layers [10,13,24]. This model is presented in Figure 8.4. Note that in Figures 8.1 and 8.2 the rotation of the magnetization vectors is presented in the figure plane.

8.2.2 The Continuous Model of a Ferromagnetic Film on the Antiferromagnetic Bilayers Substrate: Perfect Interface

In a real thin FM film, magneto-dipole interaction leads to the effective easy-plane anisotropy [(z, y) in Figure 8.4]. We assume that the magnetization is uniform in the interface plane and the model is one dimensional. The problem can be studied in the continuum approximation. All the discrete rotational angles φ_n, n being the layer number, are replaced by the continuous function $\varphi(x)$. The energy of the system in the continuum approximation is

$$E = -\frac{1}{a}\int_0^L dx\left[\frac{J}{2}\left(\frac{d\varphi}{dx}\right)^2 - \frac{\beta}{2}\cos^2\varphi - H\cos\varphi\right] - J_0\cos\varphi_0 \qquad (8.6)$$

where φ_0 is the angle of magnetization in the closest to interface FM atomic layer. Without the anisotropy, the $(\beta/2)\cos^2\varphi$ term in the energy Equation 8.6 takes the form

$$E = -\frac{1}{a}\int_0^L dx\left[\frac{J}{2}\left(\frac{d\varphi}{dx}\right)^2 - H\cos\varphi\right] - J_0\cos\varphi_0 \qquad (8.7)$$

From Equations 8.6 and 8.7, we see that, additionally, the interatomic distance and the FM film thickness parameters with length dimension appear. In the following, we introduce the magnetic length $l_0 = a\sqrt{J/\beta}$, the field characteristic $l_h = a\sqrt{J/H}$, and the characteristic of the exchange interaction $l_1 = a\,J/J_0$. In the fields of the same order as the shift magnitude, we have $l_1 \ll l_0 \ll l_h$. In some cases, the discreteness of the system is crucial and the FM layer is described by a more complicated discrete model:

$$E = -J \sum_{n=1}^{N-1} \cos\left(\varphi_n - \varphi_{n+1}\right)$$

$$-\sum_{n=1}^{N}\left(H\cos\varphi_n + \frac{\beta_1}{2}\cos^2\varphi_n \right) - J_0\cos\varphi_1 \tag{8.8}$$

We will show later that qualitatively similar magnetization dependencies of the FM/AFM bilayer may appear in the case of a uniform magnetization along the film thickness (the magnetization does not depend on x) and a nonuniform one in the interface plane.

One can also obtain the inhomogeneous magnetization in the FM film in the case of the geometrical frustration on the EB phenomenon and the hysteresis loops features. The geometrical frustration appears if the minimum of the system energy does not correspond to the minimum of all local interactions. An example of a geometrically frustrated system is a triangular lattice with the AFM interaction between each pair of magnetic moments. In this case, the frustration is observed because of the incompatibility between the local interactions and the lattice geometry. We study in Reference 29 that layered system FM/AFM consists of two monolayers on a triangular lattice with periodic boundary conditions. The cases of frozen (fixed) and nonfrozen AFM are studied.

8.2.3 Models of Inhomogeneous Ferromagnetic/Antiferromagnetic Bilayers Interface

We start with the case when the magnetization depends on z and is independent of x and y (Figure 8.5).

Inhomogeneity of the FM film magnetization in a plane parallel to the interface can be associated with various reasons. In the simplest case, the AFM can consist of the domains with different directions of the magnetization (Figure 8.5) [14,16,17] which yields the appearance of the inhomogeneous structures in the FM in the interface plane. For a fixed magnetic structure of the AFM, the effective field can be taken into account as the boundary condition for the FM. A realistic domain structure is a labyrinth, but for the low density of the domain walls, we can assume that the structure is one-dimensional. The domain walls are parallel to each other and oriented along the y-axis (Figure 8.5). Depending on the parameters, the domain walls can be thin (Figure 8.5a) or thick (Figure 8.5b). Areas with different direction of magnetization appear (Figure 8.5). When the domains have the same area, the interface is compensated and the shift of the magnetization curve is absent. Note that the appearance of the horizontal plateaus and the inclined curves parts is still possible. The energy of the discussed model is as follows:

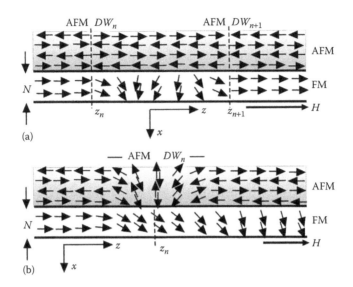

FIGURE 8.5   The model of the thin FM film on the AFM substrate with domain structure. The cases of thin (a) and thick (b) domain walls.

$$E = N \int dz \left(\frac{Ja^2}{2} \left(\frac{d\varphi}{dz} \right)^2 - \frac{\beta}{2} \cos^2 \varphi - H \cos \varphi \right)$$
$$- \int dz \left(J_0 \cos(\varphi_0 - \psi(z)) \right)$$
(8.9)

where $\psi(z)$ is the fixed magnetization distribution in the AFM surface layer. For thin domains, the AFM impact can be described by the function

$$\psi(z) = \pi \sum_n (-1)^n \theta(z - z_n)$$
(8.10)

where $\theta(z)$ is the Heaviside function, and z_n is the coordinate of the nth domain wall. For thick domain walls, the Heaviside function can be replaced by a function with the area of localization l_{AFM}, for example:

$$\psi(z) = \pi \sum_n (-1)^n \left(1 + \exp \left(\frac{(z - z_n)}{l_{AFM}} \right) \right)^{-1}$$
(8.11)

Another possibility to obtain the inhomogeneity in the FM film is to introduce a rough interface. The interface roughness in the simplest case can be described by the mono-atomic steps on the interface (Figure 8.6a).

If the thickness of the FM layer $L = aN$ is big enough, then the atomic step interface can be replaced by the perfect interface. We model the interface roughness by varying

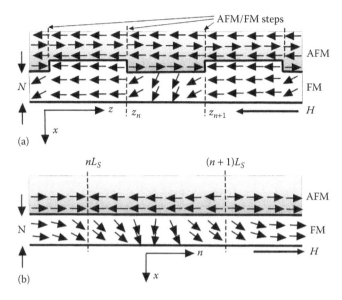

FIGURE 8.6 The model of a FM/AFM bilayer system with inhomogeneous interface (a). In the case of thick layer, the inhomogeneous interface can be described by perfect interface and varying exchange interaction (b).

the exchange interaction through the interface (Figure 8.6) and use the expression for the energy Equations 8.9 and 8.10. Using this energy, we reduce the case of an inhomogeneous interface to the case of perfect interface.

The type of the interface in experiments can be different due to the growing processes. In general, the distribution of the steps on the interface is random. For simplicity, we consider the periodically distributed steps:

$$\psi(z) = \pi \sum_n (-1)^n \theta\left(\frac{(z - nL_s)}{a}\right) \tag{8.12}$$

where L_s is the distance between the surface steps. In the case of pure periodic atomic steps in Equation 8.12, the EB is equal to zero, but we still observe the horizontal plateaus and split of the hysteresis loop. To obtain the shift of the magnetization curve, we need to modify the expression Equation 8.12 in the following way (Figure 8.7a):

$$\psi(z) = \pi \sum_n \left[\theta\left(\frac{(z - nL_s)}{a}\right) - \theta\left(\frac{(z - n(L_s - h_s))}{a}\right) \right] \tag{8.13}$$

where the areas h_s and $L_s - h_s$ are shown in Figure 8.7 with opposite directions of the magnetization on the surface of the AFM are alternating.

Next, we study the model with complicated atomic steps period structure. Qualitatively, this case is presented in Figure 8.7b. In Figure 8.7b, we use the parameters

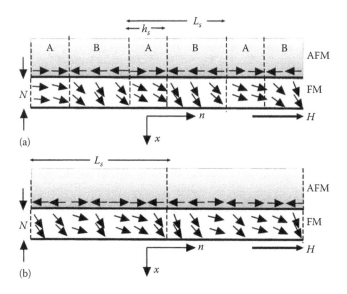

FIGURE 8.7 The model of the layered FM/AFM bilayer. The cases of uncompensated interface with different (a,b) complicated interface structures.

$L_s = 7a$ and $\langle M_s \rangle = M_0/7$, and the magnetic structure of the interface is ←←→←→→←, where the arrows denote the direction of the magnetization in the AFM domains. In Section 8.3, we will consider several types of the interface configurations and obtain information on interface roughness impact on the shape of the magnetization curves. The energy is given by the following formula:

$$E = -N\sum_n \left[J\cos\left(\varphi_n - \varphi_{n+1}\right) + \frac{\beta}{2}\cos^2\varphi_n + H\cos\varphi_n \right]$$
$$- J_s\sum_n \cos\left(\varphi_n - \psi_n\right)$$

(8.14)

The orientation of the magnetization vectors in the surface layer of the AFM is $\psi_n = 0$ or π.

The exchange interaction through the interface in experiments is so weak that it is possible to assume magnetic interaction only for small quantities of the atoms on the interface [17,22,34,35]. Here, we call these contacting atoms PMC. The scheme of the model is presented in Figure 8.8. For low density of PMC, the problem can be considered in the continuum approximation. The energy in this case takes the form

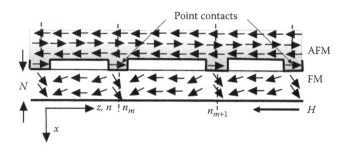

FIGURE 8.8 The model of point magnetic contacts on the FM/AFM interface. The magnetic subsytems interact only through the point magnetic contacts.

$$E = \frac{N}{a} \int dz \left(\frac{J}{2} a^2 \left(\frac{d\varphi}{dz} \right)^2 - \frac{\beta}{2} \cos \varphi^2 - H \cos \varphi \right)$$

$$- \int \frac{dz}{a} J_0 \sum_c \delta(z - z_c) \cos \varphi(z)$$

(8.15)

where $\delta(z - z_c)$ characterizes the type of the interaction through the PMC, and z_c are the coordinates of the PMC. When the PMC density increases, we need to replace the expression Equation 8.15 by its discrete analog:

$$E = -N \sum_n \left[J \cos(\varphi_n - \varphi_{n+1}) + \frac{\beta}{2} \cos^2 \varphi_n + H \cos \varphi_n \right]$$

$$- J_0 \sum_n \delta_{n,c} \cos \varphi_n$$

(8.16)

$\delta_{n,c}$ is equal to 1 when $n = c$ and zero otherwise. In this expression, the index n denotes the magnetic moments in the FM layer, and index c corresponds to the magnetic moments that interact through the interface with AFM subsystem.

All the models considered above are 1D models. The magnetization changes along the x- or z-axis. The magnetization distribution in the y-direction was uniform. In reality, the PMC distributed in a random way. We study the periodically distributed PMC in the plane (y,z). To compare the 1D and 2D models, we study the models presented in Figure 8.9. The magnetic moments position in the FM film and PMC in Figure 8.9a are like in Figure 8.8. The thick arrows correspond to the magnetization vectors at the PMC and the thin arrows to the magnetic moments of the AFM. In Figure 8.8, the rotational angles are presented in the figure plane and in Figure 8.9a in the plane (z, y). The magnetization configuration is presented for the periodically distributed along the z-axis PMC with the density $C = 1/6$ (from six magnetic moments on the interface just one interact through the interface). In Figure 8.9b, the distribution of the

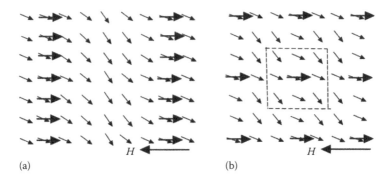

(a)   (b)

FIGURE 8.9 The structure of the FM/AFM interface in the one-dimensional (a) and two-dimensional (b) model of PMC. (Reprinted from *Superlattices and Microstructures*, Vol. 73, A. Kovalev, M. Pankratova, Field dependence of magnetization for a thin ferromagnetic film on rough antiferromagnetic surface, pp. 275–280, Copyright 2014, with permission from Elsevier.)

magnetization is periodic in the plane (z, y) with the concentration $C = 1/9$. In this 2D model, the energy of the FM layer takes the form

$$E = -N \sum_{nm} \left[J \cos\left(\varphi_{nm} - \varphi_{nm-1}\right) + \frac{\beta}{2} \cos^2 \varphi_{nm} + H \cos\varphi_{nm} \right]$$
$$- J_0 \sum_{nm} \delta_{nm,c} \cos(\varphi_{nm}) \tag{8.17}$$

where:

sum on n, m corresponds to the interaction with the closest neighbors in the FM, index c is the interaction through the point contacts, and
$\delta_{nm,c}$ is equal to 1 when $n = m = c$ and zero otherwise.

For a FM film thick enough and a rough interface, it is necessary to take into account nonuniform distribution of the FM magnetization both in the FM/AFM interface plane and in the direction perpendicular to the interface.

Near an atomic step on the interface, the nonuniform magnetization distribution appears [16]. In the absence of the external magnetic field in the film near the steps, the magnetization distribution takes the form of a magnetic vortex (Figure 8.10a). In the presence of the magnetic anisotropy, the vortex transforms to the domain wall (Figure 8.10b). For this model, in the continuum approximation, the system energy is as follows:

$$E = \frac{1}{a^2} \int_0^N dz \int dx \left(\frac{J}{2} a^2 \left(\nabla\varphi\right)^2 - \frac{\beta}{2} \cos^2\varphi - H\cos\varphi \right) - E_s \tag{8.18}$$

with

$$E_s = -\frac{1}{a} \int dx \left(J_s \cos(\varphi_s - \psi(x)) + \frac{\beta_s}{2} \cos^2\varphi_s \right) \tag{8.19}$$

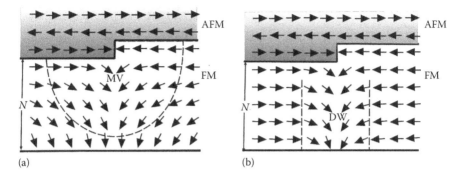

FIGURE 8.10 The magnetic structure of the FM film in contact with an AFM in the presence of the atomic step on the interface in the absence (a) and presence (b) of the magnetic anisotropy.

where $\psi(x)$ is the function that corresponds to the distribution of the magnetization of the AFM on the interface. If the FM film is thick enough so that a domain wall appears, then latter can be considered similar to a linear object [in the (x, z)]. This domain wall separates the areas with uniform magnetization distribution. One can analytically describe the shape of the domain wall [16]. Let us consider an uncompensated FM/AFM interface with atomic steps. The alternating steps form the pairs which we associate with the vortex–antivortex pairs (Figure 8.11a). From these vortex–antivortex pairs, the domain walls start. These domain walls are curvilinear because of the interaction between them. For the uncompensated interface case, the magnetization of the film is not zero. In the compensated interface case, the domain structure appears (Figure 8.11b). If the external magnetic field is strong enough, the domain walls become closed (Figure 8.11c).

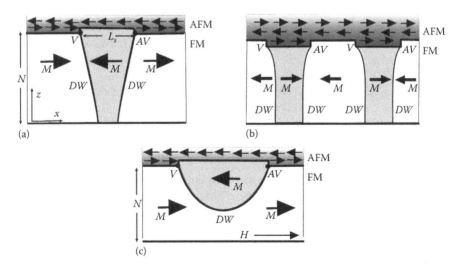

FIGURE 8.11 The magnetic structure of the FM film on the AFM substrate in the presence of the atomic steps pair (a), alternating atomic steps (b) on the interface, and the magnetic structure in the big external magnetic field (c).

8.3 The Features of the Magnetization Dependences of Ferromagnetic Film on External Field and Exchange Bias in Theoretical Models

In this section, we will study the models described in Section 8.2, analyzing the dependence of the magnetization on external magnetic field and constructing phase diagrams. We will see that the magnetization curves for perfect and inhomogeneous interface are similar. In this way, having in hand the experimentally obtained hysteresis loops, we cannot draw any conclusions about the properties of the interface and source of the EB.

In the model, presented in Figure 8.1, in the absence of the magnetic anisotropy $\beta = 0$, there is a jump of the magnetization of the FM film and thus the slope of the magnetization curve is absent. For the film that consists of N atomic layers in the uniform state, the EB decreases as $\delta H_{EB} = -J_0/N$.

This means that the EB is a surface phenomenon. For nonzero values of the magnetic anisotropy, the metastable states and hysteresis loops appear. For the monolayer FM film, the shift of the hysteresis loop is equal $\delta H_{EB} = -J_0$. The width of the hysteresis $\Delta H = \beta$.

8.3.1 The Magnetization Curves of the Ferromagnetic/ Antiferromagnetic Bilayers with Perfect Interface: Discrete Models

As all magnetization processes of the FM/AFM system can be associated with magnetically inhomogeneous states in the FM, here we do not take into account the inhomogeneity in the AFM. The generalization of the simplest model is the model with FM film made of two FM atomic layers. This model can describe the trilayer system AFM/FM1/ FM2 with the layers of thickness N_1 and N_2. To describe this trilayer system, we should change in the energy (2) $\beta_1 \to \beta_1 N_1$, $\beta_2 \to \beta_2 N_2$, and $J \to J_{12}$. The minimization of the energy Equation 8.2 leads to the system of the algebraic equations for the angles φ_1, φ_2, and the solutions $\varphi_1(H)$, $\varphi_2(H)$ give the magnetization curves $M = M(H)$.

The system is characterized by the parameters: external magnetic field H and the ratio of the exchange interactions J/J_0. The magnetization of such system is given by the expression

$$M = \cos\varphi_1 + \cos\varphi_2$$

Depending on these parameters, there are the following states in the FM/AFM system: two collinear configuration $\uparrow\uparrow$, $\downarrow\downarrow$, antiparallel $\uparrow\downarrow$, and the noncollinear states $\varphi_i \neq 0, \pi$. The areas of the existence of different configurations in the plane $(J/J_0, H)$ are presented in Figure 8.12a. In Figure 8.12b, the normalized magnetization curves of the FM film $M/2M_0 = M/2$ for various values of the exchange interaction are shown. The exchange interaction values in Figure 8.12b correspond to the dashed lines in Figure 8.12a. The value of the EB is $\delta H_{EB} = -J_0/2$.

To obtain the hysteresis loop, it is necessary to take into account the magnetic anisotropy. We start with the case $\beta_1 = \beta_2 = \beta$. The anisotropy changes essentially the number

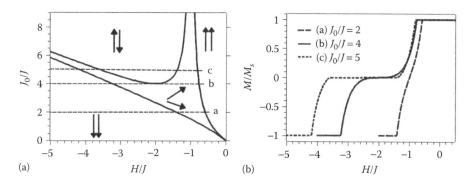

FIGURE 8.12 Phase diagram of the FM/AFM bilayer in the absence of the anisotropy (a); the magnetization curves for the various values of the exchange interaction (b). The FM film consists of two atomic layers.

of magnetization curves. Here, we observe 12 types of hysteresis loops [9]. The lines that separate the areas of existence of various phases (collinear, antiparallel, and noncollinear) are shown in Figure 8.13a. Usually, the anisotropy is small $\beta/J \sim 10^{-2}$. The case when $\beta \sim J$ can appear for the system AFM/FM1/FM2, where J is exchange interaction through the FM1/FM2 interface. The magnetization curve may contain one or two hysteresis loops. The obtained magnetization curves are presented in Figure 8.13b–f. It can be seen from the graph that with the increase of the exchange interaction through the FM/AFM interface, the hysteresis loop changes its form and splits into two loops. With further increasing of the parameter J_0 the horizontal plateau appears, between two hysteresis loops. The obtained curves are qualitatively similar to the experimental ones [35]. The most unusual hysteresis loops appear for large values of the exchange interaction through the interface. Note that in the all the obtained results, the magnetization curve is symmetric in the shifted field. But in some experiments, the symmetry is absent. We presume that the asymmetry can be associated with different values of the magnetic anisotropy in the atomic layers of the FM, that is, $\beta_1 \neq \beta_2$. In the trilayer system, it can be associated with the various properties of the FM films. In a thin FM film, different anisotropy constants can appear due to the various crystallographic neighbors of the atomic magnetic layers.

In Figure 8.14, the magnetization curves for the FM/AFM system are presented for the case of the magnetic anisotropy $\beta_1 \neq \beta_2$. In contrast to the previous case, the number of possible magnetization configurations increase. We present only two types of the magnetization curves for this model. The feature of this magnetization curves is horizontal plateaus at $M = 0$. The position of the horizontal plateau is caused by the same thickness of the FM atomic layers. For the case of three or more FM layers, or FM layers with different thickness, the horizontal plateau shifts from the position $M = 0$.

One can consider a model with different FM layers thickness N_1, N_2 on the AFM. The horizontal plateau in this case shifts from the $M = 0$ position (Figure 8.15). The results are presented for $\beta_1 \neq \beta_2$. The curves are presented for various values of the exchange interaction. The values of the parameters are presented in Figure 8.15. The dependences of the magnetization on magnetic field are presented in Figure 8.15b.

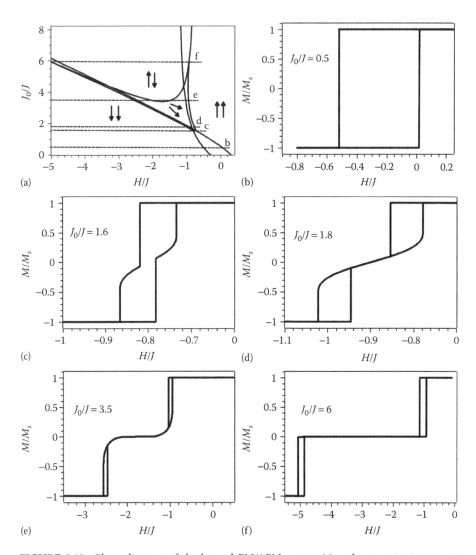

FIGURE 8.13 Phase diagram of the layered FM/AFM system (a) and magnetization curves obtained for various values of the exchange interaction (b–f).

The hysteresis loop splits into two loops. One of the loops with the width $\delta H \approx 2\beta$ is located close to $H = 0$ like in the case of FM. The other hysteresis loop with the same width is shifted along the field axis on the value J_0/N. If the magnetic anisotropy parameters are different, then the widths of the hysteresis loops are $2\beta_2$ and $2\beta_1$. We discussed above the case of two FM layers but the results can also describe a thin FM films. In this case, the magnetic state near the interface is different from the volume magnetic state. We can assume that the exchange interactions of the surface FM layer with other FM atomic layers $J_{1,2}$ and with the AFM J_0 are smaller than the exchange interaction in the volume J. These exchange interactions $J_{1,2}$ can be of the same order as

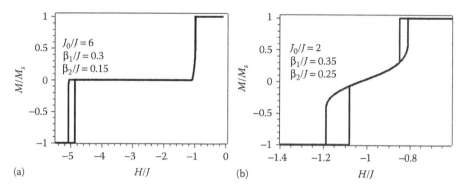

FIGURE 8.14 The magnetization curves of the FM/AFM in the case of two-layered FM film and surface anisotropy $\beta_1 \neq \beta_2$ for $J/J_0 = 6$, $\beta_1/J = 0.3$, $\beta_2/J = 0.15$ (a) and $J/J_0 = 2$, $\beta_1/J = 0.35$, $\beta_2/J = 0.25$ (b).

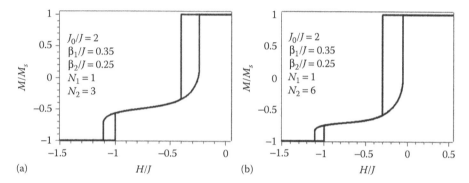

FIGURE 8.15 The magnetic hysteresis loops of the FM/AFM bilayer. The thickness of the FM layers is different $N_1 \neq N_2$, $N_1 = 1$, $N_2 = 3$ (a) and $N_1 = 1$, $N_2 = 6$ (b).

the anisotropy β. For example, for the values $J_0 = 10$, $J_{1,2} = 3$, $N_1 = 1$, $N_2 = 9$, $\beta_1 = 0.35$, $\beta_2 = 0.6$, we can obtain the magnetization curve similar to the experimental one (Figure 8.15b).

8.3.2 The Magnetization Curves of the Ferromagnetic/Antiferromagnetic Bilayers: Continuum Models

We discussed above the discrete models of the FM layer. For the FM films that are thick enough, we can use continuum approximation and study the corresponding differential equations. Let us start from the model presented in Figure 8.5 in the absence of the magnetic anisotropy (Equation 8.7). The distribution of the magnetization $\varphi = \varphi(x)$ is defined by a simple pendulum equation. A solution of the equation is given in terms of the elliptic Jacobi function and gives the magnetization $M = \int_0^L dx \cos \varphi(x)$. There is neither hysteresis loop nor horizontal plateau on the magnetization curve (see phase diagram Figure 8.16a). On the phase diagram, only the areas of existence of collinear ($\varphi = 0, \varphi = \pi$) and canted ($\varphi \neq 0, \pi$) states are presented. To obtain a horizontal plateau, we have to take into account the discreteness of the system. Note that the magnetization

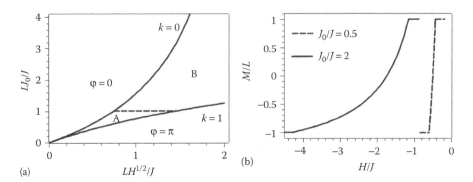

FIGURE 8.16 The phase diagram (a) and the corresponding magnetization curves (b) obtained in the continuum model of the FM film on the AFM substrate.

curve become asymmetric. The shape of the magnetization curves varies depending on the system parameters. The phase diagram for such system and the magnetization curves with the increase of the exchange interaction are presented in Figure 8.16.

Note that if the exchange interaction is fixed and we change the thickness of the FM film, the magnetization curves have different shapes depending on the system parameters. The area of the magnetization rotation of the FM film shifts with the thickness increase to the smaller fields $\delta H \sim J_0/L$. The magnetization dependence $M = M(H)$ is different for various film thicknesses. In this model, there are three length dimension parameters $l_1 = J/J_0, l_h = \sqrt{J/H}$ and L. There are two types of the hysteresis loops that correspond to the thin $L < l_1$ and thick $L > l_1$ FM film. In the thin film, the magnetization curve is shifted by $\delta H \sim J_0/L$ with the thin area of the magnetization reorientation $\Delta H \simeq 2J_s^2/3J$. For $L \sim l_s$, the qualitative change of the magnetization curve type takes place. For $L \ll l_s$, the magnetization curve becomes sufficiently asymmetric.

To obtain the hysteresis loops, we have to take into account the magnetic anisotropy in the continuum model β and return to the energy Equation 8.6. Here, in addition to the parameters l_1, l_h, and L, the magnetic length $l_0 = \sqrt{J/\beta}$ parameter appears. The anisotropy parameter leads to the appearance of the hysteresis loop. For a thin FM film, the hysteresis loop is symmetric, whereas in the case of a thick layer we obtain new types of hysteresis loops. From Equation 8.6, it can be seen that the anisotropy leads to the shift of the collinear states boundaries to the values $H \rightarrow H + \beta$ and $H \rightarrow H - \beta$ (Figure 8.17a).

The absence of the horizontal plateaus and the split of the hysteresis loops are associated with the continuum approximation. Therefore, the next step is to consider the discreteness in the thick enough FM film with the energy Equation 8.8. It is easy to study the transition from a two-layered FM film to the system with $N = 3$. The magnetization curves (Figure 8.13) are modified as is shown in Figure 8.18 [13]. In this trilayer FM system, only one antiparallel configuration is possible ↑↓↓. With the magnetic field

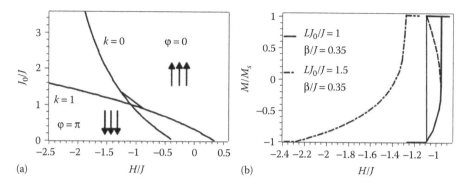

FIGURE 8.17 The phase diagram of a FM/AFM bilayer (a) and the magnetization curve (b) in continuum model with anisotropy.

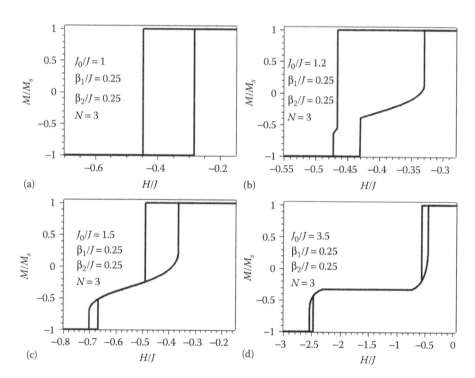

FIGURE 8.18 Magnetization curves of the FM/AFM system for the various values of the exchange interaction $J_0/J = 1$ (a), $J_0/J = 1.2$ (b), $J_0/J = 1.5$ (c), $J_0/J = 3.5$ (d).

increase, the magnetization of the second and third FM atomic layers rotate and the magnetization in the surface FM layer is fixed by the AFM.

With the increase of the FM film thickness, the magnetization curves become close to the ones obtained in the continuum approximation. The discreteness of the system manifests itself in the antiparallel states appearance. These antiparallel states lead to the appearance of a horizontal plateau and to the split of the hysteresis loop.

The magnetization curves and phase diagram are obtained also for the case when the geometrical frustration takes place in the system. It is shown in Reference 29 that the shift of the magnetization curve is observed for the fixed (frozen) AFM. If the AFM is not fixed, there is no shift of the hysteresis loop. We see however the split of the hysteresis loop and one or multiple hysteresis loops on the magnetization curve.

8.3.3 The Dependencies of Magnetization on Magnetic Field of the Ferromagnetic Film on the Antiferromagnetic Bilayers Substrate: Inhomogeneous Interface

We discussed above the nonuniform magnetic structures with the distribution of magnetization variation along the FM film thickness, but uniform along the interface. We observe the same types of the magnetic hysteresis loops in the case of magnetization inhomogeneity that is parallel to the interface. We start from the models where this inhomogeneity depends on a coordinate in the interface plane z. We assume that the ferromagnetic film is thin. The magnetization distribution in the direction perpendicular to the interface is uniform. The scheme of the model is presented in Figure 8.6, and we use the energy Equation 8.9. We start from the case when the atomic steps or the areas with the various AFM magnetization appear on the FM/AFM interface. The distance between the defects is assumed to be the same and equals L_s. For the fixed values of the exchange interactions J, J_0, and β, the shape of the hysteresis loops is dependent significantly on the interface nonuniformity. The parameter $S = L/\sqrt{l_1}$ characterizes the degree of the interface inhomogeneity. Sufficient variation of the hysteresis loops takes place for $S \sim 1$. We assume that the FM film is monolayer. The magnetic structure of the AFM is fixed like in the previous cases so in the domains A and B we have $\psi A = 0$, $\psi B = \pi$. If we take the discreteness into account, then in the system appear: the parallel states.. ↑↑↑...,.. ↓↓↓..., and antiparallel states.. ↑↑↑↓↓↓....

For simplicity, let us consider the periodic case with the domains $A = B$. In this case, the average field that acts from the AFM on the FM equals zero and the shift is absent, but some features of the magnetization curves are present. The boundaries of the existence of the parallel and antiparallel states are easy to find and the results are presented in Figure 8.19a and b in the absence of the magnetic anisotropy. The type of the magnetization curve depends on the period of the atomic steps on the interface L_s. For the weak nonuniformity of the interface $L_s > L_* = a\pi^2/2$, there is no hysteresis loop. For the strong inhomogeneity $L_s < L+$, the magnetization curves are more interesting. Namely, the hysteresis loops appear even in the absence of the magnetic anisotropy.

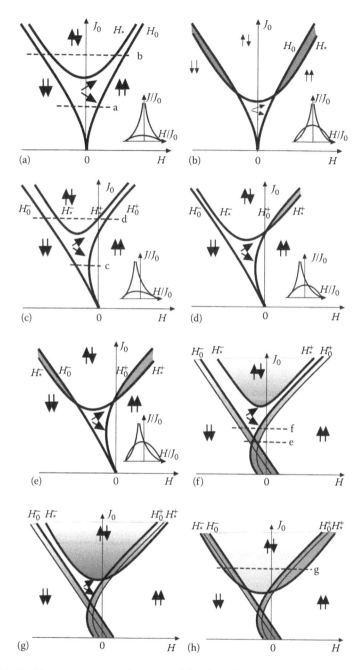

FIGURE 8.19 The schematic transformation of the phase diagrams of the FM/AFM bilayer with atomic steps on the interface. The compensated (a, b) and uncompensated (c–e) interface in the absence of the magnetic anisotropy. The uncompensated interface in the presence of the magnetic anisotropy (f–h).

For various growing processes of the FM film on the AFM substrate give uncompensated interfaces with periodically distributed areas A and B with different width h_s and $L_s - h_s$ with opposite magnetization direction in the AFM. The system behavior changes depending the period of the atomic steps: $L_s/2 < h_s < L_s - L_*(C)$, L_*, $L_s - L_* < h_s < L_s(D)$, $L_s/2 < h_s < L_s$, L_*. The corresponding curves are presented in Figure 8.19c–e. We have not taken into account the impact of the magnetic anisotropy on the magnetization curves. The anisotropy leads to the shift of the boundaries of the collinear states $H_0^+ \rightarrow H_0^+ - \beta$ and $H_0^- \rightarrow H_0^- + \beta$. The area of the existence of the antiparallel states shifts to the smaller values of the interlayer exchange interaction on the value $\delta J_0 \sim \beta$.

The results for the case of the compensated AFM interface are similar to the hysteresis loops obtained in the model of two FM atomic layers. However, the shift of the magnetization curves is absent. If the discreteness of the system is not taken into account, then the horizontal plateaus on the magnetization curves are absent. We see only the parts of the curve with small slope [14].

The results in the presence of the magnetic anisotropy for the uncompensated interface case are presented in Figure 8.20, for example. For the uncompensated FM/AFM interface, we observe the shift and asymmetry of the hysteresis loop [14].

For a more complicated period structure of the interface, we can study the model described by the energy Equation 8.14 (Figure 8.8b). The problem is studied numerically (Figure 8.21). Earlier, we studied the systems with a complicated interaction through the FM/AFM interface.

In these systems, however, all the atoms interact magnetically through the interface. Real interfaces are not perfect and due to this only several percent of the magnetic atoms interact through the interface. To study this case, we use the energy Equation 8.15 or the discrete analog Equation 8.16. The PMC are distributed periodically along the interface [17]. The distribution of the magnetization is given by the double sine-Gordon equation. This problem is easy to solve analytically. *The main result is that the magnetization curves M (H) actually coincide, after the renormalization of the parameters, with those obtained earlier for the FM/AFM system with the perfect interface.* To explain the experimental data, we have to know the dependencies of the magnetization curves shape on the system parameters. The transformation of the magnetic hysteresis loops with variation of the exchange interaction is presented in Figure 8.22a. The value of the interface inhomogeneity is fixed. The hysteresis loops have the same form as in the case of the perfect interface presented in Figure 8.17. The transformation of the magnetization curve with interface roughness is presented in Figure 8.22b. These results are obtained in the framework of the continuum model. The discrete model gives the same magnetization curves. At some values of the system parameters, the horizontal plateau is observed on the hysteresis loop. The model of the PMC has been studied also in the 2D model. The scheme of the interface is presented in Figure 8.9b. The obtained magnetization curves are very close to the ones in 1D system (Figure 8.22c). The magnetization curves have the same features and the same shape like in the 1D system, but for different values of the exchange interaction and magnetic anisotropy (Figure 8.22b).

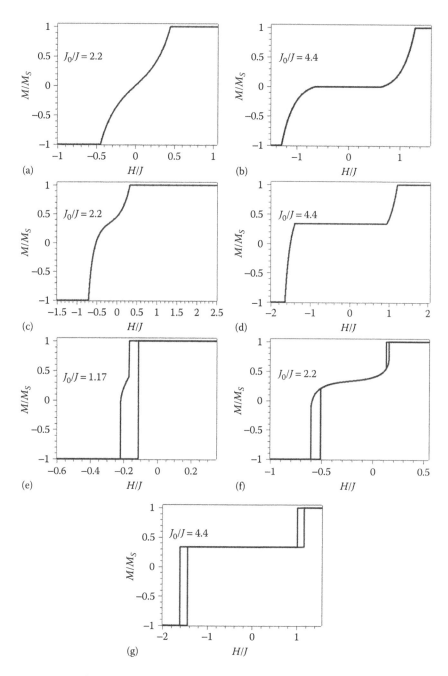

FIGURE 8.20 The magnetization curves of a FM/AFM bilayer with atomic steps on the interface. The compensated (a, b) and uncompensated (c, d) interface cases in the absence of the magnetic anisotropy. The uncompensated interface in the presence of the magnetic anisotropy (e–g). The hysteresis corresponds to the dashed lines on the phase diagram in Figure 8.19.

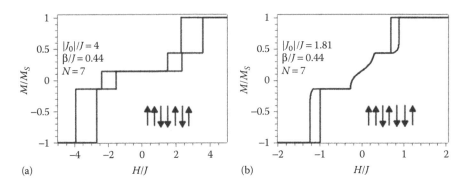

FIGURE 8.21 The magnetization curves of the FM/AFM bilayer with atomic steps on the interface for various structures of the interface and exchange interaction $J_0/J = 4$ (a), $J_0/J = 1.81$ (b). Complicated structure of the interface.

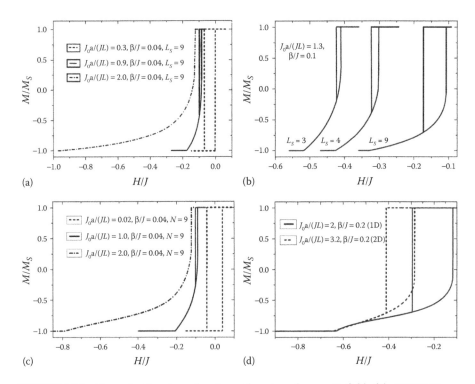

FIGURE 8.22 Magnetization dependencies on the external magnetic field of the FM/AFM system with PMC on the interface in the continuous (a, b) and the discrete (c, d) model. (Reprinted from *Superlattices ad Microstructures*, Vol. 73, A. Kovalev, M. Pankratova, Field dependence of magnetization for a thin ferro-magnetic film on rough antiferromagnetic surface, pp. 275–280, Copyright (2014), with permission from Elsevier.)

8.4 Conclusions

It has been shown that the existence of the experimentally observed features of the magnetization dependences on the external magnetic field of the layered systems FM/AFM can be explained by the nonuniform magnetization distribution in the FM film. This nonuniform magnetization distribution can be both parallel and perpendicular to the interface. The behavior of this domain walls explain the appearance of the horizontal plateaus, asymmetry, splitting of the hysteresis loops. All these magnetization curve features can be explained in the framework of classical Heisenberg models. We show that the same type of the magnetization curves can be observed both in the perfect and inhomogeneous interface cases and can be caused by various reasons.

Acknowledgment

This work was supported by the project of the National Academy of Science of Ukraine 4/16 – *H* and by the scientific program 1.4.10.26/-26-3.

The support of the Scientific Grant Agency of Ministry of Education of Slovak Republic (Grant No. 1/0331/15) and National Scholarship Programme of the Slovak Republic is gratefully acknowledged. Authors thank the journal *Superlattices and Microstructures* for permission to reproduce Figures 8.9 and 8.22.

References

1. W.H. Meiklejohn, C.P. Bean, New magnetic anisotropy, *Phys. Rev.* 102, 1413 (1956).
2. J. Nogues, I.K. Schuller, Exchange bias, *J. Magn. Magn. Mater.* 192, 203 (1999).
3. A.E. Berkowitz, K. Takano, Exchange bias—a review, *J. Magn. Mater.* 200, 552–570 (1999).
4. M. Kiwi, Exchange bias theory, *J. Magn. Mater.* 234, 584 (2001).
5. A.P. Malozemoff, Random-field model of exchange anisotropy at rough ferromagnetic-antiferromagnetic interfaces, *Phys. Rev. B* 35, 3679 (1987).
6. M.D. Stiles, R.D. McMichael, Model for exchange bias in polycrystalline ferromagnet–antiferromagnet bilayers, *Phys. Rev. B* 59, 3722 (1999).
7. D. Mauri, H.C. Siegmann, P.S. Bagus, E. Kay, Simple model for thin ferromagnetic films exchange coupled to an antiferromagnetic substrate, *J. Appl. Phys.* V.62, 3047 (1987).
8. J. Geshev, Analytical solutions for exchange bias and coercivity in ferromagnetic/antiferromagnetic bilayers, *Phys. Rev. B.* V.62, 5627 (2000).
9. A.G. Grechnev, A.S. Kovalev, M.L. Pankratova, Influence of magnetic anisotropy on hysteresis behavior in the two-spin model of a ferro/antiferromagnet bilayer with exchange bias, *Low Temp. Phys.* 38, 937 (2012).
10. A.G. Grechnev, A.S. Kovalev, M.L. Pankratova, Field dependences of the magnetization and exchange bias in ferro/antiferromagnetic systems. II. Continuum model of a ferromagnetic layer, *Low Temp. Phys.* 35, 526 (2009).

11. A.G. Grechnev, A.S. Kovalev, M.L. Pankratova, Magnetization field-dependences and the exchange bias in ferro/antiferromagnetic systems. I. Model of a bilayer ferromagnetic, *Low Temp. Phys.* 35, 476 (2009).

12. A.G. Grechnev, A.S. Kovalev, M.L. Pankratova, Influence of magnetic anisotropy on hysteresis behavior in the two-spin model of a ferro/antiferromagnet bilayer with exchange bias, *Low Temp. Phys.* 38, 937 (2012).

13. A.G. Grechnev, A.S. Kovalev, M.L. Pankratova, Effect of the exchange bias on the magnetization hysteresis of a ferromagnetic film in contact with an antiferromagnet, *Low Temp. Phys.* 39, 1060 (2013).

14. A.S. Kovalev, M.L. Pankratova, Properties of ferromagnetic film hysteresis, on the surface of a hard-magnetic antiferromagnet, with a domain structure, *Low Temp. Phys.* 40, 990 (2014).

15. M.L. Pankratova, A.S. Kovalev, Model of exchange bias in a trilayer FM/AFM/FM structure, *Low Temp. Phys.* 41, 838 (2015).

16. A.S. Kovalev, M.L. Pankratova, The magnetic structure of a thin ferromagnetic film on the rough surface of an antiferromagnet, *Low Temp. Phys.* 37, 866 (2011).

17. A.S. Kovalev, M.L. Pankratova, Field dependence of magnetization for a thin ferromagnetic film on rough antiferromagnetic surface, *Superlattices Microstruct.* 73, 275 (2014).

18. E.L. Fertman, A.V. Fedorchenko, A.V. Kotlyar, V.A. Desnenko, E.Čižmar, A. Baran, D.D. Khalyavin, A.N. Salak, V.V. Shvartsman, A. Feher, Exchange bias phenomenon in $(Nd_{1x}Y_x)_{2/3}Ca_{1/3}MnO_3(x = 0, 0.1)$ perovskites, *Low Temp. Phys.* 41, 1001 (2015).

19. W.H. Meiklejohn, C.P. Bean, New Magnetic Anisotropy, *Phys. Rev.* 105, 904 (1957).

20. W.H. Meiklejohn, *J. Appl. Phys.* 33, 1328 (1962).

21. J. Stohr, A. Scholl, T.J. Regan, S. Anders, J. Loning, M.R. Scheinfein, H.A. Padmore, R.L. White, Images of the antiferromagnetic structure of a NiO(1 0 0) surface by means of X-Ray magnetic linear dichroism spectromicroscopy, *Phys. Rev. Lett.* 83, 1862 (1999).

22. P. Kappenberger, S. Martin, Y. Pellmont, H.J. Hug, J.B. Kortright, O. Hellwig, E.E. Fullerton, Direct imaging and determination of the uncompensated spin density in exchange-biased Co/(CoPt) multilayers, *Phys. Rev. Lett.* 91, 267202 (2003).

23. W.J. Antel, Jr. Perjeru, G.R. Harp, Spin structure at the interface of exchange biased FeMn/Co bilayers, *Phys. Rev. Lett.* 83, 1439 (1999).

24. L. Neel, Ferro-antiferromagnetic coupling in thin layers, *Ann. Phys. Paris* 2, 61 (1967).

25. N.C. Koon, Calculations of exchange bias in thin films with ferromagnetic/antiferromagnetic interfaces, *Phys. Rev. Lett.* 78, 4865 (1997).

26. B. Skubic, J. Hellsvik, L. Nordstrom, O. Eriksson, Exchange coupling and exchange bias in FM/AFM bilayers for a fully compensated AFM interface, *Acta Phys. Pol. A* 115, 27–29 (2009).

27. C. Mitsumata, A. Sakuma, K. Fukamichi, *IEEE Trans. Mag.* 41, 10 (2005).

28. P. Song, G.K. Li, L. Ma, C.M. Zhen, D.L. Hou, W.H. Wang, E.K. Lui, J.L. Chen, G.H. Wu, Magnetization jumps and exchange bias induced by a partially disordered antiferromagnetic state in $(FeTiO_3)_{0.9} - (Fe_2O_3)_{0.1}$, *J. Appl. Phys.* 115, 213907 (2014).
29. M. Pankratova, M. Žukovič, Magnetization curves of geometrically frustrated exchange-biased FM/AFM bilayers, *Acta Phys. Pol. A* 131, p. 642–644 (2017).
30. L.S. Uspenskaya, Asymmetric kinetics of magnetization reversal of thin exchange-coupled ferromagnetic films, *Phys. Sol. State* 52, 2274–2280 (2010).
31. C.Y. You, H.S. Goripati, T. Furubayashi, Y.K. Takahashi, K. Hono, Exchange bias of spin valve structure with a top-pinned $Co_{40}Fe_{40}B_{20}/IrMn$, *App. Phys. Lett.* 93, 012501-1–012501-3 (2008).
32. C.L. Chien, V.S. Gornakov, V.I. Nikitenko, A.J. Shapiro, R.D. Shull, Hybrid domain walls and antiferromagnetic domains in exchangecoupled-ferromagnet/antiferromagnet bilayers, *Phys. Rev. B.* 68, 014418 (2008).
33. G. Vallejo-Fernandez, L.E. Fernandez-Outon, K. OGrady, Thermal activation of bulk and interfacial order in exchange biased systems, *J. Appl. Phys.* 103, 07C101-1–07C101-3 (2008).
34. Z. Lu, W.-Y. Lai, C.-L. Chai, The effect of microstructure and interface conditions on the exchange coupling fields of NiFe FeMn, *Thin Solid Films* 375, 224–227 (2000).
35. P.D. Kim, G.S. Patrin, D.A. Muraschenko, T.V. Rudenko, V.V. Polyakov, T.V. Kim, The investigation of magnetization reversal of free- and pinning-layers in the spin valve structures, *J. Siberian Federal Univ.* 5(2), 196204 (2012).

Exchange Bias in Core–Shell Nanowires and Nanotubes

Mariana P. Proenca
and João Ventura

9.1 Introduction

The exchange bias (EB) effect is a coupling phenomenon between two magnetic layers with different spin structures. The most commonly studied bilayers are composed of a ferromagnetic (FM) and an antiferromagnetic (AFM) material, in which the Curie temperature (T_C) of the ferromagnet is typically higher than the Néel temperature (T_N) of the antiferromagnet. The exchange interaction between the AFM and FM layers will then influence the reversal of the FM layer, shifting the switching field of the ferromagnet by an EB field (H_{EB}), as illustrated in Figure 9.1. The strength of this coupling is highly influenced by the quality of the interface between the two magnetic layers. When reducing a material's size to the nanoscale, surface effects become increasingly important. In particular, the EB coupling effect between the surfaces of different magnetic materials also increases at the nanoscale. An accurate understanding of the nanomaterial's

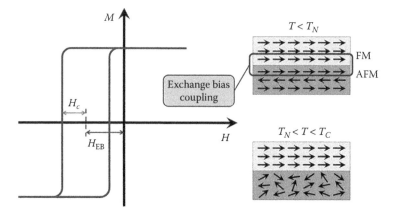

FIGURE 9.1 Exchange bias effect on the magnetic hysteresis loop of an FM layer coupled to an AFM layer, where H_c, H_{EB}, T_N, and T_C are the coercive field, exchange bias field, Néel temperature, and Curie temperature, respectively.

surface contribution is thus of crucial importance, as it will highly influence the magnetic response of the nanomaterial.

There are three main dimensional classes of nanomaterials depending on the number (one, two, or three) of dimensions reduced to the nanoscale (Figure 9.2) (Sajanlal et al., 2011; García-Calzón and Díaz-García, 2012). For example, a thin film is a nanomaterial in which only one of its three dimensions has been reduced to the nanoscale (see 2D nanostructures in Figure 9.2). Magnetic thin films and stacks of magnetic and non-magnetic layers are considered the heart of many modern magnetic devices, such as spin valves and magnetic tunnel junctions (Coey, 2009). When two dimensions of a material are reduced to the nanoscale, we fabricate nanostripes, nanowires (NWs), nanotubes (NTs), nanorods, or nanofibers (see 1D nanostructures in Figure 9.2). Owing to their high shape anisotropy, magnetic NWs are potential candidates for magnetic storage devices. Finally, when we reduce the three dimensions of a material to the nanoscale, we obtain molecular clusters or nanoparticles (see 0D nanostructures in Figure 9.2). Magnetic nanoparticles are very attractive for biomedical applications, ranging from magnetic-assisted drug delivery methods to forefront imaging techniques (Colombo et al., 2012).

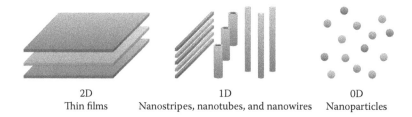

| 2D | 1D | 0D |
| Thin films | Nanostripes, nanotubes, and nanowires | Nanoparticles |

FIGURE 9.2 Illustrations of the various types of materials by size of their structural elements: 2D thin films, 1D nanostripes/nanowires/nanotubes, and 0D nanoparticles.

The EB effect was first observed in core–shell cobalt/cobalt oxide (Co/CoO) nanoparticles by Meiklejohn and Bean (1956). Since then, the EB coupling has been measured in many other different systems containing an FM/AFM interface, ranging from 2D thin films (Maat et al., 2001; Sun et al., 2003; Brems et al., 2005; Shipton et al., 2009; Suszka et al., 2012) to 1D NWs and NTs (Salabas et al., 2006; Maurer et al., 2009; Proenca et al., 2013a; Buchter et al., 2015) and other compositions of 0D nanoparticles (Respaud et al., 1998; Skumryev et al., 2003; Nogués et al., 2006; Kovylina et al., 2009; Sharma et al., 2010; Fernández-García et al., 2011; Proenca et al., 2011). In addition to the magnetic coupling reported in FM/AFM interfaces, EB and related effects have also been observed in other types of interfaces, such as AFM materials with uncompensated spins at the surface (which will be addressed in Section 9.4) and FM-multiferroic core–shell nanostructures (as will be discussed in Section 9.5).

Of all the interfacial nanosystems, EB-coupled magnetic thin films and multilayers have been the most intensively used, especially in magnetic recording devices such as read heads (Zhu and Park, 2006; Sbiaa et al., 2011). The interesting coupling behavior found between FM and AFM layers, with a few nanometers in thickness, has propelled the development of new devices used in magnetic data storage, sensors, and biomedical industries (Coey, 2009; Guimaraes, 2009). Nevertheless, with the miniaturization of devices, the search for smaller nanostructures that exhibit a similar or enhanced EB effect has increased. Most of the 0D bilayered nanosystems are constituted by spherical nanoparticles that exhibit an isotropic magnetic behavior. One of the main advantages of 2D multilayered thin films is their anisotropic shape, which provides an anisotropic magnetic behavior, enhancing the EB effect along one particular direction or plane. As in thin films, the EB in high-aspect-ratio nanoparticles (1D nanostructures) also evidences two main directions of coupling. In bilayered thin films, one usually has a stronger coupling along the in-plane direction of the film, while in the out-of-plane direction, the EB effect is usually smaller (Sun et al., 2003; Phuoc and Suzuki, 2008). In the case of NWs and NTs, the EB effect also depends on the direction of the applied magnetic field. In fact, studies have shown different EB fields when applying the magnetic field along the long axis of the NW/NT structures or perpendicular to it (De La Torre Medina et al., 2009; Proenca et al., 2013a). Therefore, high-aspect-ratio nanoparticles are potential candidates for future applications, as they are reduced in size (when compared with 2D thin films) and still allow tuning the EB effect with the direction of the applied magnetic field.

This chapter describes the EB effect in high-aspect-ratio NWs and NTs that present a core–shell structure and therefore an EB coupling between an outer shell and an inner shell with different magnetic structures. Section 9.2 summarizes some of the main fabrication methods to obtain such structures. The magnetic properties of these core–shell nanostructures are then described in more detail in the subsequent sections. Section 9.3 describes the EB effect in NWs and NTs with a bilayered core–shell FM–AFM structure. Section 9.4 presents the EB effect observed in AFM NWs and NTs with uncompensated spins at the surface. Finally, Section 9.5 introduces the EB effect in FM-multiferroic core–shell NWs and NTs. The most important applications of such EB 1D nanostructures and the main conclusions of this chapter are then summarized in Section 9.6.

9.2 Fabrication Routes of Core–Shell 1D Nanostructures

The basic approaches for the fabrication of nanomaterials can be divided into two main types: top-down and bottom-up approaches. In the top-down approach, one starts from a large-scaled material and then carves it into smaller pieces, giving them the desired final shape until a smaller-sized (nanoscaled) material is obtained. Top-down methods, such as nano-lithography techniques, are usually expensive and have a low throughput, thus hindering low-cost industrial production of nanodevices. On the other hand, bottom-up fabrication is made by assembling small particles (atoms, molecules, and/or clusters) to form nanomaterials. The bottom-up fabrication techniques are usually based on self-assembling or template-filling processes, such as sol–gel, electrodeposition, co-precipitation, and layer-by-layer deposition, which are cheaper and with a much higher throughput, ideal for the large-scale inexpensive production of nanodevices.

For the fabrication of 1D nanostructures, the most commonly used techniques are template-assisted depositions, in which the NWs/NTs are grown inside a nanoporous template (Schonenberger et al., 1997; Nielsch et al., 2000; Toimil Molares et al., 2001; Lee et al., 2005; Batista et al., 2009; Proenca et al., 2012a; Apolinário et al., 2014; Lee and Park, 2014; Arshad et al., 2016). Low-cost nanoporous templates with tuned pore diameters (ranging from 6 to 200 nm) can be obtained through a controlled anodization process of aluminum foils or by etching heavy ion tracks in a polycarbonate membrane (Masuda and Fukuda, 1995; Masuda et al., 2001; Nielsch et al., 2002; Lee et al., 2006; Sulka, 2008; Lee and Kim, 2010; Sousa et al., 2014a). Other porous templates can also be obtained by using nano-lithographic techniques; however, in this case, the production costs and efficiencies are much lower (Woldering et al., 2008). Figure 9.3 shows scanning electron microscopy (SEM) images of the most commonly used nanoporous templates, evincing the easy control of the pore diameter, length, interpore distance, vertical alignment, order of the array, and pore modulation along their length.

The most reported template-filling processes to grow arrays of NWs and/or NTs are the sol–gel and the electrodeposition methods. The sol–gel method basically consists of

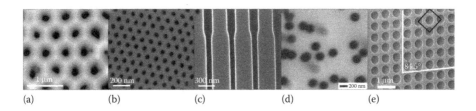

(a)   (b)   (c)   (d)   (e)

FIGURE 9.3   SEM images of (a–c) nanoporous alumina templates with different pore sizes, (d) polycarbonate porous template, and (e) nano-lithographed pores. ([c] From Salem, M. S. et al., *Nanoscale*, 5, 3941–3947, 2013. Reproduced by permission of The Royal Society of Chemistry; [d] From Arshad, M. S. et al., *J. Phys. D: Appl. Phys.*, 49, 185006, 2016, IOP Publishing, all rights reserved; [e] From Woldering, L. A. et al., *Nanotechnology*, 19, 145304, 2008, IOP Publishing, all rights reserved.)

three main steps: first, the creation of a precursor solution with the components that one wants to deposit; then, the gelation and aging of the solution until a final gel is obtained; and finally, the drying of the gel through calcination and thermal annealing processes (Brinker and Scherer, 1990; Caruso and Antonietti, 2001). This method is mostly used for the growth of nanoparticles composed of metal oxides (Meulenkamp, 1998; Lu et al., 2002; Proenca et al., 2011). However, by immersing a nanoporous template in the sol–gel solution during the first and/or second steps of the process, the solution will infiltrate the pores, creating elongated 1D nanostructures (Figure 9.4a and b) (Xu et al., 2008; Sousa et al., 2009).

For the electrodeposition of NW/NT arrays inside nanoporous templates, three different methods have been developed, aiming at its uniform growth (Sousa et al., 2014a; Proenca et al., 2015; Stepniowski and Salerno, 2015; Sharma and Kuanr, 2015): direct current (DC; potentiostatic mode) (Whitney et al., 1993; Proenca et al., 2012a, b), alternating current (AC) (Gerein and Haber, 2005; Sharma et al., 2007), or a hybrid constant potential and constant current in pulses (pulsed electrodeposition [PED]) (Nielsch et al., 2000; Sousa et al., 2011). The PED usually consists of the consecutive application of cycles with three different pulses: a deposition pulse at a constant current for a few milliseconds, a shorter pulse with opposite polarity and constant potential to discharge the insulator barrier layer capacitance (sometimes present at the bottom of nanoporous alumina templates), and a longer rest pulse to allow the repositioning of the ions at the deposition interface. Alternating current deposition is characterized by the application of an AC deposition voltage with a given deposition frequency. It is important to note that the anodization of Al leads to an insulating barrier layer at the bottom of the pores that prevents direct contact between the electrodepositing electrolyte and the conducting cathode. Therefore, for the AC and PED processes, the alumina barrier layer must first be thinned to less than 10 nm, for which hollow dendrites are created at the bottom of the pores (Sousa et al., 2011; Sousa et al., 2014b). The electrodeposition process then takes place through such thin insulating barrier layer. In DC deposition, a constant potential is applied between the working and the counter (or reference) electrodes and the current

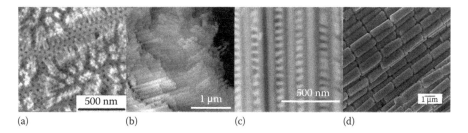

(a)   (b)   (c)   (d)

FIGURE 9.4 SEM images of 1D nanostructures fabricated inside nanoporous alumina templates by (a and b) sol–gel and (c and d) electrodeposition methods. ([a] Reprinted from *Colloids Surf. B: Biointerfaces*, 94, Sousa, C. T., et al., pH sensitive silica nanotubes as rationally designed vehicles for NSAIDs delivery, 288–295. Copyright (2012), with permission from Elsevier; [c] Reproduced with permission from Susano, M. et al., *Nanotechnology*, 27(33), 335301, 2016, IOP Publishing, all rights reserved; [d] From Liew, H. F. et al., *J. Physics: Conf. Ser.*, 266(1), 012058, 2011.)

transient monitored during the deposition process. For the growth of NW/NT arrays inside the nanopores of a template, a metallic contact is sputtered on one side of the free-standing membrane to serve as the working electrode. One of the biggest advantages of the DC method is that it allows the deposition of tubular structures with controlled wall dimensions, by tuning the thickness of the sputtered metallic contact (Proenca et al., 2015), while PED and AC methods only allow the deposition of NWs. Nevertheless, template filling using any of the three electrodeposition methods mentioned above is a low-cost and high-yield process that allows one to easily tune the NWs/NTs dimensions, composition, and crystallinity, even allowing modulations (in diameter, composition, and/or crystallographic structure) to occur along their length (Figure 9.4c and d) (Tao et al., 2006; Iglesias-Freire et al., 2015; Cantu-Valle et al., 2015; Susano et al., 2016).

Another method for the fabrication of NTs inside nanoporous templates is atomic layer deposition (ALD) (Daub et al., 2007; Bachmann et al., 2007; Lu et al., 2008; George, 2010; Zierold et al., 2011). As the deposition occurs by a layer-by-layer process, the control of the tubular thickness is easier and more accurate when compared with the sol–gel or electrodeposition methods. In addition, by changing the precursors introduced in the chamber during the deposition process, multilayered structures can be deposited in a core–shell tubular configuration, as depicted in Figure 9.5.

Although porous template-filling methods are the most reported for the fabrication of NW and NT arrays, other growth techniques can also be used for the production of NWs/NTs, such as hydrothermal processes (Wu et al., 2010), chemical self-assembling (Yu and Urban, 2007; Smith et al., 2011; Anagnostopoulou et al., 2016), electrospinning (Fu et al., 2012), and using template chemical assembling on the surface of other molecular/tubular structures (Grzelczak et al., 2007; Correa-Duarte and Salgueirino, 2010; Mohamed et al., 2012; Buchter et al., 2015).

Finally, for the fabrication of core–shell 1D nanostructures, a combination of the several methods described above can be employed. In particular, by using nanoporous template-filling methods, ordered arrays of coaxial multiwalled NWs/NTs can be easily fabricated (Crowley et al., 2005; Daly et al., 2006). The main approach is to first grow the

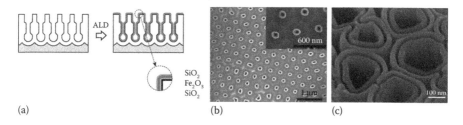

(a)   (b)   (c)

FIGURE 9.5 (a) Schematic representation of the fabrication of modulated core–shell nanotubular structures by using atomic layer deposition on nanoporous alumina templates. (From Pitzschel, K. et al., *ACS Nano*, 3(11), 3463–3468, 2009.) SEM images of, (b) single layered nanotubes. (Reprinted with permission from Daub, M. et al., *J. Appl. Physics.*, 101(9), 09J111. Copyright 2007, American Institute of Physics.), and (c) coaxial multiple-walled nanotubes grown by atomic layer deposition. (Reprinted with permission from Gu, D. et al., *ACS Nano*, 4(2), 753–758. Copyright 2010 American Chemical Society.)

NTs inside the templates by using sol–gel, ALD, or electrodeposition methods and then fill the empty core of the tubes with a different material by using electrodeposition or sol–gel techniques. Nevertheless, reports have already shown that, by tuning the deposition conditions of a single technique (such as ALD or electrodeposition), one can also obtain core–shell nanotubular structures (Wang et al., 2005; Park et al., 2007; Liu et al., 2007; Pitzschel et al., 2009; Gu et al., 2010; Chong et al., 2010; Li et al., 2010; Narayanan et al., 2010; Shi et al., 2012; Narayanan et al., 2012).

For the study of the EB effect in core–shell NWs/NTs, one has to ensure that there is a good interfacial coupling between the two (or more) coaxial layers. Therefore, one of the best methods for the fabrication of core–shell FM–AFM nanotubular structures is by thermal or natural oxidation of one of the magnetic layers (Salgueiriño-Maceira et al., 2008; Maurer et al., 2009; Hsu et al., 2012; Proenca et al., 2013a; Buchter et al., 2015). Nickel, cobalt, and iron are ferromagnetic materials that when oxidized to NiO, CoO, and α-Fe_2O_3, respectively, become antiferromagnetic (below the respective Néel temperature). As a result, most of the EB studies on FM–AFM core–shell 1D nanostuctures that will be presented in Section 9.3 thus refer to FM NWs/NTs in which a native AFM oxide layer was formed, creating a coaxial FM–AFM exchange-coupled nanostructure.

9.3 Antiferromagnetic–Ferromagnetic Bilayered Nanowires and Nanotubes

One of the first reports on EB in elongated FM/AFM 1D nanostructures was made by Kazakova et al. (2006) in cobalt and magnetite (Fe_3O_4)-based nanocables with a CoO thin layer naturally formed at their interfaces. The coaxial nanocables with diameters and lengths of around 100 nm and 60 µm, respectively, were synthesized within the pores of anodic alumina membranes by using a supercritical fluid inclusion-phase technique (Daly et al., 2006). The core–shell nature of the obtained coaxial nanocables was clearly seen by transmission electron microscope (TEM), as illustrated in Figure 9.6a and b. Single-phased cobalt and magnetite NWs were also prepared in an identical way, in order to compare magnetic properties. Magnetic hysteresis loops measured at low temperatures after field-cooled (FC) in 1 kOe (1 Oe = $10^3/4\pi$ A/m) revealed loop shifts (Figure 9.6c), confirming the EB effect due to the formation of CoO thin layers at the nanocables' interfaces. It was also shown that the manifestation and strength of the EB effect depended on the thickness and oxidation state of the Co layer. In addition, the combination of the EB coupling between magnetic layers and the high geometrical aspect ratio of the nanocables was found to significantly stabilize the magnetization of the NWs at low temperatures, thus making them prospective candidates for future spintronic devices.

In 2008, Salgueiriño-Maceira et al. reported EB in unique wires of Ni/NiO synthesized by using carbon nanotubes (CNTs) as substrates. After the functionalization of the CNTs, Pt nanoparticles were attached to their surface, providing the necessary catalytic activity for the further deposition of a very uniform and homogeneous layer of Ni (Grzelczak et al., 2007). During the Ni deposition process, the CNT@Pt/Ni nanocomposites were exposed to an oxygen-rich environment, promoting the formation of stable NiO outer shells, thus favoring the appearance of interfacial FM–AFM exchange

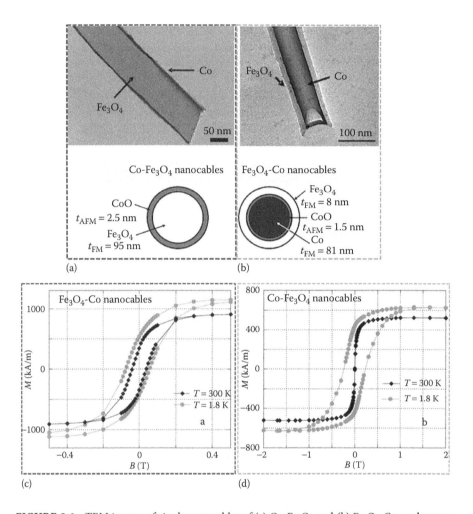

FIGURE 9.6 TEM images of single nanocables of (a) Co-Fe₃O₄ and (b) Fe₃O₄-Co, and respective schematic representations of their cross-sections. (c) Magnetization curves for Fe₃O₄-Co and (d) Co-Fe₃O₄ nanocables measured at 300 K and 1.8 K (after field cooling in 1 kOe). Note the different field scales. (Reprinted with permission from Kazakova, O. et al., *Phys. Rev. B*, 74, 184413. Copyright 2006 by the American Physical Society.)

anisotropy. Each NW was then considered a heterogeneous magnetic system consisting of FM Ni nanocrystals surrounded by AFM NiO shells. The magnetic properties of the CNT@Pt/Ni/NiO nanocomposites were then measured in the solid state, as nanopowders. Figure 9.7a shows the magnetic hysteresis loops measured at 25, 100, 200, and 300 K, after zero-field-cooled (ZFC). The large values of the coercive field (H_c), which increase at lower temperatures (Figure 9.7b), indicate that the magnetic moments are in a blocked state in the Ni/NiO nanostructures. When cooling the system under an applied magnetic field of $H_{FC} = 5$ kOe, the hysteresis loops are shifted with respect to the

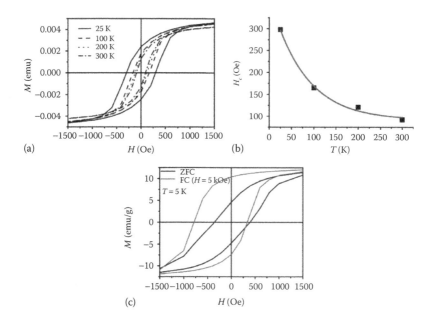

FIGURE 9.7 (a) Hysteresis loops of Ni/NiO-coated CNTs/Pt nanocomposites measured at different temperatures after ZFC, (b) temperature dependence of the coercivity (H_c), and (c) magnified view of the magnetic hysteresis loops measured at 5 K after ZFC and FC under an applied magnetic field of 5 kOe. (Salgueiriño-Maceira, V. et al., *Adv. Funct. Mater.*, 18, 616–621, 2008. Copyright Wiley-VCH Verlag GmbH & Co. KGaA. Reproduced with permission.)

field axis (Figure 9.7c), confirming the presence of a unidirectional anisotropy due to the EB coupling at the interface of the FM–AFM materials.

A year later, EB in Co/CoO NWs was reported in two different systems: (1) Co/CoO NW arrays electrodeposited in polycarbonate porous templates (De La Torre Medina et al., 2009) and (2) Co/CoO dispersed NWs synthesized by chemical reduction (Maurer et al., 2009). In the first case, Co NWs were fabricated by a three-probe electrodeposition technique inside 30-nm-diameter pores of homemade 21-μm-thick track-etched polycarbonate templates (De La Torre Medina et al., 2009). By keeping the samples in air for long periods of time, an AFM oxide layer was formed between the wires and the membrane, creating a FM–AFM cylindrical interface at the wires' surface. Magnetic hysteresis loops were then measured at 10 K after field cooling the samples at −15 kOe from a temperature $T > T_N$ (T_N = 291 K for bulk CoO). Figure 9.8a shows the hysteresis loops obtained for non-oxidized Co NW arrays (carried out just after the electrodeposition process) and oxidized Co/CoO NW arrays (measured 804 days after electrodeposition). The magnetic field H was applied parallel to the long axis of NWs. The hysteresis loop of the non-oxidized Co NWs is centered at $H = 0$, thus indicating the absence of oxidation, since no EB effect is measured. On the other hand, the hysteresis loop measured after 804 days shows an increase in the coercive field H_c and the presence of an EB field

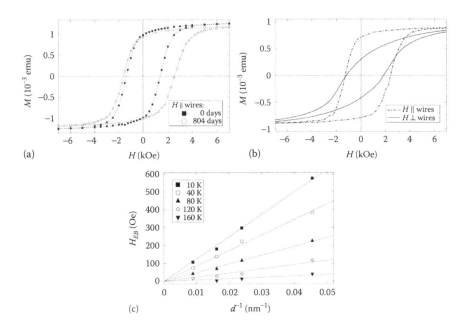

FIGURE 9.8 Magnetic hysteresis loops measured for Co NW arrays embedded in polycarbonate templates. Measurements were carried out in (a) a freshly prepared sample measured with the field applied along the wires at 0 (filled squares) and 804 (empty squares) days after fabrication of the samples, (b) a 5-year-old sample, with the field applied parallel (dash-dotted line) and perpendicular (continuous line) to the wires, and (c) EB field as a function of the inverse of the wire's magnetic core diameter, d, for 5-year-old Co NW arrays measured at different temperatures. The dotted lines are linear fits to the data. (Reprinted with permission from De La Torre Medina, J. et al., *J. Appl. Phys.*, 106, 023921. Copyright 2009, American Institute of Physics.)

(H_{EB}) of 455 Oe in the opposite direction to the cooling field ($H_{FC} = -15$ kOe). The wires' oxidation is thus identified from the hysteresis loop shift or H_{EB} measured after field cooling. In fact, H_{EB} can be used to characterize the oxidation state of the samples as a function of time, as it is usually an increasing function of the AFM thickness (Nogués and Schuller, 1999).

Figure 9.8b shows the hysteresis loops measured along the parallel and perpendicular directions in a 5-year-old sample. The H_{EB} obtained along the parallel direction of applied magnetic field is 578 Oe and thus higher than the H_{EB} measured for the 804-day-old sample. This means that the oxidation rate of these nanocomposite magnetic systems, under standard atmospheric conditions, is delayed in time, and a steady oxidation state is reached only after a few years, creating an approximately 4-nm-thick CoO layer. Moreover, Figure 9.8b also illustrates a larger H_{EB} when H is applied parallel to the wires, indicating that a stronger EB coupling is taking place and that the FM–AFM interface lies preferentially along the wires' axis. Five-year-old samples with different Co NW diameters d were also analyzed. Figure 9.8c shows the linear behavior of H_{EB} with d^{-1} at different temperatures, evidencing the interfacial nature of the EB coupling for a fixed oxide thickness. In addition, as T approached 160 K, the slope variation of H_{EB} with d^{-1} became very small, suggesting that the blocking temperature T_B of the AFM layer was around 200 K and thus lower than the T_N of the bulk CoO. This was mainly attributed to the thin nature of the oxide layer formed at the wire/polymer cylindrical interface.

The other work on the EB effect in Co/CoO core–shell NWs studied bundles of dispersed wires (Figure 9.9a) with smaller dimensions (average diameters and lengths of 15 and 130 nm, respectively). The Co NWs were synthesized by reduction in liquid polyol (Ung et al., 2005; Soumare et al., 2008; 2009). By keeping in their polyol solution, NWs were well preserved from oxidation (non-oxidized Co wires, used as reference samples). However, when the Co powders were dried and exposed to air, the wires oxidized at their surface, reaching a stable magnetic state after a few weeks through a passivation mechanism (Figure 9.9b) (Gangopadhyay et al., 1993; Gallant and Simard, 2005). Figure 9.9c shows the evolution of the coercive ($\mu_0 H_c$) and EB ($\mu_0 H_{EB}$) fields for both systems (oxidized and non-oxidized Co NWs), as a function of temperature. The hysteresis loops were measured after field cooling the samples under 50 kOe. In the case of non-oxidized Co NWs, no EB is observed and H_c decreases monotonously with increasing temperature, as expected. However, for the Co/CoO NWs, an EB field appears below $T_B = 100$ K, reaching 1 kOe at low temperatures. The most striking feature is the non-monotonous behavior of H_c with increasing temperature, decreasing to a local minimum at T_B, then increasing and reaching a maximum at about $T_N = 200$ K, and finally decreasing again until reaching room temperature (open circles in Figure 9.9c). The unexpected increase of H_c between T_B and T_N was explained by the presence of superparamagnetic fluctuations of the AFM CoO grains. Below T_N, the AFM moment

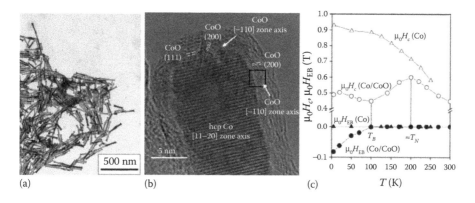

(a)   (b)   (c) T (K)

FIGURE 9.9 (a) TEM image of Co NWs, (b) high-resolution TEM image of the tip of a Co wire showing the local structure of the Co wire surrounded by a CoO shell (1.2 nm in thickness), and (c) temperature dependence of the coercive field $\mu_0 H_c$ (open symbols) and the exchange-biased field $\mu_0 H_{EB}$ (filled symbols) for non-oxidized Co (triangles) and oxidized Co/CoO (circles) nanowires. ([a] Soumare, Y. et al., *Adv. Funct. Mater.*, 19(12), 1971–1977, 2009. Copyright Wiley-VCH Verlag GmbH & Co. KGaA. Reproduced with permission; [b] and [c] Reprinted with permission from Maurer, T. et al., *Phys. Rev. B*, 80(6), 064427, 2009. Copyright 2009 by the American Physical Society.)

fluctuations of the CoO freeze progressively as the temperature decreases (Scarani et al., 2000), leading to a maximum of the magnetic viscosity below T_B. The physical origin of the superparamagnetism could be attributed to a small fraction of uncompensated spins at the FM–AFM interface. Figure 9.10 summarizes the proposed scenario for the magnetization and relaxation processes occurring in oxidized NWs, in which three different regimes can be identified: (1) below the blocking temperature ($T < T_B$), the CoO particles are blocked and a finite H_{EB} appears, (2) for $T_B < T < T_N$, the CoO particles are antiferromagnetically ordered but subjected to superparamagnetic fluctuations, and (3) for $T > T_N$, the CoO shell is not magnetically ordered and there is no effective interaction between the FM NW core and the AFM shell.

A few years later, EB effects in Ni/NiO NWs and nanorods were again reported (Maaz et al., 2011; Hsu et al., 2012). Similar to the study performed by De La Torre Medina et al. (2009), Maaz et al. (2011) also used the H_{EB} to investigate the oxidation rate at the surface of Ni NWs embedded in a polycarbonate template. The Ni NWs with diameters and lengths of around 14 nm and 24 µm, respectively, were electrodeposited at the pores of homemade etched ion-track polycarbonate templates. Magnetic hysteresis loops were then measured as a function of time (starting from the as-fabricated to 2-year-old samples). The measurements were made at 10 K after field cooling at 10 kOe, with the

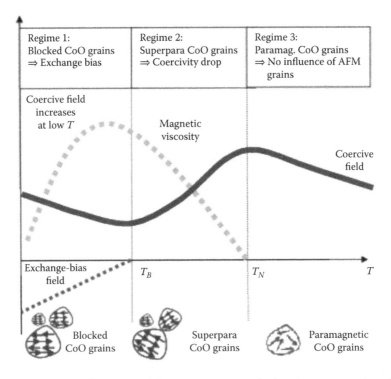

FIGURE 9.10 Proposed scenario of the magnetization and relaxation processes in oxidized nanowires. (Reprinted with permission from Maurer, T. et al., *Phys. Rev. B*, 80(6), 064427, 2009. Copyright 2009 by the American Physical Society.)

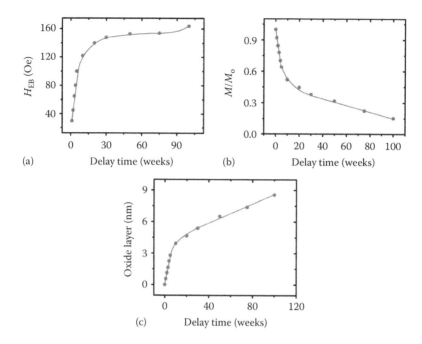

FIGURE 9.11 (a) Variation of the exchange bias field H_{EB} as a function of the delay time after the Ni NWs' fabrication. (b) Decrease in the normalized magnetization at saturation with time. (c) Increase in the NiO layer thickness with the age of the wires. (Reprinted with permission from Maaz, K., S. et al., *J. Appl. Phys.*, 110, 013908. Copyright 2011, American Institute of Physics.)

magnetic field applied parallel to the wires' axis. After only a few weeks, a shift in the hysteresis loops was already observed and an EB field was measured. Figure 9.11a shows the increase of the H_{EB} with the delay time. A rapid increase of H_{EB} within the first 15–20 weeks can be observed, indicating a rapid oxidation of the wires in the initial weeks. After that, the increase in H_{EB} becomes slower and almost constant, indicating a steady state oxidation, as the outer layers prevent the rapid oxidation of Ni at the center of the NWs. By measuring the normalized magnetization at saturation as a function of time (Figure 9.11b), a continuous decrease has been recorded due to the natural oxidation of Ni into NiO at the surface of the wires. From the drop of the magnetization, the thickness of the oxide layer was then estimated at different stages and plotted in Figure 9.11c as a function of the delay time.

On the other hand, Hsu et al. (2012) studied the effect of frozen spins at the Ni/NiO interface of free-standing Ni/NiO core–shell nanorods on an Si substrate. The Ni nanorods with 70 nm in diameter and 350 nm in length were prepared on an Si substrate by template-assisted electroless-plating, using nanoporous alumina membranes (Figure 9.12a). For the subsequent controlled oxidation of the Ni nanorods' surface, the alumina templates were removed and the Ni/Si samples were annealed at 623 K in air, for 15 and 30 min, creating Ni/NiO core–shell structures with NiO thicknesses of approximately 6 and 10 nm, respectively. Non-oxidized Ni nanorods were also used as

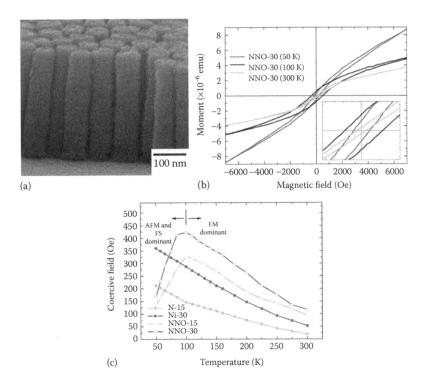

FIGURE 9.12 (a) SEM image of the Ni nanorods standing freely on an Si substrate. (b) Selected temperature-dependent magnetic hysteresis curves of the Ni/NiO nanorods oxidized for 30 min. (c) Temperature-dependent H_c of the reference non-oxidized Ni nanorods (N-15 and Ni-30, heat-treated for 15 and 30 min, respectively) and the oxidized Ni/NiO nanorods (NNO-15 and NNO-30, annealed for 15 and 30 min, respectively). Vertical line on top of NNO-30 marks the boundary for the AFM (below 100 K) and FM (above 100 K)-dominant areas. (Reprinted with permission from Hsu, H. et al., *J. Appl. Phys.*, 111(6), 063919. Copyright 2012, American Institute of Physics.)

reference samples, in which, to prevent the formation of NiO, the alumina template was not removed and similar heat treatments were performed, but this was done in a nitrogen atmosphere. Magnetic hysteresis loops measured at 50 K after field cooling the Ni/NiO nanorod arrays at ±30 kOe illustrated a vertical shift in magnetization, signaling the existence of frozen spins. These were likely the result of Ni spins locked to the NiO at the Ni/NiO interface. Exchange bias was also measured as a function of temperature. The EB field was found to decrease with temperature, following the same trend as the NiO AFM layer thickness. Figure 9.12b shows selected magnetic hysteresis loops at different temperatures, in which a non-monotonous temperature dependence of coercivity can be observed. For the non-oxidized samples, H_c decreases with temperature, as expected. However, for the Ni/NiO nanorods, H_c first increases until 100 K and then decreases up to room temperature (Figure 9.12c), as similarly reported by Maurer et al. (2009) in Co/CoO NWs. In this work, the suppression of H_c at 50 K, which corresponds to the largest values of H_{EB} and frozen spins, was explained by the dominance of NiO. When reaching 100 K, the highest values of H_c and saturation magnetization are observed, indicating the restoration of the FM phase, as, at this temperature (and above), Ni dominates again the properties of the rods. In addition, the disappearance of H_{EB} at approximately 100 K could also be linked to the vanishing of the frozen spins and the NiO pinning effect on Ni, yielding magnetic frustration at the core–shell interface and thus enhancing H_c.

Finally, in 2013, the first studies on EB effects in ordered arrays of core–shell FM/ AFM NTs were reported (Proenca et al., 2013a, b). The Co NTs and NWs were electrodeposited inside homemade nanoporous alumina templates with hexagonally ordered arrays of 40-nm-diameter pores. Nanotubular FM/AFM bilayered structures were then obtained after the natural oxidation of the inner Co NT walls, forming Co/CoO

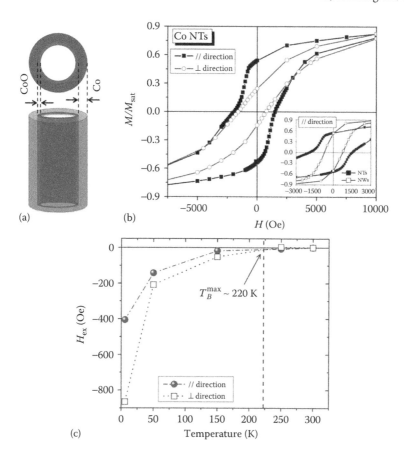

FIGURE 9.13 (a) Schematic representation of a Co/CoO NT. (b) Magnetic hysteresis loops of Co/CoO NT arrays measured at 6 K along both directions of the applied magnetic field. Inset shows a comparison between the hysteresis loops of Co/CoO NT and Co NW arrays, evidencing the absence of loop shift in the NW arrays. (c) Temperature dependence of the EB field when applying the magnetic field parallel (filled symbols) and perpendicular (open symbols) to the tube long axis. (Reprinted with permission from Proenca, M. P. et al., *Phys. Rev. B*, 87(13), 134404, 2013a. Copyright 2013 by the American Physical Society.)

NTs (Figure 9.13a). Temperature-dependent magnetic measurements evidenced the existence of EB coupling between the Co FM outer wall and the CoO AFM inner wall of the NTs. The magnetic hysteresis loops were measured at different temperatures by field cooling the samples in 50 kOe, with the magnetic field applied parallel and perpendicular to the wire/tube axis. Figure 9.13b shows the hysteresis loops measured at 6 K, illustrating a loop shift along both directions of applied magnetic field only for the tubular sample. This evidences the partial oxidation of the walls of Co NTs, forming an FM–AFM bilayered nanostructure with EB coupling, while no oxidation is observed in the Co NW arrays, owing to the protection of the NW walls by the alumina template. The temperature dependence of the EB field allowed the estimation of a maximum blocking temperature T_B^{max} of around 220 K, similar to the values reported by De La Torre Medina et al. (2009) and Maurer et al. (2009) for Co/CoO NWs (Figure 9.13c). An interesting behavior illustrated in Figure 9.13c is the higher EB field value obtained when applying the magnetic field perpendicular to the NTs' long axis. Most of the previous works on the study of longitudinal and perpendicular EB in FM–AFM multilayers report higher H_{EB} values when both the cooling and measurement fields are applied along the easy axis of magnetization of the FM layer (Maat et al., 2001; Sun et al., 2003; Phuoc and Suzuki, 2008; De La Torre Medina et al., 2009; Shipton et al., 2009). In this case, in which the easy axis of magnetization of the Co NTs is in the parallel direction, the EB field along the parallel direction is smaller than that along the perpendicular direction. This difference was attributed to the microstructure of the distinct crystalline orientations of both the Co NT walls and their respective oxidized layers (Maat et al., 2001; Phuoc and Suzuki, 2008).

The training effect of the EB field, that is, the reduction of the magnitude of the EB field with the number of hysteretic cycles performed, was also studied as a function of temperature, after field cooling the Co/CoO NT arrays in 50 kOe along both the parallel and perpendicular directions of the applied magnetic field (Proença et al., 2013b). The variation of the EB field with the number of magnetic cycles, n, is represented in Figure 9.14a. A monotonic decrease of the EB effect, when cycling the magnetic field through consecutive loops, is seen, corresponding to the training effect. This was fitted by using the recursive Binek formula (Binek, 2004):

$$H_{EB}^{n+1} - H_{EB}^{n} = -\gamma \left(H_{EB}^{n} - H_{EB}^{\infty} \right)^3 \tag{9.1}$$

where H_{EB}^{n} (H_{EB}^{∞}) is the EB field at the nth cycle (in the limit of an infinite number of cycles) and γ describes the characteristic decay rate of the training behavior. The γ values obtained are one to two orders of magnitude smaller than those previously reported for Co/CoO bilayers (Ali et al., 2012). This can be interpreted as the existence of a stronger coupling between the AFM and FM layers when shaped into a core–shell nanotubular structure as compared with spherical core–shell or multilayer nanostructures.

Bimodal T_B distributions were also measured along both the parallel and perpendicular directions (more details can be found in Proença et al., 2013a). Two peaks were identified: one at very low T and the other centered at $T \sim 87$ K and extending to higher

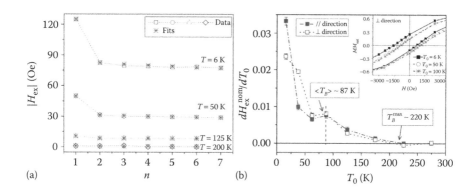

FIGURE 9.14 (a) Dependence of the EB field on the number of magnetic hysteresis cycles (n), measured at different temperatures after field cooling in 50 kOe, by applying the magnetic field perpendicular to the tube axis, and respective fits using Equation 9.1. (Reprinted with permission from Proenca, M. P. et al., *J. Appl. Phys.*, 114, 043914, 2013b. Copyright 2013, American Institute of Physics.) (b) Representation of the T_B distribution of the CoO AFM grains formed at the inner Co NT walls, by illustrating the dependence on the intermediate temperature (T_0) of the derivative of normalized EB field, along both the parallel (filled symbols) and perpendicular (open symbols) directions. Inset shows the magnetic hysteresis loops measured at 6 K along the perpendicular direction for $T_0 = 6$, 50, and 100 K. (Reprinted with permission from Proenca, M. P. et al., *Phys. Rev. B*, 87(13), 134404, 2013a. Copyright 2013 by the American Physical Society.)

temperatures (Figure 9.14b). The low-T peak was ascribed to interfacial spin-glass phases, while the high-T peak was attributed to the thermal reversal of AFM grains (Proenca et al., 2013a).

Exchange bias in alloyed NWs has also been recently reported by Chandra et al. (2013), Buchter et al. (2015), and Liébana-Viñas et al. (2015). Chandra et al. (2013) studied the EB effect in hydrothermally synthesized randomly distributed single-crystalline $La_{0.5}Sr_{0.5}MnO_3$ NWs with diameters of 20–50 nm and lengths between 1 and 10 µm. The core of the NWs was phase-separated, creating a double exchange-driven ferromagnetism in the AFM matrix, while the surface was found to be composed of disordered magnetic spins. The NWs underwent a paramagnetic-FM transition at $T_C = 310$ K, followed by an FM–AFM transition at $T_N = 210$ K. Usually, the development of EB in nanostructures can be described by considering a core–shell formation of different magnetic phases, such as the FM–AFM interface found in these NWs. However, the NWs were found to exhibit no EB above 42 K, even though the phase coexistence persisted up to T_N. In this case, the EB effect was attributed to a coupling between the FM core and the spin-glass shell, which thus only appeared below the surface spin freezing temperature of around 42 K.

Buchter et al. (2015), on the other hand, studied the EB effect in individual permalloy (Py) NTs (Figure 9.15a), using a nanometer-scaled superconducting interference device magnetometer and a cantilever torque sensor. The NTs were fabricated by thermally evaporating 30 nm of polycrystalline Py onto GaAs NWs with a hexagonal

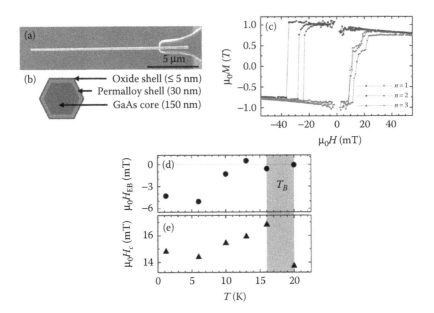

FIGURE 9.15 (a) SEM image of an individual Py NT on a cantilever. (b) Schematic cross-section of a Py NT grown on a GaAs NW. (c) Magnetic hysteresis loops measured at the 1st, 2nd, and 10th cycle. Temperature dependence of (d) the EB field and (e) the coercivity, indicating the range of T_B by the gray shading. (Reprinted with permission from Buchter, A. et al., *Phys. Rev. B*, 92, 214432, 2015. Copyright 2015 by the American Physical Society.)

cross-section, as depicted in Figure 9.15b. The GaAs NWs were grown by molecular beam epitaxy, having lengths of 10–20 μm and widths of around 150 nm at their widest point. Magnetic hysteresis loops were measured at 3.4 K after field cooling the sample in a magnetic field of 2 kOe applied parallel to the tube axis. Loop shifts were observed at a low temperature, indicating the presence of an EB coupling between the FM Py tubular shell and an AFM outer shell of native oxide. Training effect was also detected when measuring consecutive cycles (Figure 9.15c). The temperature dependence of the EB and coercive fields were also studied, from which a blocking temperature of around 18 K was determined (Figure 9.15d and e). This very low value of T_B suggested that the oxide layer formed was very thin (3–5 nm) and with a grainy and non-homogeneous structure.

Liébana-Viñas et al. (2015) studied the EB coupling in oxidized $Co_{80}Ni_{20}$ nanorods with mean diameters and lengths of 6.5 and 52.5 nm, respectively. The NWs were prepared by a polyol reduction of mixed cobalt and nickel acetates, and these were then exposed to the air. Figure 9.16 shows a high-resolution TEM image of the tip of an oxidized CoNi NW, illustrating the core–shell structure of a single-crystalline metallic core covered with a polycrystalline oxide shell. Temperature-dependent magnetic hysteresis loops of the dried powder of oxidized $Co_{80}Ni_{20}$ nanorods were measured after field cooling the samples in 45 kOe. Horizontal and vertical loop shifts and enhanced coercivity values were found at low temperatures, indicating

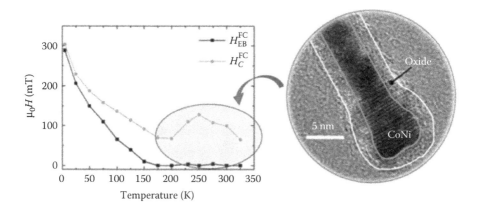

FIGURE 9.16 Temperature dependence of the coercive and EB fields of the CoNi nanorods after field cooling. High-resolution TEM image of an oxidized CoNi nanorod. (Reprinted with permission from Liébana-Viñas, S. et al., *Chem. Mater.*, 27, 4015–4022, 2015. Copyright 2015 American Chemical Society.)

the presence of FM–AFM interfaces exhibiting EB coupling. The EB field was found to decrease with increasing temperature, disappearing at around 175 K (Figure 9.16). A temperature-dependent local maximum of the coercive field at $T = 250$ K was identified (Figure 9.16), which originated from non-collinear spin orientations in the FM CoNi alloy core and the AFM Co-rich oxide shell. This effect had also been found by Maurer et al. (2009) in Co/CoO NWs and was explained by a three-regime model (Figure 9.10), which considered superparamagnetic fluctuations of the AFM grains in the Co-rich oxide shell. Note that such behavior is only expected when the magnetic easy axes of the FM core and the AFM shell are not collinear. Therefore, 1D core–shell nanostructures are ideal systems for the investigation of the influence of superparamagnetic fluctuations in EB.

To conclude this section on EB in FM/AFM 1D nanostructures, a recent work on Co/CoO coaxial core–shell NWs by Salazar-Alvarez et al. (2016) demonstrated a distinct way to manipulate the high-field magnetization in exchange-biased systems. The Co NWs with around 5-μm length were deposited by pulsed electrodeposition into the pores of anodic alumina templates with diameters of around 30 nm. The template was then removed by chemical etching in an NaOH aqueous solution, resulting in the formation of a CoO passivation layer on the Co NWs' surface with approximately 4-nm thickness. Temperature-dependent magnetic measurements of Co/CoO NWs, investigated with cooling and applied fields perpendicular to the wire axis, illustrated unexpected EB effects. In particular, the strong FM–AFM coupling resulted in striking differences between the FC and ZFC loops at high fields and a high-field irreversibility at large fields in ZFC loops (Figure 9.17). This work thus illustrates the tunability of the high-field magnetization by the EB coupling in core–shell 1D nanostructures when applying the magnetic field perpendicular to the long axis, providing additional degrees of freedom for the design of new spintronic devices.

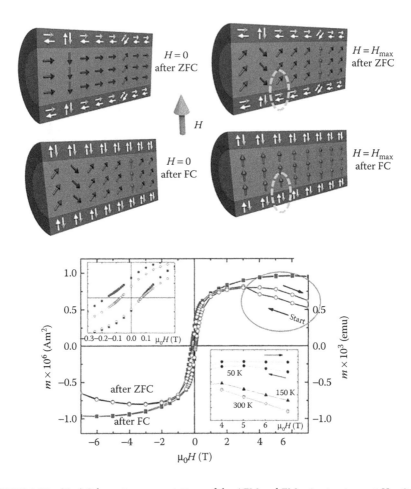

FIGURE 9.17 (Top) Schematic representations of the AFM and FM spin structures at $H = 0$ and H_{max} (where H_{max} is the maximum measuring field) after ZFC and FC with the magnetic field applied perpendicular to the wire axis. (Bottom) Out-of-plane hysteresis loops for free-standing Co/CoO core–shell NWs measured at 10 K after either ZFC or FC in 50 kOe from 300 K to 10 K. The top inset shows a magnification of the low field region of the loops. The bottom inset shows the high-field magnetization measured at 50, 150, and 300 K, after ZFC. No diamagnetic correction has been performed to any of the loops. The lines are guides to the eye. (Reprinted with permission from Salazar-Alvarez, G. et al., *ACS Appl. Mater. Interfaces*, 8(34), 22477–22483. Copyright 2016 American Chemical Society.)

9.4 Antiferromagnetic Nanowires with Uncompensated Spins at the Surface

Exchange bias in AFM nanoparticles is an already-established research topic, with a large number of systems and conditions studied. In this case, magnetic properties are dominated by surface effects, and EB is usually associated with the presence of uncompensated surface spins. However, the study of EB in AFM NWs and NTs is still emerging. The AFM NWs and NTs also exhibit quite interesting magnetic properties, owing to their inherent shape anisotropy and large surface-to-volume ratio.

Arrays of AFM Co_3O_4 NWs were synthesized following a nanocasting route that uses mesoporous silica as a mold to grow the intended nanostructures (Salabas et al., 2006). The fabricated sample comprised randomly oriented arrays of Co_3O_4 NWs with lengths up to 100 nm and an average diameter of 8 nm (Figure 9.18a and b). The ZFC and FC magnetization curves confirmed the presence of the AFM phase of Co_3O_4 with a Néel temperature of approximately 30 K (near the bulk value of 40 K). A large increase in the magnetization at low temperatures was also observed, which was ascribed to the presence of uncompensated spins at the surface of the NWs that become frozen at low temperatures. Furthermore, hysteretic cycles performed at 2 K after FC showed a horizontal sift of the magnetization in the measured hysteretic cycles (H_{EB} = 240 Oe). A vertical shift from the origin toward negative fields and enhanced coercivity were also observed (Figure 9.18c). A training effect was also measured in such Co_3O_4 NWs (Figure 9.18d). As usually appears in this effect, the largest reduction occurred between the first and second cycles, which was followed by a smother behavior. Furthermore, with increasing cooling field (the field applied during cooling through the Néel temperature), the EB field also increased (Figure 9.18e). All these results suggest the existence of an exchange coupling between AFM core spins and FM-like surface spins, with properties similar to those of spin-glass systems with multiple stable configurations (see also open loop in Figure 9.18c). Such surface spins can be frozen uncompensated spins that originate EB (and thus the horizontal loop shift) or reversible spins that are aligned by the magnetic field and contribute to the coercivity enhancement. With cycling, or by applying different cooling fields, it is possible to change the magnetic configuration of the surface spins of the AFM NWs owing to the instability of the spin structure and thus to tune the EB effect. Finally, temperature-dependent measurements showed the disappearance of EB just below the NWs' Néel temperature (Figure 9.18f). A core–shell magnetic model, including intracore, intraplane, core–plane, and shell-shell interactions, was recently developed based on first-principles calculations to explain the properties of Co_3O_4 NWs (Chen et al., 2013). This model was able to account for the spin-glass behavior and the EB effect experimentally encountered in this type of structures.

Exchange bias was also seen below a blocking temperature of 19 K in single-crystalline CuO NWs (65 nm in diameter and 5.5 µm in length) obtained by thermal oxidation of Cu (Díaz-Guerra et al., 2010). Magnetization measurements suggested a spin-glass-like behavior at low temperatures arising from uncompensated spins at the surface of the CuO NWs. After field cooling, the hysteretic loops showed both vertical and horizontal shifts, revealing the presence of EB (H_{EB} = 230 Oe at 2 K and decreases to 0 above the blocking temperature). Open loops, training effect, and a strong dependence of the EB on the cooling field were also found and explained as signatures of the spin-glass-like nature of the AFM NW system, with a spin-glass-like shell of uncompensated surface spins having a net magnetic moment embedded in an AFM core.

Bundles of AFM goethite (α-FeOOH) NWs with 118-nm length and 12-nm diameter were prepared by using a modified Massart method, based on the co-precipitation of ferrous and ferric ion solutions (Mariño-Fernández et al., 2011). Goethite is one of the most

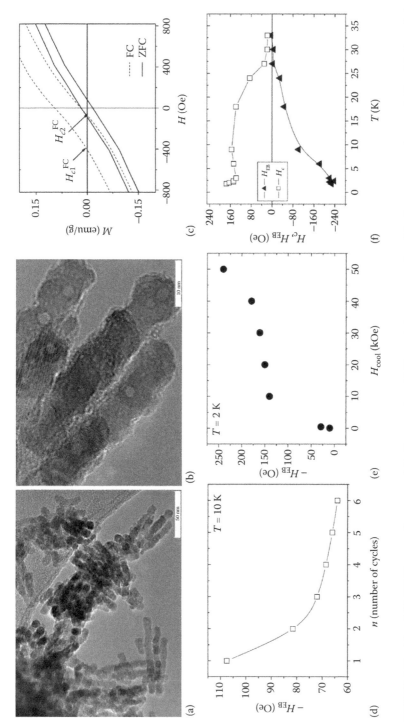

FIGURE 9.18 (a) and (b) TEM images of nanocasted Co_3O_4 nanowires. (c) Magnetic hysteretic cycles of Co_3O_4 nanowires after zero field and field-cooled conditions. (d) Cycling, (e) cooling field, and (f) temperature dependences of the exchange bias field. (Reprinted with permission from Salabas, E. L. et al., *Nano Lett.*, 6(12), 2977–2981. Copyright 2006 American Chemical Society.)

stable iron oxides at room temperature. It is an AFM material with a Néel temperature of approximately 400 K. Mariño-Fernández et al. (2011) found the presence of a blocking temperature of about 40 K in α-FeOOH NWs associated with the paramagnetic behavior of surface spins arising from structural disorder superimposed to an AFM component. An EB field of 30 Oe at 6 K after field cooling under 50 kOe was observed in these structures due to the presence of an exchange coupling interaction between uncompensated surface spins and the AFM core. Pure goethite single-crystalline AFM NWs were also prepared inside polycarbonate membrane (175 nm in pore diameter and 6 μm in length) by using electrodeposition (Llavona et al., 2013). Magnetic hysteretic cycles measured at low temperatures showed the presence of a strong magnetic anisotropy and the EB effect ascribed to the intrinsic properties of goethite NWs due to the absence of other iron oxide phases. The EB effect then results from the interaction between the AFM core of the NWs and the uncompensated spins from the surface.

Akaganeite (β-FeOOH) AFM NWs (80 nm in length and 15 nm in diameter) with a Néel temperature of 259 K were also produced by using a microwave process (Tadic et al., 2015). Magnetization cycles revealed an EB effect and clear hysteresis below T_N, attributed to the presence of a 1-nm-thick disordered shell at the surface of the akaganeite NWs and its coupling with the AFM core.

Arrangements of AFM pure α-MnO₂ NWs (20 nm in diameter and 1 μm in length) were fabricated (Figure 9.19a and b) by using a hydrothermal method, and spin-glass-like behavior was confirmed in this type of structures (Li et al., 2014). Temperature-dependent magnetization measurements set the AFM behavior of the α-MnO₂ NWs with $T_N = 13$ K. At 5 K and after FC under 10 kOe, the measured magnetization loops revealed the presence of hysteresis and EB ($H_{EB} = 456$ Oe; Figure 9.19c), associated with the coupling of a surface spin-glass shell and the α-MnO₂ core. The training effect (Figure 9.19d) was also analyzed in detail by using different models. The obtained training effect data were initially fitted to a known phenomenological model (Binek, 2004; Ventura et al., 2008):

$$\mu_0 H_{EB}^n - \mu_0 H_{EB}^\infty = \frac{k}{\sqrt{n}} \tag{9.2}$$

where:
   n is the number of cycles
   k is a system-dependent constant

This model is valid for $n \geq 2$, as the EB field reduction from the first to the second field is usually associated with a different mechanism (Hoffmann, 2004) than the subsequent evolution. This can be seen in Figure 9.19d, where fits were performed by using Equation 9.2, considering $n \geq 1$ and $n \geq 2$, and only the latter resulted in a good correlation with the experimental data. To account for the variation of the EB in the whole range, Li et al. (2014) used a two-term exponential model. Although the monotonic decrease of H_{EB} with n is related with changes in the interfacial spin structure between the AFM core and its FM shell, the rearrangement of AFM domains should also be considered (Li et al., 2014):

$$\mu_0 H_{EB}^n = \mu_0 H_{EB}^\infty + A_f e^{-n/P_f} + A_i e^{-n/P_i} \tag{9.3}$$

FIGURE 9.19 (a) SEM and (b) TEM images of α-MnO$_2$ nanowires, (c) hysteretic cycles of α-MnO$_2$ nanowires displaying exchange bias, and (d) cycling dependence of the exchange bias and coercive fields. (Reprinted by permission from Macmillan Publishers Ltd: Scientific Reports, Li, W. et al., *Scientific Reports*, 4, 6641, 2014. Copyright 2014.)

where the parameters A_f (A_i) and P_f (P_i) are associated with changes in the frozen AFM spins (interfacial disordered spins). The experimental data were well fitted by using this model for the whole number of cycles (Figure 9.19d). Note that, in the case of α-MnO$_2$ NWs, the uncompensated spins at the surface are constituted by uncoupled spins in the 3d^3 orbitals of the Mn^{4+} ions in the MnO$_6$ octahedra.

Finally, in the case of AFM NTs, EB was observed in polycrystalline Co$_3$O$_4$ NTs with 50–60 nm outer diameter, a wall thickness of 10 nm, and a length of 300 nm, fabricated by using a gas-bubble-induced self-assembly technique (Figure 9.20a–c) (Tong et al., 2013). A unidirectional anisotropy associated with the interaction between FM and AFM components was also obtained in this type of nanostructures, with a maximum H_{EB} of 145 Oe at 10 K, disappearing above the Néel temperature of 25 K (Figure 9.20d). The obtained EB field value was higher than those observed in the NW case, owing to the shape and internal stress of the fabricated vertical array of Co$_3$O$_4$ NTs. The Co$_3$O$_4$ NTs showing ferrimagnetic behavior were also fabricated inside nanoporous alumina templates by using chemical vapor deposition (Shen et al., 2008).

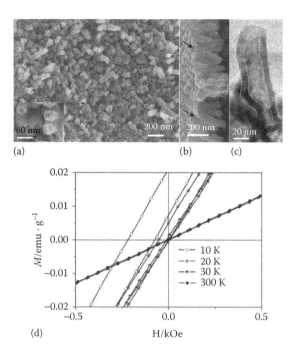

FIGURE 9.20 (a) Top and (b) tilted SEM images of Co_3O_4 nanotubes, (c) corresponding high-magnification field emission SEM image, and (d) hysteretic cycles of Co_3O_4 nanotubes for different temperatures. (Tong, G. et al., *Adv. Funct. Mater.*, 23, 2406–2414, 2013. Copyright Wiley-VCH Verlag GmbH & Co. KGaA. Reproduced with permission.)

9.5 Ferromagnetic–Multiferroic Core–Shell 1D Nanostructures

The coexistence of electric and magnetic orders in a so-called multiferroic material offers large prospects for their application in novel memories, logic devices, transducers, and field sensors (Wang et al., 2003; Hur et al., 2004; Eerenstein et al., 2006; Zhao et al., 2006; Chu et al., 2008). However, the number of known multiferroic materials is still very scarce, and alternative routes to control magnetism by using electric fields, or vice-verse, are also being followed. In that respect, a more general phenomenon is the magnetoelectric coupling of magnetic and electrical properties that can even arise when materials with different ferroic properties are put into close contact.

The fabrication of ferromagnetic–multiferroic hybrid core–shell nanostructures displaying EB is an extremely recent topic, with only a small number of reports in the literature. In fact, the actual control of EB by an electrical field in this type of structures has not been demonstrated so far, and the obtained EB fields are usually very small. Given the broad classification of multiferroic materials, including antiferroic order (Eerenstein et al., 2006), one can combine materials, such as $BiFeO_3$ (also known as BFO), displaying

antiferromagnetism and ferroelectricity, to fabricate core–shell nanostructures displaying EB. The BFO is a commensurate ferroelectric (T_C = 1100 K) and an incommensurate
antiferromagnet (T_N = 640 K) at room temperature. It is thus one of the most promising
multiferroic materials being studied and, given its antiferromagnetism, is often used in
a two-phase system, in combination with a ferromagnetic material (Dho et al., 2006). Shi
et al. (2014) used nanoporous alumina templates with pore diameters of 120 and 300 nm
to report the first fabrication of core–shell ferromagnetic–multiferroic Ni-BFO coaxial
nanostructures. Single-phase perovskite BFO NTs with wall thicknesses of approximately 20 nm were prepared by dipping the nanoporous templates for 15 minutes in
a sol–gel solution constituted by high-purity $Bi(NO_3)_3 \cdot 5H_2O$ and $Fe(NO_3)_3 \cdot 9H_2O$ with
a molar ratio of 1.15:1 dissolved in 2-methoxyethanol with a concentration of 0.3 M.
The samples were then annealed at 873 K, with both X-ray diffraction and polarization
hysteresis loops confirming the presence of ferroelectric BFO. The Ni NWs or NTs were
then electrodeposited inside the BFO NTs. Figure 9.21 shows SEM and TEM images of
the BFO NTs with Ni NTs (Figure 9.21a and b) and NWs (Figure 9.21c and d) inside the
porous template with 300-nm-diameter pores.

The magnetic properties revealed interesting differences between the fabricated samples.
The hysteretic cycles of the Ni(NTs)/BFO structures with an outer diameter of 300 nm

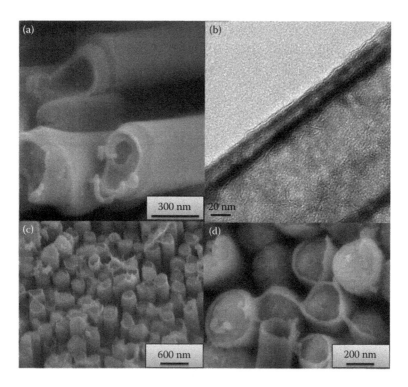

FIGURE 9.21 (a) SEM and (b) TEM images of Ni(nanotubes)/BFO 1D structures. (c) and (d)
SEM images of Ni(nanowires)/BFO 1D structures. (Shi, D.-W. et al., *Nanoscale*, 6, 7215–7220, 2014.
Reproduced by permission of The Royal Society of Chemistry.)

revealed a magnetic easy axis perpendicular to the long axis of the NTs (Figure 9.22a). Furthermore, hysteretic cycles measured under magnetic fields both parallel and perpendicular to the long axis revealed a loop shift, denoting the existence of an EB coupling between the AFM shell and the FM core, without the need for a field-cooling process.

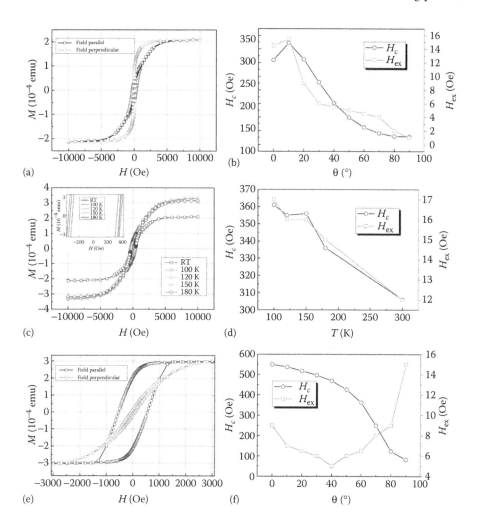

FIGURE 9.22 (a) Magnetization versus field hysteretic curves of Ni(nanotubes)/BFO 1D structures on 300-nm pores for magnetic fields parallel and perpendicular to the NTs long axis. (b) Corresponding angular dependence of the coercive and EB fields (with respect to the NTs axis). (c) Magnetization versus field curves of Ni(nanotubes)/BFO on 300-nm pores at different temperatures. (d) Corresponding dependence of the coercive and EB fields. (e) Magnetization versus field curves of Ni(nanowires)/BFO 1D structures on 120-nm pores for magnetic fields parallel and perpendicular to the NTs long axis. (f) Corresponding angular dependence of the coercive and EB fields. (Shi, D.-W. et al., *Nanoscale*, 6, 7215–7220, 2014. Reproduced by permission of The Royal Society of Chemistry.)

Such effect is related with the presence of uncompensated spin at the interface between the FM core (Ni) and the AFM shell (BFO). Angular dependent measurements of both the coercivity and EB field revealed a peak at $\theta = 10°$ with respect to the long axis of the core–shell structure (Figure 9.22b), which was explained as a transition between different reversal modes and the impact of the Ni and BFO crystallinity on the exchange interaction. Temperature-dependence measurements revealed the usual increase of coercive and EB fields with decreasing temperature (Figure 9.22c and d). On the other hand, magnetic measurements on the Ni(NWs)/BFO structures on the 300-nm-diameter porous template did not show any EB effect due to the large (~280 nm) thickness of the FM core. However, EB was again observed for Ni(NWs)/BFO on 120-nm-diameter pores, having an easy axis along the long NW axis (Figure 9.22e). In this case, a monotonous decrease of the coercivity with θ was found, associated with domain wall nucleation-driven magnetization reversal (Figure 9.22f). For the EB field, a minimum was observed for $\theta = 40°$ and was related with the spin structure at the Ni/BFO interface.

A similar fabrication route was followed to obtain 1D core–shell structures composed of CoPt NWs surrounded by a $Bi_{0.87}La_{0.13}FeO_3$ (also called BLFO) shell (Ali et al., 2015). To circumvent the formation of BFO parasitic phases, Ali et al. (2015) synthesized BLFO NTs with 20-nm wall thickness inside 120-nm-diameter pores of nanoporous alumina templates by using a sol–gel process. This was followed by the electrodeposition of the CoPt NWs core and an annealing at 973 K to remove impurity phases. Magnetic measurements showed a peak in the angular dependence of the coercivity for $\theta = 50°$, associated with a curling (coherent) reversal mechanism below (above) this critical angle (Figure 9.23a). On the other hand, the EB field showed a maximum for $\theta = 0°$ (corresponding to the easy magnetization axis), followed by a monotonous decrease with increasing θ (Figure 9.23a). Temperature-dependent measurements again showed an increase of both the coercive and EB fields with decreasing temperature, with the maximum H_{EB} value of 77 Oe obtained for 5 K (Figure 9.23b). The presence of intrinsic ferroelectricity in the fabricated BLFO shell structures was also confirmed by polarization measurements (Figure 9.23c).

The same group reported the fabrication of hybrid Co-BFO core–shell nanostructures on 120-nm-diameter pores, using similar methods to the one described previously (Ali et al., 2016). The fabricated structures, showing a well-defined easy axis along the NW's long axis due to the dominance of shape anisotropy, revealed increasing EB field values for increasing θ. Thus, a maximum EB was obtained for 90° (the hard axis), revealing an out-of-plane exchange coupling between the FM and AFM materials, without the need for a field-cooling process. The microscopic mechanisms behind this effect were still not revealed.

The substitution of Fe^{3+} ions by Co^{2+} is known to enhance the magnetic properties of BFO (Naganuma et al., 2008; Xu et al., 2009). For that reason, Javed et al. (2015) studied $Ni_{80}Fe_{20}$ (NWs/NTs)/BFO and $Ni_{80}Fe_{20}$ (NWs/NTs)/$BiFe_{0.95}Co_{0.05}O_3$ (also called BFC) core–shell nanostructures fabricated inside nanoporous alumina templates with pore diameters of 100 and 300 nm. Confirming the report of Shi et al. (2014), the NWs grown on the largest pores (300 nm) showed little influence of the 24-nm thick BFO shell. However, when grown on the smallest pores, a clear enhancement of the coercive field (Figure 9.24a and b) and an EB effect (Figure 9.24c and d) could be observed. Note that the maximum values of the coercive and EB fields were higher for all the structures fabricated with the

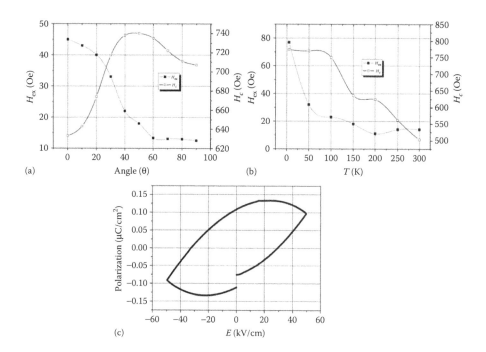

FIGURE 9.23 (a) Angular and (b) temperature dependences of the coercive and EB fields of CoPt/BFLO 1D structures on 120-nm pores, and (c) room temperature polarization versus electric field hysteresis loop of the reported BLFO NTs. (Ali, S. S. et al., *Nanoscale*, 7, 13398, 2015. Reproduced by permission of The Royal Society of Chemistry.)

BFC shell. For the case of the NiFe NWs, the decrease of coercivity with increasing θ shows that, in this case also, magnetization reversal is dominated by coherent rotation (Figure 9.24a). Regarding EB (Figure 9.24c), both BFO and BFC show a minimum for $\theta \approx 30°–40°$. However, while NiFe-BFO NWs have their maximum H_{EB} value when the magnetic field is applied along the easy axis ($\theta = 0°$), for NiFe-BFC, maximum H_{EB} occurs for H along the hard axis ($\theta = 90°$). This can be explained by a change in the spin structure of the interface between the Py ferromagnetic wires and the BFO/BFC NTs.

On the other hand, the NiFe (NTs)/BFC core–shell structure (grown on 300-nm pores) showed an easy axis perpendicular to the NTs' long axis. The angular dependence of the coercivity revealed a maximum value for both the BFO (at $\theta = 20°$) and BFC (at $\theta = 30°$) shells, demonstrating curling (coherent reversal) below (above) the critical angle (Figure 9.24b). A contrary behavior was seen for the angular dependence of the EB field for the BFO and BFC structures (Figure 9.24d). While for BFO, H_{EB} increases up to an angle of 50°, for BFC, the EB field decreases up to 60° and then increases slightly. The enhanced H_{EB} values obtained in the case of the NiFe NTs likely resulted from the smaller core thickness when compared with the NW structures.

Finally, note that hybrid multiferroic core–shell nanostructures constituted by a ferroelectric barium titanate ($BaTiO_3$) shell and a Co FM core demonstrated

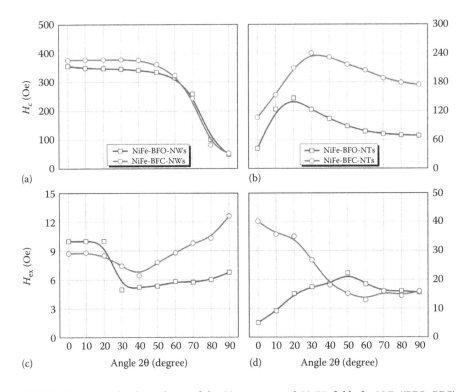

FIGURE 9.24 Angular dependence of the (a) coercive and (c) EB fields for NiFe/(BFO, BFC) nanowires. Angular dependence of the (b) coercive and (d) EB fields for NiFe/(BFO, BFC) nanotubes. (From Javed, K. et al., *Scientific Reports*, 5, 18203, 2015.)

magnetoelectric coupling and the possibility to tune the dielectric permittivity with an external magnetic field (Narayanan et al., 2012). However, since no AFM material was present, no EB was measured in this system.

9.6 Novel Applications and Conclusions

Exchange bias is one of the most important effects in spintronic applications. In thin films, it allows to pin the magnetization of a ferromagnet adjacent to an AFM layer and thus control the magnetic state (parallel or antiparallel) of spin valve and magnetic tunnel junction devices. Its first use in spin valves (Dieny et al., 1991) was behind the increase of areal density in hard disk drives in the 1990s. Its study and optimization in isotropic 1D nanostructures could lead to new applications in the fields of magnetic data storage, sensors, and biomedicine. Just as an example of one of the opened possibilities, a recent report used $Ir_{17}Mn_{83}$ (10 nm)/$Co_{70}Fe_{30}$ (7.5 nm) NTs to achieve deterministic transport of superparamagnetic beads (Ueltzhöffer et al., 2016). The NTs, with lengths of 300 μm, inner diameters of 10 μm, and wall thickness of 350 nm, were fabricated by using sacrificial germanium and pre-strained titanium layers (Figure 9.25a–c).

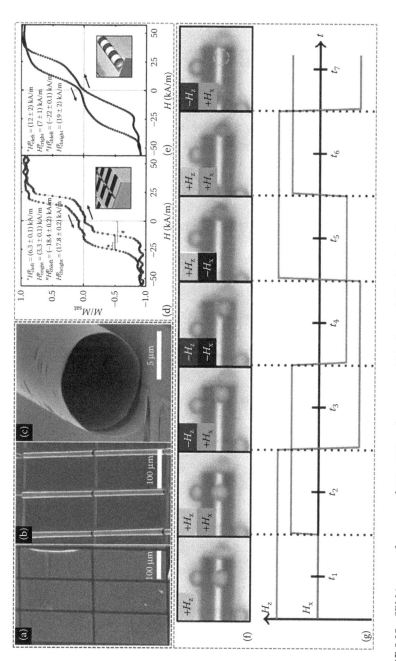

FIGURE 9.25 SEM images of an array of $300 \times 100\ \mu m^2$ structures (a) before and (b) after the rolling-up process. (c) SEM image of a rolled-over tube. Exchange bias measurements of the (d) thin films and (e) nanotubes after rolling up. (f) Motion of a superparamagnetic bead on top, at the edge, and inside the rolled-over nanotubes. (g) Corresponding magnetic field pulses applied. (Reprinted with permission from Ueltzhöffer, T. et al., *ACS Nano*, 10(9), 8491–8498. Copyright 2016 American Chemical Society.)

This allowed the rolling over of the AFM-FM bilayer and the fabrication of NTs with three AFM/FM sheets. Magnetic characterization using the magneto-optic Kerr effect (MOKE) showed that the EB properties are left mostly unchanged after the rolling-up process (Figure 9.25d–e). Using an appropriate magnetic field sequence (Figure 9.25g), it was then possible to move a micron-sized superparamagnetic particle (with velocities up to 220 µm/s) on top of the fabricated tube, place it at the tube entrance, and then move it inside the NT (Figure 9.25f). This proof of concept opens new applications in biotechnology and lab-on-chip systems for 3D particle transport.

To conclude, EB effects have been observed in a number of isotropic 1D nanostructures, including AFM-FM core–shell NWs and NTs, AFM NWs and NTs, and FM-multiferroic systems. In AFM/FM core–shell NWs and NTs, typical EB behaviors are observed, including temperature, thickness and cooling field dependencies, training effects, and the presence of bimodal blocking temperature distributions. However, the particular morphology of the high-aspect-ratio 1D nanostructures provides an enhanced tool to probe isotropic and surface effects. In that respect, the influence of FM locked spins at the FM–AFM interface or of the applied magnetic field direction on EB was thoroughly studied. In the case of AFM NWs and NTs, it is the presence of frozen uncompensated spins at the surface having properties similar to those of spin-glass systems that leads to the appearance of an EB-like effect. Finally, in FM-multiferroic structures, EB appears due to the AFM nature of the multiferroic material used. Although the modulation of EB with an electrical field in these systems was still not observed, it was already possible to observe the typical properties of exchange-biased multilayers in FM-multiferroic structures.

Acknowledgments

Mariana Proenca is thankful to FCT for post-doctoral grant SFRH/BPD/84948/2012 supported by funding POPH/FSE. João Ventura acknowledges financial support through FSE/POPH and project PTDC/CTM-NAN/3146/2014. The authors acknowledge funding from FCT through the Associated Laboratory IN, Institute of Nanoscience and Nanotechnology (POCI-01-0145-FEDER-016623).

References

Ali, S. R., M. R. Ghadimi, M. Fecioru-Morariu, B. Beschoten, and G. Guntherodt. Training effect of the exchange bias in Co/CoO bilayers originates from the irreversible thermoremanent magnetization of the magnetically diluted antiferromagnet. *Physical Review B*, 85:012404, 2012. doi:10.1103/PhysRevB.85.012404.

Ali, S. S., W. J. Li, K. Javed, D. W. Shi, S. Riaz, G. J. Zhai, and X. F. Han. Exchange bias in two-step artificially grown one-dimensional hybrid Co–BiFeO$_3$ core–shell nanostructures. *Nanotechnology*, 27:045708, 2016. doi:10.1088/0957-4484/27/4/045708.

Ali, S. S., W. J. Li, K. Javed, D. W. Shi, S. Riaz, Y. Liu, Y. G. Zhao, G. J. Zhai, and X. F. Han. Utilizing the anti-ferromagnetic functionality of a multiferroic shell to study exchange bias in hybrid core–shell nanostructures. *Nanoscale*, 7:13398, 2015. doi:10.1039/c5nr02977e.

Anagnostopoulou, E., B. Grindi, L.-M. Lacroix, F. Ott, I. Panagiotopoulos, and G. Viau. Dense arrays of cobalt nanorods as rare-earth free permanent magnets. *Nanoscale*, 8:4020–4029, 2016. doi:10.1039/c5nr07143g.

Apolinário, A., P. Quitério, C. T. Sousa, M. P. Proença, J. Azevedo, M. Susano, S. Moraes, P. Lopes, J. Ventura, and J. P. Araújo. Bottom-up nanofabrication using self-organized porous templates. *Journal of Physics: Conference Series*, 534:012001, 2014. doi:10.1088/1742-6596/534/1/012001.

Arshad, M. S., M. P. Proenca, S. Trafela, V. Neu, U. Wolff, S. Stienen, M. Vazquez, S. Kobe, and K. Z. Rozman. The role of the crystal orientation (c-axis) on switching field distribution and the magnetic domain configuration in electrodeposited hcp Co–Pt nanowires. *Journal of Physics D: Applied Physics*, 49:185006, 2016. doi:10.1088/0022-3727/49/18/185006.

Bachmann, J., J. Jing, M. Knez, S. Barth, H. Shen, S. Mathur, U. Gosele, and K. Nielsch. Ordered iron oxide nanotube arrays of controlled geometry and tunable magnetism by atomic layer deposition. *Journal of the American Chemical Society*, 129(31):9554–9555, 2007. doi:10.1021/ja072465w.

Batista, E. A., D. P. dos Santos, G. F. S. Andrade, A. C. Sant'Ana, A. G. Brolo, and M. L. A. Temperini. Using polycarbonate membranes as templates for the preparation of Au nanostructures for surface-enhanced Raman scattering. *Journal of Nanoscience and Nanotechnology*, 9:3233–3238, 2009. doi:10.1166/jnn.2009.209.

Binek, C. Training of the exchange-bias effect: A simple analytic approach. *Physical Review B*, 70:014421, 2004. doi:10.1103/PhysRevB.70.014421.

Brems, S., D. Buntinx, K. Temst, C. Van Haesendonck, F. Radu, and H. Zabel. Reversing the training effect in exchange biased CoO/Co bilayers. *Physical Review Letters*, 95(15):157202, 2005. doi:10.1103/PhysRevLett.95.157202.

Brinker, C. J. and G. W. Scherer. *Sol–gel Science: The Physics and Chemistry of Sol–gel Processing*. Academic Press, NY, 1990.

Buchter, A., R. Wolbing, M. Wyss, O. F. Kieler, T. Weimann, J. Kohlmann, A. B. Zorin, D. Ruffer, F. Matteini, G. Tutuncuoglu, F. Heimbach, A. Kleibert, A. Fontcuberta i Morral, D. Grundler, R. Kleiner, D. Koelle, and M. Poggio. Magnetization reversal of an individual exchange-biased permalloy nanotube. *Physical Review B*, 92:214432, 2015. doi:10.1103/PhysRevB.92.214432.

Cantu-Valle, J., E. D. Barriga-Castro, V. Vega, J. García, R. Mendoza-Reséndez, C. Luna, V. M. Prida, K. Nielsch, F. Mendoza-Santoyo, M. Jose-Yacaman, and A. Ponce. Quantitative magnetometry analysis and structural characterization of multisegmented cobalt–nickel nanowires. *Journal of Magnetism and Magnetic Materials*, 379:294–299, 2015. doi:10.1016/j.jmmm.2014.12.022.

Caruso, R. A., and M. Antonietti. Sol–gel nanocoating: An approach to the preparation of structured materials. *Chemistry of Materials*, 13:3272–3282, 2001. doi:10.1021/cm001257z.

Chandra, S., A. Biswas, S. Datta, B. Ghosh, A. K. Raychaudhuri, and H. Srikanth. Inverse magnetocaloric and exchange bias effects in single crystalline $La_{0.5}Sr_{0.5}MnO_3$ nanowires. *Nanotechnology*, 24:505712, 2013. doi:10.1088/0957-4484/24/50/505712.

Chen, L., Y. Yang, and X. Meng. Core–shell magnetic exchange model for Co_3O_4 nanowires. *Applied Physics Letters*, 102:203102, 2013. doi:10.1063/1.4807393.

Chong, Y., T. D. Gorlitz, S. Martens, M. Y. E. Yau, S. Allende, J. Bachmann, and K. Nielsch. Multilayered core/shell nanowires displaying two distinct magnetic switching events. *Advanced Materials*, 22(22):2435–2439, 2010. doi:10.1002/adma.200904321.

Chu, Y.-H., L. W. Martin, M. B. Holcomb, M. Gajek, S.-J. Han, Q. He, N. Balke, C.-H. Yang, D. Lee, W. Hu, Q. Zhan, P.-L. Yang, A. Fraile-Rodríguez, A. Scholl, S. X. Wang, and R. Ramesh. Electric-field control of local ferromagnetism using a magnetoelectric multiferroic. *Nature Materials*, 7:478–482, 2008. doi:10.1038/nmat2184.

Coey, J. M. D. *Magnetism and Magnetic Materials*. Cambridge University Press, New York, 2009. ISBN 9780521816144.

Colombo, M., S. Carregal-Romero, M. F. Casula, L. Gutierrez, M. P. Morales, I. B. Bohm, J. T. Heverhagen, D. Prosperi, and W. J. Parak. Biological applications of magnetic nanoparticles. *Chemical Society Reviews*, 41:4306–4334, 2012. doi:10.1039/C2CS15337H.

Correa-Duarte, M. A. and V. Salgueirino. Magnetic properties of nanowires guided by carbon nanotubes. *Nanowires Science and Technology*. Edited by N. Lupu, INTECH, Croatia, Balkans, pp. 84–112, 2010.

Crowley, T. A., B. Daly, M. A. Morris, D. Erts, O. Kazakova, J. J. Boland, B. Wue, and J. D. Holmes. Probing the magnetic properties of cobalt–germanium nanocable arrays. *Journal of Materials Chemistry*, 15:2408–2413, 2005. doi:10.1039/b502155c.

Daly, B., D. C. Arnold, J. S. Kulkarni, O. Kazakova, M. T. Shaw, S. Nikitenko, D. Erts, M. A. Morris, and J. D. Holmes. Synthesis and characterization of highly ordered cobalt–magnetite nanocable arrays. *Small*, 2(11):1299–1307, 2006. doi:10.1002/smll.200600167.

Daub, M., M. Knez, U. Goesele, and K. Nielsch. Ferromagnetic nanotubes by atomic layer deposition in anodic alumina membranes. *Journal of Applied Physics*, 101(9):09J111, 2007. doi:10.1063/1.2712057.

De La Torre Medina, J., M. Darques, and L. Piraux. Exchange bias anisotropy in Co nanowires electrodeposited into polycarbonate membranes. *Journal of Applied Physics*, 106:023921, 2009. doi:10.1063/1.3183949.

Dho, J., X. Qi, H. Kim, J. L. MacManus-Driscoll, and M. G. Blamire. Large electric polarization and exchange bias in multiferroic $BiFeO_3$. *Advanced Materials*, 18(11):1445–1448, 2006. doi:10.1002/adma.200502622.

Díaz-Guerra, C., M. Vila, and J. Piqueras. Exchange bias in single-crystalline CuO nanowires. *Applied Physics Letters*, 96:193105, 2010. doi:10.1063/1.3428658.

Dieny, B., V. S. Speriosu, S. S. P. Parkin, B. A. Gurney, D. R. Wilhoit, and D. Mauri. Giant magnetoresistance in soft ferromagnetic multilayers. *Physical Review B*, 43(1):1297–1300, 1991. doi:10.1103/PhysRevB.43.1297.

Eerenstein, W., N. D. Mathur, and J. F. Scott. Multiferroic and magnetoelectric materials. *Nature*, 442:759–765, 2006. doi:10.1038/nature05023.

Fernández-García, M. P., P. Gorria, M. Sevilla, A. B. Fuertes, R. Boada, J. Chaboy, G. Aquilanti, and J. A. Blanco. Co nanoparticles inserted into a porous carbon amorphous matrix: the role of cooling field and temperature on the exchange bias effect. *Physical Chemistry Chemical Physics*, 13:927–932, 2011. doi:10.1039/c0cp00396d.

Fu, J., J. Zhang, Y. Peng, J. Zhao, G. Tan, N. J. Mellors, E. Xie, and W. Han. Unique magnetic properties and magnetization reversal process of $CoFe_2O_4$ nanotubes fabricated by electrospinning. *Nanoscale*, 4:3932, 2012. doi:10.1039/c2nr30487b.

Gallant, D. and S. Simard. A study on the localized corrosion of cobalt in bicarbonate solutions containing halide ions. *Corrosion Science*, 47(7):1810–1838, 2005. doi:10.1016/j.corsci.2004.08.008.

Gangopadhyay, S., G. C. Hadjipanayis, C. M. Sorensen, and K. J. Klabunde. Exchange anisotropy in oxide passivated Co fine particles. *Journal of Applied Physics*, 73:6964, 1993. doi:10.1063/1.352398.

García-Calzón, J. A. and M. E. Díaz-García. Synthesis and analytical potential of silica nanotubes. *Trends in Analytical Chemistry*, 35:27–38, 2012. doi:10.1016/j.trac.2012.01.003.

George, S. M. Atomic layer deposition: An overview. *Chemical Reviews*, 110:111–131, 2010. doi:10.1021/cr900056b.

Gerein, N. J. and J. A. Haber. Effect of ac electrodeposition conditions on the growth of high aspect ratio copper nanowires in porous aluminum oxide templates. *The Journal of Physical Chemistry B*, 109(37):17372–17385, 2005. doi:10.1021/jp051320d.

Grzelczak, M., M. A. Correa-Duarte, V. Salgueiriño-Maceira, B. Rodríguez-González, J. Rivas, and L. M. Liz-Marzán. Pt-Catalyzed formation of Ni nanoshells on carbon nanotubes. *Angewandte Chemie (International ed. in English)*, 46:7026–7030, 2007. doi:10.1002/anie.200701671.

Gu, D., H. Baumgart, T. M. Abdel-Fattah, and G. Namkoong. Synthesis of nested coaxial multiple-walled nanotubes by atomic layer deposition. *ACS Nano*, 4(2):753–758, 2010. doi:10.1021/nn901250w.

Guimaraes, A. P. *Principles of Nanomagnetism*. Springer-Verlag, Berlin, Heidelberg, 2009.

Hoffmann, A. Symmetry driven irreversibilities at ferromagnetic-antiferromagnetic interfaces. *Physical Review Letters*, 93:097203, 2004. doi:10.1103/PhysRevLett.93.097203.

Hsu, H.-C., C.-C. Lo, and Y.-C. Tseng. Competing magnetic interactions and interfacial frozen-spins in Ni-NiO coreshell nano-rods. *Journal of Applied Physics*, 111(6):063919, 2012. doi:10.1063/1.3699039.

Hur, N., S. Park, P. A. Sharma, J. S. Ahn, S. Guha, and S.-W. Cheong. Electric polarization reversal and memory in a multiferroic material induced by magnetic fields. *Nature*, 429:392–395, 2004. doi:10.1038/nature02572.

Iglesias-Freire, O., C. Bran, E. Berganza, I. Mínguez-Bacho, C. Magén, M. Vázquez, and A. Asenjo. Spin configuration in isolated FeCoCu nanowires modulated in diameter. *Nanotechnology*, 26:395702, 2015. doi:10.1088/0957-4484/26/39/395702.

Javed, K., W. J. Li, S. S. Ali, D. W. Shi, U. Khan, S. Riaz, and X. F. Han. Enhanced exchange bias and improved ferromagnetic properties in Permalloy–$BiFe_{0.95}Co_{0.05}O_3$ core–shell nanostructures. *Scientific Reports*, 5:18203, 2015. doi:10.1038/srep18203.

Kazakova, O., B. Daly, and J. D. Holmes. Tunable magnetic properties of metal/metal oxide nanoscale coaxial cables. *Physical Review B*, 74:184413, 2006. doi:10.1103/PhysRevB.74.184413.

Kovylina, M., M. García del Muro, Z. Konstantinovic, M. Varela, O. Iglesias, A. Labarta, and X. Batlle. Controlling exchange bias in Co–CoO$_x$ nanoparticles by oxygen content. *Nanotechnology*, 20:175702, 2009. doi:10.1088/0957-4484/20/17/175702.

Lee, W. and J.-C. Kim. Highly ordered porous alumina with tailor-made pore structures fabricated by pulse anodization. *Nanotechnology*, 21:485304, 2010. doi:10.1088/0957-4484/21/48/485304.

Lee, W. and S.-J. Park. Porous anodic aluminum oxide: Anodization and templated synthesis of functional nanostructures. *Chemical Reviews*, 114 (15):7487–7556, 2014. doi:10.1021/cr500002z.

Lee, W., R. Ji, U. Gosele, and K. Nielsch. Fast fabrication of long-range ordered porous alumina membranes by hard anodization. *Nature Materials*, 5(9):741–747, 2006. doi:10.1038/nmat1717.

Lee, W., R. Scholz, K. Nielsch, and U. Gosele. A template-based electrochemical method for the synthesis of multisegmented metallic nanotubes. *Angewandte Chemie (International ed. in English)*, 44(37):6050–6054, 2005. doi:10.1002/anie.200501341.

Li, W., R. Zeng, Z. Sun, D. Tian, and S. Dou. Uncoupled surface spin induced exchange bias in μ-MnO$_2$ nanowires. *Scientific Reports*, 4:6641, 2014. doi:10.1038/srep06641.

Li, X., Y. Wang, G. Song, Z. Peng, Y. Yu, X. She, J. Sun, J. Li, P. Li, Z. Wang, and X. Duan. Fabrication and magnetic properties of Ni/Cu Shell/Core nanocable arrays. *Journal of Physical Chemistry C*, 114:6914–6916, 2010. doi:10.1021/jp910979g.

Liébana-Viñas, S., U. Wiedwald, A. Elsukova, J. Perl, B. Zingsem, A. S. Semisalova, V. Salgueiriño, M. Spasova, and M. Farle. Structure-correlated exchange anisotropy in oxidized Co$_{80}$Ni$_{20}$ nanorods. *Chemistry of Materials*, 27:4015–4022, 2015. doi:10.1021/acs.chemmater.5b00976.

Liew, H. F., S. C. Low, and W. S. Lew. Fabrication of constricted compositionally-modulated Ni$_x$Fe$_{1-x}$ nanowires. *Journal of Physics: Conference Series*, 266(1):012058, 2011. doi:10.1088/1742-6596/266/1/012058.

Liu, M., X. Li, H. Imrane, Y. Chen, T. Goodrich, Z. Cai, K. S. Ziemer, J. Y. Huang, and N. X. Sun. Synthesis of ordered arrays of multiferroic NiFe$_2$O$_4$-Pb(Zr$_{0.52}$Ti$_{0.48}$)O$_3$ core-shell nanowires. *Applied Physics Letters*, 90:152501, 2007. doi:10.1063/1.2722043.

Llavona, A., A. Prados, V. Velasco, P. Crespo, M. C. Sánchez, and L. Pérez. Electrochemical synthesis and magnetic properties of goethite single crystal nanowires. *CrystEngComm*, 15:4905–4909, 2013. doi:10.1039/C3CE26772E.

Lu, H. L., G. Scarel, X. L. Li, and M. Fanciulli. Thin MnO and NiO films grown using atomic layer deposition from ethylcyclopentadienyl type of precursors. *Journal of Crystal Growth*, 310:5464–5468, 2008. doi:10.1016/j.jcrysgro.2008.08.031.

Lu, Y., Y. Yin, B. T. Mayers, and Y. Xia. Modifying the surface properties of superparamagnetic iron oxide nanoparticles through a Sol–Gel approach. *Nano Letters*, 2(3):183–186, 2002. doi:10.1021/nl015681q.

Maat, S., K. Takano, S. S. P. Parkin, and Eric E. Fullerton. Perpendicular exchange bias of Co/Pt multilayers. *Physical Review Letters*, 87(8):087202, 2001. doi:10.1103/PhysRevLett.87.087202.

Maaz, K., S. Ishrat, S. Karim, and G.-H. Kim. Effect of aging on the magnetic characteristics of nickel nanowires embedded in polycarbonate. *Journal of Applied Physics*, 110:013908, 2011. doi:10.1063/1.3603006.

Mariño-Fernández, R., S. H. Masunaga, N. Fontaíña-Troitiño, M. P. Morales, J. Rivas, and V. Salgueirino. Goethite (μ-FeOOH) nanorods as suitable antiferromagnetic substrates. *Journal of Physical Chemistry C*, 115:13991–13999, 2011. doi:10.1021/jp201490j.

Masuda, H. and K. Fukuda. Ordered metal nanohole arrays made by a two-step replication of honeycomb structures of anodic alumina. *Science*, 268(5216):1466–1468, 1995.

Masuda, H., H. Asoh, M. Watanabe, K. Nishio, M. Nakao, and T. Tamamura. Square and triangular nanohole array architectures in anodic alumina. *Advanced Materials*, 13(3):189–192, 2001. doi:10.1002/1521-4095(200102)13:3 <189::AID-ADMA189> 3.0.CO;2-Z.

Maurer, T., F. Zighem, F. Ott, G. Chaboussant, G. Andre, Y. Soumare, J.-Y. Piquemal, G. Viau, and C. Gatel. Exchange bias in Co/CoO core–shell nanowires: Role of anti-ferromagnetic superparamagnetic fluctuations. *Physical Review B*, 80(6):064427, 2009. doi:10.1103/PhysRevB.80.064427.

Meiklejohn, W. H. and C. P. Bean. New magnetic anisotropy. *Physical Review*, 102:1413, 1956. doi:10.1103/PhysRev.102.1413

Meulenkamp, E. A. Synthesis and growth of ZnO nanoparticles. *Journal of Physical Chemistry B*, 102(29):5566–5572, 1998. doi:10.1021/jp980730h.

Mohamed, H. D. A., S. M. D. Watson, B. R. Horrocks, and A. Houlton. Magnetic and conductive magnetite nanowires by DNA-templating. *Nanoscale*, 4:5936–5945, 2012. doi:10.1039/C2NR31559A.

Naganuma, H., J. Miura, and S. Okamura. Ferroelectric, electrical and magnetic prop-erties of Cr, Mn, Co, Ni, Cu added polycrystalline $BiFeO_3$ films. *Applied Physics Letters*, 93:052901, 2008. doi:10.1063/1.2965799.

Narayanan, T. N., B. P. Mandal, A. K. Tyagi, A. Kumarasiri, X. Zhan, M. G. Hahm, M. R. Anantharaman, G. Lawes, and P. M. Ajayan. Hybrid multiferroic nano-structure with magnetic–dielectric coupling, *Nano Letters*, 12:3025–3030, 2012. doi:10.1021/nl300849u.

Narayanan, T. N., M. M. Shaijumon, P. M. Ajayan, and M. R. Anantharaman. Synthesis of high coercivity core–shell nanorods based on nickel and cobalt and their magnetic properties. *Nanoscale Research Letters*, 5:164–168, 2010. doi:10.1007/s11671-009-9459-7.

Nielsch, K., F. Muller, A.-P. Li, and U. Gosele. Uniform nickel deposition into ordered alumina pores by pulsed electrodeposition. *Advanced Materials*, 12(8):582–586, 2000.

Nielsch, K., J. Choi, K. Schwirn, R. B. Wehrspohn, and U. Gosele. Self-ordering regimes of porous alumina: The 10% porosity rule. *Nano Letters*, 2(7):677–680, 2002. doi:10.1021/nl025537k.

Nogués, J. and I. K. Schuller. Exchange bias. *Journal of Magnetism and Magnetic Materials*, 192(2):203–232, 1999. doi:10.1016/S0304-8853(98)00266-2.

Nogués, J., V. Skumryev, J. Sort, S. Stoyanov, and D. Givord. Shell-driven magnetic stability in core–shell nanoparticles. *Physical Review Letters*, 97:157203 (2006). doi:10.1103/PhysRevLett.97.157203.

Park, D. H., Y. B. Lee, M. Y. Cho, B. H. Kim, S. H. Lee, Y. K. Hong, J. Joo, H. C. Cheong, and S. R. Lee. Fabrication and magnetic characteristics of hybrid double walled nanotube of ferromagnetic nickel encapsulated conducting polypyrrole. *Applied Physics Letters*, 90(9):093122, 2007. doi:10.1063/1.2710748.

Phuoc, N. N. and T. Suzuki. In-plane and out-of-plane exchange biases in epitaxial FePt-FeMn multilayers with different crystalline orientations. *IEEE Transactions on Magnetics*, 44(11):2828–2831, 2008. doi:10.1109/TMAG.2008.2002199.

Pitzschel, K., J. M. Montero-Moreno, J. Escrig, O. Albrecht, K. Nielsch, and J. Bachmann. Controlled introduction of diameter modulations in arrayed magnetic iron oxide nanotubes. *ACS Nano*, 3(11):3463–3468, 2009. doi:10.1021/nn900909q.

Proenca, M. P., C. T. Sousa, A. M. Pereira, P. B. Tavares, J. Ventura, M. Vazquez, and J. P. Araujo. Size and surface effects on the magnetic properties of NiO nanoparticles. *Physical Chemistry Chemical Physics*, 13:9561–9567, 2011. doi:10.1039/c1cp00036e.

Proenca, M. P., C. T. Sousa, J. Ventura, and J. P. Araujo. Electrochemical synthesis and magnetism of magnetic nanotubes. *Magnetic Nano- and Microwires: Design, Synthesis, Properties and Applications*. Edited by M. Vazquez, Woodhead Publishing – Elsevier, Cambridge, UK, pp. 727–781, 2015. doi:10.1016/B978-0-08-100164-6.00024-2.

Proenca, M. P., C. T. Sousa, J. Ventura, M. Vazquez, and J. P. Araujo. Ni growth inside ordered arrays of alumina nanopores: Enhancing the deposition rate. *Electrochimica Acta*, 72:215–221, 2012a. doi:10.1016/j.electacta.2012.04.036.

Proenca, M. P., C. T. Sousa, J. Ventura, M. Vazquez, and J. P. Araujo. Distinguishing nanowire and nanotube formation by the deposition current transients. *Nanoscale Research Letters*, 7(1):280, 2012b. doi:10.1186/1556-276X-7-280.

Proenca, M. P., J. Ventura, C. T. Sousa, M. Vazquez, and J. P. Araujo. Exchange bias, training effect, and bimodal distribution of blocking temperatures in electrode-posited core–shell nanotubes. *Physical Review B*, 87(13):134404, 2013a. doi:10.1103/PhysRevB.87.134404.

Proenca, M. P., J. Ventura, C. T. Sousa, M. Vazquez, and J. P. Araujo. Temperature dependence of the training effect in electrodeposited Co/CoO nanotubes. *Journal of Applied Physics*, 114:043914, 2013b. doi:10.1063/1.4816696.

Respaud, M., J. M. Broto, H. Rakoto, A. R. Fert, L. Thomas, B. Barbara, M. Verelst, E. Snoeck, P. Lecante, A. Mosset, J. Osuna, T. Ould Ely, C. Amiens, and B. Chaudret. Surface effects on the magnetic properties of ultrafine cobalt particles. *Physical Review B*, 57(5): 2925–2935, 1998. doi:10.1103/PhysRevB.57.2925.

Sajanlal, P. R., T. S. Sreeprasad, A. K. Samal, and T. Pradeep. Anisotropic nanomaterials: Structure, growth, assembly, and functions. *Nano Reviews*, 2:5883, 2011. doi:10.3402/nano.v2i0.5883.

Salabas, E. L., A. Rumplecker, F. Kleitz, F. Radu, and F. Schuth. Exchange anisotropy in nanocasted Co_3O_4 nanowires. *Nano Letters*, 6(12):2977–2981, 2006. doi:10.1021/nl060528n.

Salazar-Alvarez, G., J. Geshev, S. Agramut-Puig, C. Navau, A. Sanchez, J. Sort, and J. Nogués. Tunable high-field magnetization in strongly exchange-coupled free-standing Co/CoO core/shell coaxial nanowires. *ACS Applied Materials and Interfaces*, 2016. doi:10.1021/acsami.6b05588.

Salem, M. S., P. Sergelius, R. M. Corona, J. Escrig, D. Görlitz, and K. Nielsch. Magnetic properties of cylindrical diameter modulated $Ni_{80}Fe_{20}$ nanowires: Interaction and coercive fields. *Nanoscale*, 5:3941–3947, 2013. doi:10.1039/C3NR00633F.

Salgueiriño-Maceira, V., M. A. Correa-Duarte, M. Bañobre-López, M. Grzelczak, M. Farle, L. M. Liz-Marzán, and J. Rivas. Magnetic properties of Ni/NiO nanowires deposited onto CNT/Pt nanocomposites. *Advanced Functional Materials*, 18:616–621, 2008. doi:10.1002/adfm.200700846.

Sbiaa, R., H. Meng, and S. N. Piramanayagam. Materials with perpendicular magnetic anisotropy for magnetic random access memory. *Physica Status Solidi—Rapid Research Letters*, 5(12):413–419, 2011. doi:10.1002/pssr.201105420.

Scarani, V., H. De Riedmatten, and J.-Ph. Ansermet. ^{59}Co nuclear magnetic resonance studies of magnetic excitations in ferromagnetic nanowires. *Applied Physics Letters*, 76(7):903, 2000. doi:10.1063/1.125624.

Schonenberger, C., B. M. I. van der Zande, L. G. J. Fokkink, M. Henny, C. Schmid, M. Kruger, A. Bachtold, R. Huber, H. Birk, and U. Staufer. Template synthesis of nanowires in porous polycarbonate membranes: Electrochemistry and morphology. *Journal of Physical Chemistry B*, 101:5497–5505, 1997. doi:10.1021/jp963938g.

Sharma, G., M. V. Pishko, and C. A. Grimes. Fabrication of metallic nanowire arrays by electrodeposition into nanoporous alumina membranes: Effect of barrier layer. *Journal of Materials Science*, 42(13):4738–4744, 2007. doi:10.1007/s10853-006-0769-1.

Sharma, M., and B. K. Kuanr. Electrodeposition of ferromagnetic nanostructures. *Electroplating of Nanostructures*. Edited by M. Aliofkhazraei, InTech, 2015. doi:10.5772/61226.

Sharma, S. K., J. M. Vargas, E. De Biasi, F. Béron, M. Knobel, K. R. Pirota, C. T. Meneses, S. Kumar, C. G. Lee, P. G. Pagliuso, and C. Rettori. The nature and enhancement of magnetic surface contribution in model NiO nanoparticles. *Nanotechnology*, 21:035602, 2010. doi:10.1088/0957-4484/21/3/035602.

Shen, X.-P., H.-J. Miao, H. Zhao, and Z. Xu. Synthesis, characterization and magnetic properties of Co$_3$O$_4$ nanotubes. *Applied Physics A*, 91(1):47–51, 2008. doi:10.1007/s00339-007-4361-6.

Shi, D., J. Chen, S. Riaz, W. Zhou, and X. Han. Controlled nanostructuring of multiphase core–shell nanowires by a template-assisted electrodeposition approach. *Nanotechnology*, 23:305601, 2012. doi:10.1088/0957-4484/23/30/305601.

Shi, D.-W., K. Javed, S. S. Ali, J.-Y. Chen, P.-S. Li, Y.-G. Zhao, and X.-F. Han. Exchange-biased hybrid ferromagnetic–multiferroic core–shell nanostructures. *Nanoscale*, 6:7215–7220, 2014. doi:10.1039/C4NR00393D.

Shipton, E., K. Chan, T. Hauet, O. Hellwig, and E. E. Fullerton. Suppression of the perpendicular anisotropy at the CoO Néel temperature in exchange-biased CoO/(Co/Pt) multilayers. *Applied Physics Letters*, 95:132509, 2009. doi:10.1063/1.3240402.

Skumryev, V., S. Stoyanov, Y. Zhang, G. Hadjipanayis, D. Givord, and J. Nogués. Beating the superparamagnetic limit with exchange bias. *Nature*, 423:850–853, 2003. doi:10.1038/nature01687.

Smith, K. H., E. Tejeda-Montes, M. Poch, and A. Mata. Integrating top-down and self-assembly in the fabrication of peptide and protein-based biomedical materials. *Chemical Society Reviews*, 40:4563–4577, 2011. doi:10.1039/c1cs15064b.

Soumare, Y., C. Garcia, T. Maurer, G. Chaboussant, F. Ott, J.-Y. Piquemal, and G. Viau. Kinetically controlled synthesis of hexagonally close-packed cobalt nanorods with high magnetic coercivity. *Advanced Functional Materials*, 19(12):1971–1977, 2009. doi:10.1002/adfm.200800822.

Soumare, Y., J.-Y. Piquemal, T. Maurer, F. Ott, G. Chaboussant, A. Falqui, and G. Viau. Oriented magnetic nanowires with high coercivity. *Journal of Materials Chemistry*, 18(46):5696–5702, 2008. doi:10.1039/B810943E.

Sousa, C. T., A. Apolinário, M. P. Proença, D. C. Leitão, J. Azevedo, J. Ventura, and J. P. Araújo. Dendritic nanostructures grown in hierarchical branched pores. *Advances in Nanotechnology—Vol. 12*. Edited by Z. Bartul and J. Trenor, Nova Science Publishers, Hauppauge, New York, 2014b.

Sousa, C. T., A. M. L. Lopes, M. P. Proenca, D. C. Leitao, J. G. Correia, and J. P. Araujo. Rapid synthesis of ordered manganite nanotubes by microwave irradiation in alumina templates. *Journal of Nanoscience and Nanotechnology*, 9(10):6084–6088, 2009. doi:10.1166/jnn.2009.1572.

Sousa, C. T., C. Nunes, M. P. Proença, D. C. Leitão, J. L. F. C. Lima, S. Reis, J. P. Araújo, and M. Lúcio. pH sensitive silica nanotubes as rationally designed vehicles for NSAIDs delivery. *Colloids and Surfaces B: Biointerfaces*, 94:288–295, 2012. doi:10.1016/j.colsurfb.2012.02.003.

Sousa, C. T., D. C. Leitao, M. P. Proenca, A. Apolinario, J. G. Correia, J. Ventura, and J. P. Araujo. Tunning pore filling of anodic alumina templates by accurate control of the bottom barrier layer thickness. *Nanotechnology*, 22(31):315602, 2011. doi:10.1088/0957-4484/22/31/315602.

Sousa, C. T., D. C. Leitao, M. P. Proenca, J. Ventura, A. M. Pereira, and J. P. Araujo. Nanoporous alumina as templates for multifunctional applications. *Applied Physics Reviews*, 1:031102, 2014a. doi:10.1063/1.4893546.

Stepniowski, W. J. and M. Salerno. Fabrication of nanowires and nanotubes by anodic alumina template-assisted electrodeposition. *Manufacturing Nanostructures*. Edited by W. Ahmed and N. Ali, One Central Press, Manchester, UK, pp. 321–357, 2015.

Sulka, G. D. Highly ordered anodic porous alumina formation by self-organized anodizing. *Nanostructured Materials in Electrochemistry*. Edited by Ali Eftekhari, Wiley-VCH Verlag GmbH & Co. KGaA, Weinheim. 2008. ISBN: 978-3-527-31876-6.

Sun, L., S. M. Zhou, P. C. Searson, and C. L. Chien. Longitudinal and perpendicular exchange bias in FeMn/(FeNi/FeMn)$_n$ multilayers. *Journal of Applied Physics*, 93(10): 6841–6843, 2003. doi:10.1063/1.1544447.

Susano, M., M. P. Proenca, S. Moraes, C. T. Sousa, and J. P. Araújo. Tuning the magnetic properties of multisegmented Ni/Cu electrodeposited nanowires with controllable Ni lengths. *Nanotechnology*, 27(33):335301, 2016. doi:10.1088/0957-4484/27/33/335301.

Suszka, A. K., O. Idigoras, E. Nikulina, A. Chuvilin, and A. Berger. Crystallography-driven positive exchange bias in Co/CoO bilayers. *Physical Review Letters*, 109:177205, 2012. doi:10.1103/PhysRevLett.109.177205.

Tadic, M., I. Milosevic, S. Kralj, M.-L. Saboungi, and L. Motte. Ferromagnetic behavior and exchange bias effect in akaganeite nanorods. *Applied Physics Letters*, 106:183706, 2015. doi:10.1063/1.4918930.

Tao, F., M. Guan, Y. Jiang, J. Zhu, Z. Xu, and Z. Xue. An easy way to construct an ordered array of nickel nanotubes: The triblock-copolymer-assisted hard-template method. *Advanced Materials*, 18(16):2161–2164, 2006. doi:10.1002/adma.200600275.

Toimil Molares, M. E., J. Brotz, V. Bushmann, D. Dobrev, R. Neumann, R. Scholz, I. U. Shuchert, C. Trautmann, and J. Vetter. Etched heavy ion tracks in polycarbonate as template for copper nanowires. *Nuclear Instruments and Methods in Physics Research B*, 185:192–197, 2001. doi:10.1016/S0168-583X(01)00755-8.

Tong, G., J. Guan, and Q. Zhang. In situ generated gas bubble-directed self-assembly: Synthesis, and peculiar magnetic and electrochemical properties of vertically aligned arrays of high-density Co_3O_4 nanotubes. *Advanced Functional Materials*, 23:2406–2414, 2013. doi:10.1002/adfm.201202747.

Ueltzhöffer, T., R. Streubel, I. Koch, D. Holzinger, D. Makarov, O. G. Schmidt, and A. Ehresmann. Magnetically patterned rolled-up exchange bias tubes: A paternoster for superparamagnetic beads. *ACS Nano*, Article ASAP, 2016. doi:10.1021/acsnano.6b03566.

Ung, D., G. Viau, C. Ricolleau, F. Warmont, P. Gredin, and F. Fiévet. CoNi nanowires synthesized by heterogeneous nucleation in liquid polyol. *Advanced Materials*, 17(3):338–344, 2005. doi:10.1002/adma.200400915.

Ventura, J., J. P. Araujo, J. B. Sousa, A. Veloso, and P. P. Freitas. Training effect in specular spin valves. *Physical Review B*, 77:184404, 2008. doi:10.1103/PhysRevB.77.184404.

Wang, J., J. B. Neaton, H. Zheng, V. Nagarajan, S. B. Ogale, B. Liu, D. Viehland, V. Vaithyanathan, D. G. Schlom, U. V. Waghmare, N. A. Spaldin, K. M. Rabe, M. Wuttig, R. Ramesh. Epitaxial $BiFeO_3$ multiferroic thin film heterostructures. *Science*, 299(5613):1719–1722, 2003. doi:10.1126/science.1080615.

Wang, Q., G. Wang, X. Han, X. Wang, and J. G. Hou. Controllable template synthesis of Ni/Cu nanocable and Ni nanotube arrays: A one-step co-electrodeposition and electrochemical etching method. *The Journal of Physical Chemistry B*, 109(49):23326–23329, 2005. doi:10.1021/jp0530202.

Whitney, T., P. Searson, J. Jiang, and C. Chien. Fabrication and magnetic properties of arrays of metallic nanowires. *Science*, 261(5126):1316–1319, 1993.

Woldering, L. A., R. W. Tjerkstra, H. V. Jansen, I. D. Setija, and W. L. Vos. Periodic arrays of deep nanopores made in silicon with reactive ion etching and deep UV lithography. *Nanotechnology*, 19:145304, 2008. doi:10.1088/0957-4484/19/14/145304.

Wu, W., X. Xiao, S. Zhang, J. Zhou, L. Fan, F. Ren, and C. Jiang. Large-scale and controlled synthesis of iron oxide magnetic short nanotubes: Shape evolution, growth mechanism, and magnetic properties. *Journal of Physical Chemistry C*, 114:16092–16103, 2010. doi:10.1021/jp1010154.

Xu, Q., H. Zai, D. Wu, T. Qiu, and M. X. Xu. The magnetic properties of $Bi(Fe_{0.95}Co_{0.05})O_3$ ceramics. *Applied Physics Letters*, 95:112510, 2009. doi:10.1063/1.3233944.

Xu, Y., J. Wei, J. Yao, J. Fu, and D. Xue. Synthesis of $CoFe_2O_4$ nanotube arrays through an improved sol–gel template approach. *Materials Letters*, 62:1403–1405, 2008. doi:10.1016/j.matlet.2007.08.066.

Yu, M. and M. W. Urban. Formation of concentric ferromagnetic nanotubes from biologically active phospholipids. *Journal of Materials Chemistry*, 17:4644–4646, 2007. doi:10.1039/b714093m.

Zhao, T., A. Scholl, F. Zavaliche, K. Lee, M. Barry, A. Doran, M. P. Cruz, Y. H. Chu, C. Ederer, N. A. Spaldin, R. R. Das, D. M. Kim, S. H. Baek, C. B. Eom, and R. Ramesh. Electrical control of antiferromagnetic domains in multiferroic BiFeO$_3$ films at room temperature. *Nature Materials*, 5:823–829, 2006. doi:10.1038/nmat1731.

Zhu, J.-G. and C. Park. magnetic tunnel junctions. *Materials Today*, 9(11):36–45, 2006. doi:10.1016/S1369-7021(06)71693-5.

Zierold, R., Z. Wu, J. Biskupek, U. Kaiser, J. Bachmann, C. E. Krill, and K. Nielsch. Magnetic, multilayered nanotubes of low aspect ratios for liquid suspensions. *Advanced Functional Materials*, 21(2):226–232, 2011. doi:10.1002/adfm.201001395.

10

Exchange-Bias Effect in Bulk Perovskite Manganites

A. Wisniewski,
I. Fita, R. Puzniak,
and V. Markovich

10.1 Introduction

The primary interest to study ferromagnetic mixed-valence perovskite manganites, $Ln_{1-x}M_x MnO_3$, where Ln is a rare earth atom (most often La, Pr, Nd, and Sm) and M is an alkaline earth element (most often Sr, Ca, and Ba), originated from the observation that the application of an external magnetic field, H, strongly enhances the electrical conductivity near the metal–insulator (MI) transition temperature T_{MI} (coupled to the Curie temperature, T_C), giving rise to the so-called colossal magnetoresistance (CMR) effect; see numerous review articles and books: Coey et al. (2009), Dagotto et al. (2003), Dorr (2006), Goodenough (2003), Tokura (2006), and Ziese (2002). It was also recognized that these compounds exhibit very rich phase diagrams, since several phases differing in magnetic, structural, and electronic properties may coexist in the same sample.

However, the most relevant feature leading to an appearance of exchange-bias (EB) effect is the intrinsic phase separation (PS) that occurs in manganites. The first evidence of the EB effect in mixed-valent manganites with perovskite structure was reported in $Pr_{1/3}Ca_{2/3}MnO_3$ by Niebieskikwiat and Salamon (2005). In order to correlate the relation between the observed field axis and magnetization axis shifts of hysteresis loop, they proposed a simplified exchange interaction model, in which single-domain ferromagnetic (FM) clusters or droplets are embedded in an antiferromagnetic (AFM) matrix. The EB field, H_{EB}, was introduced as a factor accounting for asymmetry in switching of the magnetization direction over the anisotropy barrier, $KV \pm \mu H_{EB}$, where K is anisotropy constant, V is the volume of FM particles, and μ is their magnetic moment. A simple linear relationship between the remanence asymmetry, M_{EB}, saturation magnetization, M_S, and the H_{EB} was found: $M_{EB}/M_S \propto -H_{EB}$ was obtained for $\mu H_{EB} < k_B T$, suggesting an equivalence of both parameters. They also proposed a formula, describing the dependence of the exchange field H_{EB} on the cooling field H_{cool} for systems with single-domain FM clusters embedded in an AFM matrix:

$$H_{EB} \propto J[(\frac{J\mu_0}{(g\mu_B)^2})L(\frac{\mu H_{cool}}{k_B T_f}) + H_{cool}]$$ (10.1)

where:
  J is the interface exchange constant
  $g \approx 2$ is the gyromagnetic factor
  μ_B is the Bohr magnetron
  L is the Langevin function
  $\mu_0 \approx 3\mu_B$ is the magnetic moment of the Mn core
  $\mu = N\mu_0$ is the magnetic moment of FM clusters
  N is number of spins in a cluster
  T_f is the temperature below which the coercivity and EB appear

Equation 10.1, which is frequently used for estimation of the average size of short-range FM clusters in phase-separated manganites and cobaltites, shows that, at relatively small cooling field, the first term dominates and H_{EB} depends on J^2, while for large cooling fields, the second term ($J H_{cool}$) becomes more important, and for $J < 0$, the absolute value of H_{EB} decreases and could even change the sign.

Later, the EB has been observed in other phase-separated bulk materials, core-shell nanoparticles, and inhomogeneous materials without clearly defined AFM/FM interfaces. It became clear that the appearance of EB requires the interfaces or inhomogeneities. As it is well known, the EB effect manifests itself in shifts of hysteresis loop (M vs. H) relative to the origin, depending on measurement conditions. For the shift in the left direction relative to the origin, the effect is called a negative EB, and for the shift in the right direction, the effect is called a positive EB. When both negative and positive EBs are observed for the same sample, depending on temperature, magnetic field, or cooling conditions, one can say that the EB effect is tunable.

It should be stressed that proper estimation of magnetization curve parameters is fundamental in studying the EB effect, discovered by Meiklejohn and Bean (1956).

The effect has been intensively studied in the last decades because of its importance for the magnetic information storage technologies. In an extensive review on the EB in nanostructures, Nogués et al. (2005) pointed out that minor loops may exhibit shifts characteristic of all unsaturated FM materials that have nothing common with EB, and later on, several papers on this subject have been published (Klein, 2006; Geshev, 2008a,b, 2009a,b, Geshev et al., 2008; Kumar et al., 2014; Harres et al., 2015). In some reports, it was shown that unlike the conventional EB, the shifts strongly depend on maximum applied magnetic field, H_{max}, and are completely eliminated if this filed is high enough, confirming that these shifts are due to unsaturated magnetization. However, it should be noted that the EB effect in bulk manganites appears when the overwhelming magnetic volume of the studied samples is occupied by AFM matrix and hysteresis loops are not saturated at all. Geshev et al. (Geshev, 2008a, b; Geshev et al., 2008) suggested that the presence of *true* EB may be concluded from *effectively saturated* hysteresis loops. Here, a system is considered effectively saturated, as the ascending and descending branches of the hysteresis loop coincide for fields higher than the anisotropy field. Very recently, Harres et al. (2016) suggested a few independent criteria for discriminating nonsaturated (minor) from saturated (major) hysteresis loops. These criteria are based on the analysis of derivatives of descending and ascending branches of the loop, of the remanent magnetization, and of zero-field-cooled/field-cooled thermomagnetic curves.

A recent review devoted to EB effect, with particular attention to the EB effect in perovskite oxides and some their nanostructures, was published by Giri et al. (2011). Hence, in this chapter, mainly the papers that were published during the last 5–6 years are reviewed.

10.1.1 Basic Magnetic Properties of Perovskite Manganites

It was recognized that the rich diversity of phase diagrams—see, for example, in Figure 10.1, the phase diagram of classic $La_{1-x}Sr_xMnO_3$ system—and a variety of magnetic ground states observed in perovskite manganites result from a competition between FM interactions mediated by itinerant charge carriers—the double-exchange (DE) mechanism (Zener, 1951; Anderson and Hasegawa, 1955; De Gennes, 1960) and AFM superexchange (SE) interactions between localized spins of manganese ions, which are combined with lattice and orbital degrees of freedom (Maezono et al., 1998).

The balance between competing interactions and different structural and magnetic phases can be effectively tuned via the choice of trivalent *Ln* and divalent *M* elements in the perovskite *A* sites of ABO_3 perovskite lattice and their ratio, as well as by substituting the Mn sites by elements with different valences (Ziese, 2002; Dagotto et al., 2003; Goodenough, 2003; Dorr, 2006; Tokura, 2006; Coey et al., 2009). The magnetic properties of manganites are mainly determined by transfer of electrons between the manganese and the oxygen orbitals. The direct overlap between Mn orbitals is small; hence, the magnetic interactions are mediated by O 2p electrons. The physical properties of manganites are determined by several factors, mainly by doping level, *x* (which affects the ratio between the number of Mn^{4+} and Mn^{3+} ions), the average size of the cation at the *A* site (site occupied by rare-earth or alkaline-earth atoms), the degree of a disorder at the *A* site, and distortion of the crystallographic structure.

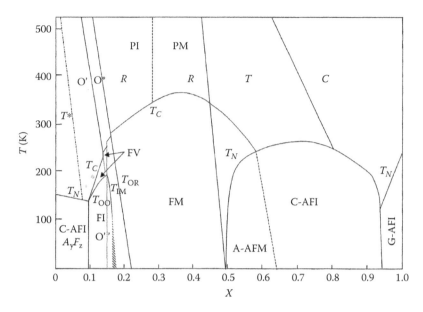

FIGURE 10.1 Phase diagram of La$_{1-x}$Sr$_x$MnO$_3$. C-AFI denotes canted-spin antiferromagnetic insulator, FI—FM insulator, FM—FM metal, AFI—AFM insulator, PI—PM insulator, PM— PM metal. (Reprinted from *Handbook on the Physics and Chemistry of Rare Earths,* Vol. 33, Goodenough, J.B. Rare earths—Manganese perovskites, 249–351, Copyright (2003), with permission from Elsevier.)

10.1.2 Phase Separation—Driving Force for Exchange-Bias Effect in Manganites

A very relevant feature of manganites is their intrinsic electronic inhomogeneity (see, Dagotto et al., 2003). They exhibit phase separation—a coexistence of different *phases* (not in the thermodynamic meaning) having different electronic, structural, and magnetic properties. The examples are ferromagnetic metallic (FMM) droplets in an insulating AFM background or insulating droplets in an FMM matrix, the so-called charge stripes. The length scales of inhomogeneities range from nanometers to micrometers. At low-doping level, phase separation between hole-poor AFM regions and hole-rich FM regions is energetically more favorable than the homogeneous canted AFM phase. The energy of the charge carriers is lower in the FM phase. Hence, if a density of carriers is too low to establish the FM order in the entire volume, the carriers form droplets or stripes that become FM regions inside the insulating AFM matrix.

The direct experimental evidences of electronic phase separation in manganites are scarce. The most convincing are the results of neutron scattering and nuclear magnetic resonance (NMR) studies of La$_{1-x}$Ca$_x$MnO$_3$ (Allodi et al., 1998; Hennion et al., 1998). The phase coexistence in Bi$_{1-x}$Ca$_x$MnO$_3$ was evidenced by using scanning tunneling microscopy by Renner et al. (2002). It was shown that charge ordering and phase separation can be resolved in real space with atomic-scale resolution. Ahn et al. (2004), in their model, assumed that strong coupling between the electronic and elastic degrees of freedom is a leading mechanism of phase coexistence. Such coupling results in local energetically favorable configuration and provides a natural mechanism for the self-organized inhomogeneities, both in the nanometer and micrometer scales. The model predicts that the phase with short- and long-wavelength distortions is insulating, and the phase without lattice distortions is metallic. Magnetic force microscopy studies performed by Wu et al. (2006) on a $(La,Pr,Ca)MnO_3$ twinned single crystal support this model. The experimental data also suggest that accommodation strain is important in the kinetics of the phase transition.

10.2 Tuning of Exchange-Bias Effect by Chemical Substitutions

Positive and negative EB effects were found in the simple perovskite $NdMnO_3$ at 30 K (Hong et al., 2012). In this compound, the Mn^{3+} ions demonstrate canted A-type AFM ordering below 79 K, and the Nd^{3+} ions show short-range ordering below the Mn^{3+} ordering temperature and long-range ferromagnetic ordering below 13 K. The Nd^{3+} ordering is induced by the Mn^{3+} ferromagnetic component, and both these ions are antiferromagnetically coupled with each other, which provides the conditions for the EB effect to occur. At 30 K, the Nd^{3+} ions exhibit only the short-range ordering, and significant negative EB effect occurs when the cooling field, H_{cool}, is relatively small, reaching for the EB field a value of $H_{EB} = -2500$ Oe when $\mu_0 H_{cool} = 1$ T. With the increasing cooling field, the EB field changes sign and reaches positive value of 1800 Oe at $\mu_0 H_{cool} = 10$ T, as one can see in Figure 10.2. In contrast, the opposite EB dependence on cooling field is observed at 8 K. The EB field reaches 130 Oe for 2 T cooling field and −120 Oe for 10 T.

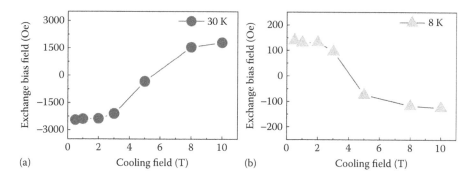

FIGURE 10.2 (a) Cooling field dependence of the exchange-bias field at 30 K (solid circles), and (b) cooling field dependence at 8 K (triangles) for $NdMnO_3$. (Reprinted with the permission from Hong, F. et al., *Appl. Phys. Lett.*, 101, 102411. Copyright 2012, American Institute of Physics.)

At 30 K, a low H_{cool} could lead to negative EB, because Nd^{3+} spins are always oriented in an opposite way with respect to the direction of H_{cool}. When the H_{cool} is high enough to align both Nd^{3+} and Mn^{3+} spins, positive EB occurs, because the Nd^{3+} spins are always along H_{cool}. For moderate values of H_{cool}, the Nd^{3+} spin orientations are distributed, total pinning force is small, and the EB effect is suppressed. It was evinced that the sign of the EB field depends both on the coupling strength between the Nd^{3+} and the Mn^{3+} ordering and on the initial states of the system.

Fertman et al. (2015) have studied EB phenomenon in $Nd_{2/3}Ca_{1/3}MnO_3$, with partial substitution of Nd by Y. It has been found that yttrium doping leads to the growth of the ferromagnetic phase fraction—saturation magnetization of the FM phase in the doped compound $(Nd_{0.9}Y_{0.1})_{2/3}Ca_{1/3}MnO_3$ is about twice larger than that in the parent one. It was found that the FM phase of the substituted compound is magnetically softer, which is evidenced by a lower coercive field H_C. The values of the EB field, H_{EB}, and H_C were found to be dependent on the cooling magnetic field. For sufficiently large $H_{cool} > 5$ kOe, the value of H_{EB} for the doped compound is about twice smaller than that for the parent compound. This difference was attributed to a lower exchange coupling and higher saturation magnetization of the FM phase of $(Nd_{0.9}Y_{0.1})_{2/3}Ca_{1/3}MnO_3$ at low temperatures.

Tunable EB was observed by Markovich et al. (2015a) in basically AFM electron-doped $Sm_{0.1}Ca_{0.6}Sr_{0.3}MnO_3$. The major part, almost 95% of the volume of this compound, undergoes first-order phase transition from paramagnetic to the C-type AFM phase at the Néel temperature $T_{N-C} \approx 150$ K. The transition is accompanied by a structural transition from $Pnma$ to $P2_1/m$ space group. The remaining part ($\sim 5\%$ of the total volume) undergoes transition to the G-type AFM phase at the Néel temperature $T_{N-G} \approx 70$ K. This phase exhibits very weak spontaneous magnetization, $M_0 \approx 0.19$ emu/g at $T = 10$ K. It was found that the phase separation, into two different AFM phases and an FM-like phase at the temperatures below T_{N-G}, leads to unusual magnetic properties, such as narrowing of magnetic hysteresis loops in field cooling process; unconventional EB effect associated with spontaneous magnetization at temperatures below T_{N-C}; strong magnetic field dependence of the negative EB at fields lower that 100 Oe, turning into practically field-independent one for fields above 0.5 kOe; and significant shift of EB with temperature, with a change of the sign from negative at 10 K to positive above 40 K. Temperature variations and change in the sign of the EB effect on increasing the temperature were ascribed to a competition between two mechanisms. According to the first one, strong negative EB appears in phase-separated G-type AFM regions, with $Pnma$ structure containing FM clusters. According to the second one, positive EB appears due to the formation of domain walls at disordered interface between structurally and magnetically different AFM phases (C and G types) during field cooling. Alternatively, positive EB may be induced by FM/AFM exchange coupling between FM clusters in the G-type AFM phase, with surface adjacent to the surrounding C-type AFM phase. The competition between the two mechanisms may be responsible for non-monotonic temperature variation of the hysteresis loop parameters and excessive evaluated size of FM droplets.

Markovich et al. (2015b) have investigated the EB effect in $CaMn_{0.9}Nb_{0.1}O_3$. With increasing Nb doping, the ground state of $CaMn_{1-x}Nb_xO_3$ ($0 \leq x \leq 0.1$) evolves from the

G-type AFM state with a weak FM component (for $x = 0$–0.08) to mostly C-type AFM state associated with charge ordering ($x = 0.1$) (see, Markovich et al., 2015c). For Nb content $x = 0.1$, a pronounced EB effect occurs. It was found that both vertical and horizontal shifts in magnetic hysteresis loop of field-cooled sample decrease monotonously with increasing temperature and vanish above 70 K, while the coercivity disappears only above 90 K, on approaching the Néel temperature. An exponential dependence, typical for various systems with frustrated interactions, successfully describes temperature variations of $H_{EB}(T)$ and $M_{EB}(T)$. The observed EB features can be understood in the framework of a scenario in which interface exchange coupling between FM domains and the G-type AFM phase is responsible for the EB effect. Moreover, the presence of different interfaces between coexisting magnetic phases may also contribute to the EB effect. The EB effect in $CaMn_{0.9}Nb_{0.1}O_3$ may be induced by the Dzyaloshinskii–Moriya interaction across the interface between FM clusters and the surrounding G-type AFM phase. Measurements of the hysteresis loops up to $H_{max} = \pm 90$ kOe demonstrated that the magnetization at $H > 20$ kOe increases in a non-linear way as a signature of a metamagnetic-like transition. It was found that such field-induced phase transformation in $CaMn_{0.9}Nb_{0.1}O_3$ completely suppresses the EB effect.

The EB effect was also reported for single crystals. Xia et al. (2009) have found that doping induced EB effect in hexagonal $Y_{0.95}Eu_{0.05}MnO_3$ single crystal. The X-ray diffraction (XRD) of the studied single crystal showed that Eu doping resulted in an expansion of the MnO_5 bipyramids along the c-axis and the contraction of the Mn-O bonds, forming trimerization in the ab-plane, which is the intrinsic mechanism of the change of the magnetic anisotropy. Magnetization measurements showed that Eu doping causes a spin-glass (SG) behavior and large zero-field-cooled EB effect, $H_{EB} = 735$ Oe. According to the authors, SG phase results from the intrinsic magnetic frustration and the disorder caused by Eu doping, and the coupling between the FM SG phase and the frustrated AFM phase gives rise to the EB effect.

Giri et al. (2014) found that in polycrystalline $Sm_{0.35}Pr_{0.15}Sr_{0.5}MnO_3$ manganite, the EB effect occurs along with tunneling magnetoresistance (MR). They observed a shift of the MR-H curve when the sample is cooled in static dc magnetic field down to 5 K, analogous to the shift in the magnetic hysteresis loop along the H-axis. The values of EB field, coercivity, remanence asymmetry, and magnetic coercivity were found to be strongly depending on H_{cool} (see Figure 10.3 for dependence of EB field [denoted in the figure as H_E] and remanence asymmetry [denoted in the figure as M_E] on H_{cool}). Figure 10.4 presents training effect for MR-H curve.

A decrease of the H_{EB} when the sample was successively field-cycled at a particular temperature (magnetic training effect is usually considered a hallmark of the EB effect, see, e.g., Nogués et al., 2005; Giri et al., 2014) have shown that $Sm_{0.35}Pr_{0.15}Sr_{0.5}MnO_3$ exhibits training effect in both independent experiments as shift of magnetic hysteresis loops as well as the shift of the MR-H curves.

Belik (2014) has studied magnetic properties of polycrystalline hexagonal $ScMnO_3$, $InMnO_3$, h-$YMnO_3$, $4H$-$SrMnO_3$, and $6H$-$SrMnO_3$ and perovskite o-$YMnO_3$, with a focus on the EB phenomenon. In these compounds, there are no magnetic ions at the A site; there is only one type of magnetic ions at the B site. All these compounds are very weak canted AFMs or pure AFMs. Negative EB was observed for all samples.

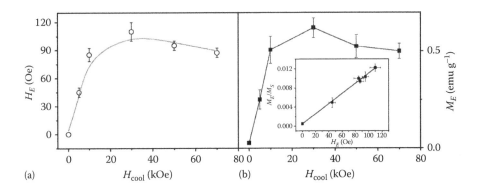

FIGURE 10.3 $Sm_{0.35}Pr_{0.15}Sr_{0.5}MnO_3$ at 5 K: (a) H_E versus cooling field (H_{cool}), (b) M_E versus cooling field. Inset shows the plot of scaled vertical shift with horizontal shift described in the text. Solid straight line is the guide to the eye for linear fit. (From Giri, S.K. et al., *IEEE. Trans. Magn.*, 50, 2000504 © 2014 IEEE.)

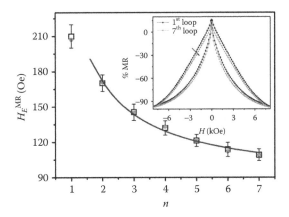

FIGURE 10.4 The n dependence of H_E^{MR} for $Sm_{0.35}Pr_{0.15}Sr_{0.5}MnO_3$ at 5 K after FC in 80 kOe. Solid line shows the best fitting with the power law to the data for $n \geq 2$. Inset shows training effect in MR-H curves up to seven successive cycles ($n = 7$). (From Giri, S.K. et al., *IEEE. Trans. Magn.*, 50, 2000504 © 2014 IEEE.)

Magnetic measurements (alternating and direct current) showed the presence of very small amount of Mn_3O_4 in $ScMnO_3$, h-$YMnO_3$, $4H$-$SrMnO_3$, and $6H$-$SrMnO_3$ and other magnetic impurities in $InMnO_3$. Hence, the EB was suggested to originate from pinned spins that exist at the interfaces of main phases and magnetic-impurity phases. Detailed studies of EB performed for mechanical mixtures of h-$YMnO_3$ and Mn_3O_4 showed apparent correlation between the amount of added impurity phase of Mn_3O_4 and the EB parameters, confirming the impurity origin of EB in the studied materials.

10.3 Exchange-Bias Effect and Magnetization-Reversal Phenomenon

Recently, several authors (e.g., Ren et al., 2000; Tung et al., 2007; Mandal et al., 2010; Mao et al., 2011; Dasari et al., 2012; Belik, 2013, 2014) discussed the relationship between magnetization-reversal phenomenon and tunable EB effect. The magnetization reversal consists of an alignment of spins in the direction opposite to that of an applied magnetic field. The magnetization-reversal effect was originally predicted and observed in some ferrimagnets with several magnetic sublattices. Although the magnetization-reversal effect was recently also observed in many perovskites, the origin of the effect in these compounds is still a matter of debates. Some researchers suppose that the origin of magnetization-reversal effect is intrinsic. Within this approach, ideal and homogeneous chemical composition of material without defects is assumed. Following this assumption, if ABO_3-type compound contains magnetic rare earth ions at the A site, the influence of magnetic ordering in the B sublattice on the A sublattice is postulated (through a negative exchange interaction) and coupling of A and B cations is considered. If there are no magnetic ions at the A site, but there are different ions at the B site, the magnetization-reversal effect is explained by the existence of different competing magnetic interactions between different ions at the B site. If there are no magnetic ions at the A site and there is only one type of cations at the B site, a competition between single-ion magnetic anisotropy and Dzyaloshinskii–Moriya interactions is the leading mechanism. The second approach consists, in assumption, that different inhomogeneities of a sample and uncompensated spins are responsible for magnetization reversal. Belik (2013) pointed out that when detailed studies are performed, the tunable EB effect is always found in materials exhibiting the magnetization-reversal effect. Hence, both phenomena should be closely related. Below, several recent results concerning this issue will be discussed briefly.

It was demonstrated (Belik, 2013) that the magnetization-reversal effect in the canted AFM $BiFe_{0.7}Mn_{0.3}O_3$ and $BiFe_{0.6}Mn_{0.4}O_3$ compounds is dependent on the magnetic history of the samples and measurement protocols. Magnetization reversal was not observed in the virgin sample measurements at temperature below T_N. The effect appeared only if the samples were magnetized or if they were cooled in small magnetic fields from temperatures above T_N. For both samples, magnetic measurements showed the existence of the FM component (*impurities*) with T_C above T_N, and the tunable EB effect was noticed. According to Belik, single-ion magnetic anisotropy and Dzyaloshinskii–Moriya interactions play roles in the formation of canted AFM states. However, the tunable EB effect and the magnetization-reversal effect, as a result, appear due to other reasons, such as the presence of magnetic impurities or sample inhomogeneities. These results confirm that tunable EB and magnetization-reversal phenomena are closely related.

Bora and Ravi (2013) have studied the sign reversal of magnetization and the EB field in $LaCr_{0.85}Mn_{0.15}O_3$. The $LaCrO_3$ is a perovskite compound with the G-type AFM structure, with T_N close to room temperature. It exhibits magnetization reversal when

La is partially or completely replaced by magnetic rare earth ions such as Pr, Ce, Yb, and Gd (see, e.g., Manna et al., 2010; Su et al., 2010; Yoshii, 2011). The $LaCr_{0.85}Mn_{0.15}O_3$ exhibits the sign reversal of magnetization and the tunable H_{EB}. As shown in Figure 10.5, both magnetization and H_{EB} values were found to undergo sign reversal at a compensation temperature at around 100 K.

The authors attribute magnetization-reversal effect to the competition between the paramagnetic component of doped Mn^{3+} ions under the influence of negative internal field due to AFM ordering of Cr^{3+} ions and the canted FM component of Cr^{3+} ions. The sign reversal of H_{EB} is also explained to occur due to the coupling between these two components. In the studied material, the EB is observed in a wide temperature range and can be tuned from negative to positive values by varying the temperature.

Huang et al. (2011) have studied the impact of Ca content on intrinsic inhomogeneity-induced EB effect in $La_{1-x}Ca_xMnO_3$ (0.55 $\leq$ x $\leq$ 0.95). They found that EB shows a maximum value of about 2200 Oe around x = 0.60, which is mainly affected by the AFM anisotropy and exchange energy at the FM/AFM interface. For $x \geq 0.85$, the spin-canted G-type AFM phase emerges, and a competition between the reduced C-type and increased G-type AFM fractions results in another EB-effect peak near x = 0.90.

Karmakar et al. have studied the EB effect in $Pr_{0.5}Eu_{0.5}MnO_3$ (Karmakar et al., 2011a) and $Eu_{0.5}Sm_{0.5}MnO_3$ (Karmakar et al., 2011b). For $Pr_{0.5}Eu_{0.5}MnO_3$ polycrystalline manganite, the Rietveld refinement of XRD pattern at room temperature showed significant Jahn–Teller distortion due to ionic size mismatch of the isovalent ions. This structural distortion strongly influences the magnetic properties of the compound. Alternating (AC) and direct current (DC) magnetization measurements revealed a disordered AFM state with glassy magnetic behavior at low temperature. Apparent EB effect ($H_{EB} \approx$ 900 Oe at 5 K, for H_{cool} = 5 kOe) was observed; however, the absence of training effect was noticed. The authors suggested a novel low-temperature phase separation between a disordered AFM phase induced by the Mn moments and another highly anisotropic phase involving ordering related to the rare earth moments. According to them, this unique magnetic ground state exhibiting EB is a consequence of the structural distortion inherent to the system.

The $Eu_{0.5}Sm_{0.5}MnO_3$ manganite shows strong orthorhombic distortion (Karmakar, et al. 2011b). The static Jahn–Teller effect modifies the low-temperature magnetic properties, creating a complicated anisotropic magnetic ground state. A low-temperature magnetic phase separation between an A-type AFM phase and a low-temperature highly anisotropic phase originating from another ordering at the rare earth sublattice leads, also in this case, to strong EB effect. The authors pointed out that the detailed structural analysis ruled out any structural inhomogeneity in the studied compound. Hence, the observed phase separation, in this structurally single-phase compound, is strictly magnetic in nature. Cooling field dependence of EB at high H_{cool} and the absence of the training and ageing effects indicate a highly stable magnetic moment at the interface.

Xu et al. (2013) studied the impact of tiny difference in the radius of isovalent ions substituted at the A site in $Tb_{0.4}Dy_{0.6}MnO_3$ manganite. Evidence of the EB effect ($H_{EB} \approx$ 230 Oe), which originates from interfacial exchange interactions between the AFM and FM phases, was obtained at low temperatures. When Dy content is close to 0.6 (≈ 0.59),

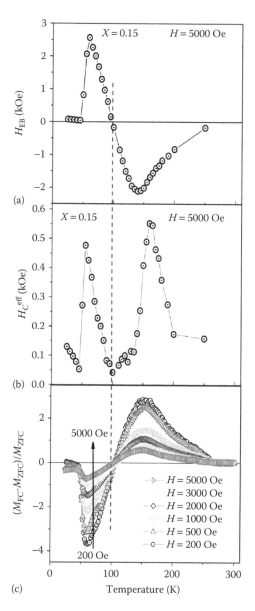

FIGURE 10.5 Temperature variations of (a) H_{EB}, (b) H_C^{eff}, and (c) irreversible magnetization for LaCr$_{0.85}$Mn$_{0.15}$O$_3$. (Reprinted with permission from Bora, T., and Ravi, S., *J. Appl. Phys.*, 114, 183902. Copyright 2013, American Institute of Physics.)

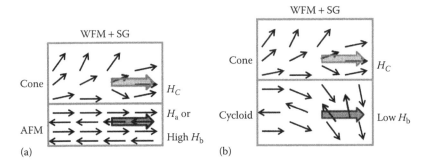

FIGURE 10.6 Schematic diagram of the spin configuration of (a) AFM/WFM and (b) WFM/cycloid AFM interfaces after field cooling below Néel temperature, T_N, in $Eu_{0.75}Y_{0.25}MnO_3$. (Reprinted from *J. Magn. Magn. Mater.*, 324, Yan, L.Q. et al., Exchange bias effect in multiferroic $Eu_{0.75}Y_{0.25}MnO_3$, 2579–2582, Copyright 2012, with permission from Elsevier.)

the Mn spin configuration changes from sinusoidal AFM to a transverse-spiral one. In addition, there is a complex interplay of $3d$ and $4f$ magnetism in this manganite. Both factors lead to the formation of an FM order at the rare earth sublattice. For Dy content close to 0.6, there is a strong competition between the FM and AFM interactions, and the EB effect appears.

Yan et al. (2012) have reported the EB effect in multiferroic $Eu_{0.75}Y_{0.25}MnO_3$ manganite. In this compound, at a microscopic scale, magnetic field changes a cycloid spin structure to a cone spin structure. The SG phase appears due to weak pinning energy at the cycloid/weak ferromagnetic (WFM) domain interface and site disorder. The EB effect is observed below the SG freezing temperature of about 16 K. The EB field (of about 1500 Oe at low temperatures), coercivity field, and remanent magnetization increase with increasing H_{cool}. The EB effect was ascribed to the frozen uncompensated spins at the AFM/WFM interfaces in the SG-like phase. A new type of EB system based on H, modifying a cycloid spin structure to a cone spin structure at a microscopic scale, is demonstrated in multiferroic $Eu_{0.75}Y_{0.25}MnO_3$. The authors suggested that a lot of uncompensated spins form an SG-like phase at the AFM/WFM phase interfaces due to weak pinning energy of the AFM layer and strong uniaxial anisotropy along the c-axis. Hence, as illustrated in Figure 10.6, the SG-like phase should partly stem from the weak exchange interaction of uncompensated cycloid AFM to weak FM cluster induced by cooling field.

10.4 Exchange-Bias Effect in Double Perovskites

The $A_2BB'O_6$ double perovskites exhibit intriguing properties, such as half metallicity, high-temperature ferrimagnetism, and a rich variety of magnetic interactions, that cannot be realized in simpler materials. The EB effect was also found in these compounds. Murthy et al. (2013) have studied phase-separated $La_{1.5}Sr_{0.5}CoMnO_6$. They reported a spontaneous hysteresis loop shift after zero-field cooling. This effect, called spontaneous

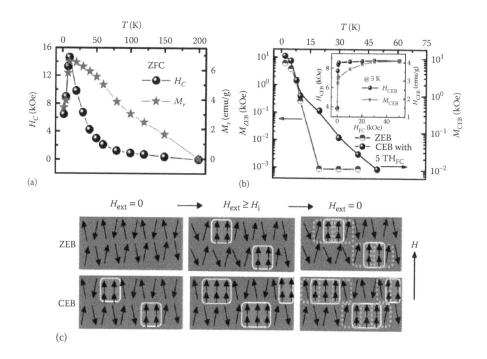

FIGURE 10.7 Temperature dependence of (a) H_C and M_r for $La_{1.5}Sr_{0.5}CoMnO_6$, (b) H_{ZEB} and H_{CEB}; the inset show the H_{FC} dependence of conventional H_{CEB} and M_{CEB} at 5 K. (c) Schematic diagram of isothermal magnetization process with spin configurations for zero-field-cooled (ZEB) and field-cooled (CEB) cases at $T < T_{CAF}$. (Reprinted with the permission from Murthy, J.K., and Venimadhav, A., *Appl. Phys. Lett.*, 103, 252410. Copyright 2013, American Institute of Physics.)

exchange-bias (SEB) or zero-field-cooled exchange-bias (ZEB) effect, was found to be very large (~ 0.65 T) in $La_{1.5}Sr_{0.5}CoMnO_6$. Magnetic study revealed a re-entrant SG at around 90 K, phase separation to SG and FM phases below 50 K, and a transition to canted AFM structure at about 10 K. They observed a small conventional exchange bias (CEB) due to spontaneous phase separation down to 10 K. Giant SEB and enhanced CEB effects were found only below 10 K (Figure 10.7) and were attributed to the large unidirectional anisotropy at the interface of isothermally field-induced FM phase and canted AFM background.

Recently, Giri et al. (2016) have investigated the EB effect in another bulk double perovskite, $Sm_{1.5}Ca_{0.5}CoMnO_6$. As one can see in Figure 10.8, they found both CEB and the enormously large value of the SEB (~ 0.51 T) in this compound after zero-field cooling from an unmagnetized state. Magnetization studies (ac and dc) showed a super-spin-glass-like (SSG) state below 19 K. The EB effects were observed below the glassy transition temperature. The CEB and SEB effects increased with a decreasing temperature and showed monotonic variation. According to the authors, both effects occur due to the multimagnetic state of this double perovskite. An SSG interface with an AFM core forms a local spin interface. This sets the unidirectional anisotropy along a particular direction at the interface.

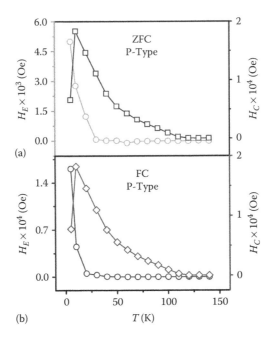

FIGURE 10.8 $Sm_{1.5}Ca_{0.5}CoMnO_6$: (a) Temperature-dependent value of EB field H_E (circles) and coercive field H_C (squares) in the ZFC case and (b) temperature-dependent H_E (circles) and H_C (squares) after FC at $H = 30$ kOe. (Reproduced with permission from Giri, S.K. et al., *J. Phys. D: Appl. Phys.*, 49, 165002, 2016, Copyright 2016, American Institute of Physics.)

Another example of SEB in double perovskites was recently reported by Murthy et al. (2016), who studied $La_{2-x}Sr_xCoMnO_6$ ($0 \leq x \leq 1$). According to the authors, the effect is driven by an antisite disorder resulting from hole carrier (Sr^{2+}) doping. An increase in disorder with an increase of Sr content up to $x = 0.5$, and next, a decrease of doping ranging from $x = 0.5$ to 1 were evinced by XRD and Raman spectroscopy. The antisite disorder at the B site interrupts the long-range FM order by introducing various magnetic interactions and initiates re-entrant glassy dynamics, phase separation, and canted AFM behavior with decreasing temperature. The authors have concluded that this leads to a novel magnetic microstructure with unidirectional anisotropy that causes SEB effect that can be tuned with the amount of antisite disorder.

It should be mentioned that, at first, the SEB effect was considered as an experimental artifact. However, several observations of this effect in various compounds, for example, in $NiMnIn_{13}$ and Mn_2PtGa Heusler alloys (Nayak et al., 2013), $BiFeO_3$-Bi_2Fe4O_9 nanocomposites (Maity et al., 2013), and $La_{0.67}Sr_{0.33}MnO_3$/$PbZr_{0.8}Ti_{0.2}O_3$/$La_{0.67}Sr_{0.33}MnO_3$ heterostructure (Mao et al., 2013), provided firm confirmation of the existence of effect. In the last case of the sandwich system, the authors suggested that the strain in the heterostructure is responsible for the SEB effect.

10.5 Impact of Hydrostatic Pressure on Exchange-Bias Effect

Hydrostatic pressure turns out to be a useful tool in studies of the nature of the EB effect in several compounds. Often, quite moderate pressure (order of 10 kbar) affects magnetic properties of manganites, including parameters characterizing EB, and in some cases, it may even be used to significantly tune these parameters.

Fita et al. (2012) observed strong enhancement of EB effect under pressure in phase-separated $CaMn_{0.9}Ru_{0.1}O_3$ manganite. The ruthenium (Ru)-doped manganites belong to an interesting class of materials with strongly pressure-dependent ferromagnetism (Markovich et al., 2004). The most interesting example of such material is low-doped $CaMn_{0.9}Ru_{0.1}O_3$, comprising of both the G-type AFM and FM phases in the same *Pnma* crystallographic structure. In this compound, saturated magnetization is suppressed by one order of magnitude at a modest pressure of 10 kbar (Figure 10.9).

The intricate behavior is caused by an exotic pressure-induced valence transition from $(Mn^{3+}-Ru^{5+})$ pair to $(Mn^{4+}-Ru^{4+})$ pair, leading to a diminishing of the number of Mn^{3+} species, controlling the FM double-exchange interaction in the system and, therefore, leading to a suppression of the volume of the FM phase.

In $CaMn_{0.9}Ru_{0.1}O_3$, a significant pressure-induced EB effect was observed. As one can see in Figure 10.10, both EB field, H_{EB}, and remanence asymmetry, M_{EB}/M_S, monotonically increase with pressure, while the coercive field, H_C, shows a distinct non-monotonic variation.

The $H_C(P)$ dependence is well described, considering a system of FM size-variable particles. A strong increase of H_{EB} under pressure is attributed to the suppression of the FM phase, that is, to the decrease in size of the FM clusters embedded in the AFM matrix. The results showed that in phase-separated FM/AFM systems, the EB effect is highly sensitive to an applied pressure because of pressure-dependent size of FM clusters. One may say that, to some extent, the morphology of phase-separated $CaMn_{0.9}Ru_{0.1}O_3$ under pressure resembles the behavior of size-variable FM nanoparticles embedded in the AFM matrix. For such system, the strong reduction in saturated FM moment, associated with an increase in surface-to-volume ratio, favors the EB effect in accordance with intuitive Meiklejohn-Bean (MB) model (Nogués et al., 2005).

In later studies, Fita et al. (2013a) have investigated the pressure effect (up to 11 kbar) on EB and coercive fields in phase-separated FM/AFM $CaMn_{1-x}Ru_xO_3$ manganites in broader Ru-doping range, with $0.06 \leq x \leq 0.15$. They found that both H_{EB} and H_C exhibit intriguing dependence on Ru doping and on applied pressure. It was found that H_{EB} is apparent only at $x \leq 0.1$ and decreases progressively with increasing doping, while H_C depends nonmonotonically on Ru content, reaching a maximum at around $x = 0.09$. The H_{EB} was found to increase strongly under applied pressure within the doping range, $0.06 \leq x \leq 0.1$, while H_C exhibits quite irregular behavior—it increases with increasing pressure for $x = 0.15$, changes nonmonotonically for $x = 0.1$, decreases for $x = 0.08$, and is almost invariable for $x = 0.06$. Complex pressure and Ru-doping effects on H_{EB} and H_C were explained within a model involving size-variable nanoscale FM regions (droplets) embedded in an AFM matrix. The enhancement of EB with increasing pressure was

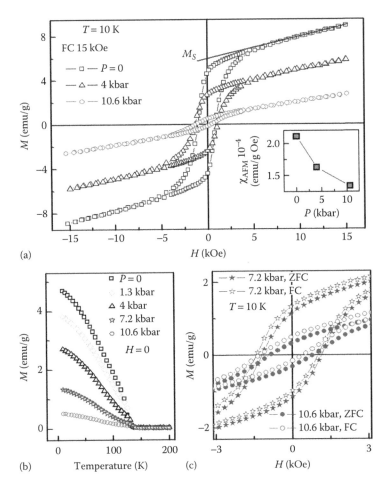

FIGURE 10.9 (a) Hysteresis loops of the FC magnetization of $CaMn_{0.9}Ru_{0.1}O_3$ measured after cooling in $H_{cool} = 15$ kOe at 10 K under several pressures. Inset shows changes in the slope of high-field magnetization, an indicative of strengthening of AFM phase with pressure, (b) thermoremanent magnetization curves at various pressures recorded at $H = 0$ after FC at 15 kOe, and (c) symmetric ZFC (closed symbols) and asymmetric FC (open symbols) hysteresis loops at two different pressures. (Reprinted with permission from Fita, I. et al., *J. Appl. Phys.*, 111, 113908. Copyright 2012, American Institute of Physics.)

attributed to the reduction in the FM droplet size, evinced by both pressure dependence of spontaneous FM moment and H_{EB} dependence on cooling field, H_{cool}. The impact of FM droplet size on the EB was further evinced by the magnetic field effect, which, in contrast to the pressure effect, leads to a growth of the FM droplets. The intricate H_C dependence on both pressure and Ru content may be explained by assuming the transition from the multidomain state to the single-domain state, induced by droplet size decrease with increasing pressure or with lowering doping. It turns out that external

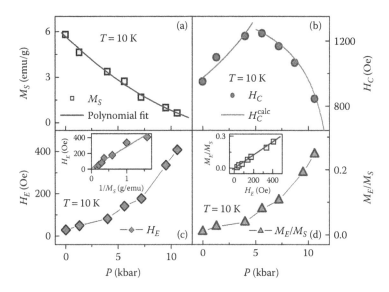

FIGURE 10.10 (a) Pressure dependence of the saturated magnetization, M_S, of $CaMn_{0.9}Ru_{0.1}O_3$ at 10 K. (b) Pressure dependence of the coercive field, H_C, at 10 K. Solid lines are calculated for both different states: for multidomain FM clusters (below 5 kbar) and for single-domain FM clusters (above 5 kbar). (c) Pressure dependence of the EB field, H_E, at 10 K. Inset shows a linear fit of H_E versus $1/M_S$. (d) Pressure dependence of the normalized remanence asymmetry, M_E/M_S, at 10 K. Inset shows a linear correlation between M_E/M_S and H_E. (Reprinted with permission from Fita, I. et al., *J. Appl. Phys.*, 111, 113908. Copyright 2012, American Institute of Physics.)

pressure appears to be an effective tool for controlling the EB and the coercivity in Ru-doped $CaMnO_3$ perovskites.

It was found that pressure strongly affects magnetic properties of $CaMn_{1-x}Nb_xO_3$ (Markovich et al., 2015c). Substitution of Nb^{5+} ion for the Mn^{4+} site of the parent matrix causes one-electron doping, accompanied by a monotonous increase of the lattice parameters, unit-cell volume, and average Mn–O bond distance and a decrease in Mn–O–Mn bond angle with increasing x. Spontaneous magnetization increases sharply with increasing doping level, approaches a maximal value of 4.1 emu/g at $T = 10$ K for $x = 0.08$, and then decreases rapidly to reach a very small value of 0.2 emu/g for $x = 0.1$. Anomalous negative magnetization behavior below the magnetic transition temperature has been observed for the compound with $x = 0.04$ in the field-cooled magnetization and remanent dc magnetization measurements. Vertical and horizontal shifts of the hysteresis loop of the field-cooled sample have been observed for $CaMn_{0.9}Nb_{0.1}O_3$ as possible signatures of the EB effect. The effect of hydrostatic pressure on dc magnetization for the sample with $x > 0.02$ revealed a significant increase in the FM phase volume under pressure, linked to progressive suppression of a negative magnetization in $x = 0.04$ sample. Application of the hydrostatic pressure resulted in a significant increase of the magnetization. Correspondingly, this leads to both progressive suppression of negative magnetization in $x = 0.04$ sample and the EB effect in $CaMn_{0.9}Nb_{0.1}O_3$. It was suggested that such behavior is related to the pressure-induced increase in the size of FM droplets embedded in AFM matrix.

10.6 Comparison of Exchange-Bias Effect in Manganites and in Other Perovskites

In FM perovskite cobaltites, $La_{1-x}M_xCoO_3$ (M = Ca, Sr, Ba), because of strong competition between the crystal-field splitting energy, Δ_{cf}, and the intra-atomic exchange interaction, J_{ex}, Co^{3+} ion exhibits three alternative spin configurations: the nonmagnetic low-spin (LS) ($S = 0$), intermediate-spin (IS) ($S = 1$), and high-spin (HS) ($S = 2$) states. Ions with different spin states may coexist in crystals, and a switching between them depends strongly on doping, temperature, and changes in crystal structure. A connection between the variable spin state and the lattice is an additional source for phase separation in cobaltites. Usually, they exhibit a ground state with coexisting hole-reach FM regions, SG regions, and hole-poor LS regions (Kuhns et al., 2003). The EB effect has been carefully investigated in low-doped $La_{1-x}Sr_xCoO_3$ ($x \leq 0.18$) cobaltites, in which it was ascribed to the pinning effect at the FM/SG interface (Tang et al., 2006; Huang et al., 2008). The existence of the EB effect was also reported for $La_{1-x}Ba_xCoO_3$ in a wide doping range, $0.18 \leq x \leq 0.5$ (Luo and Wang, 2007); however, it was further argued that the observed effect is not a true EB, but rather, it is an effect of minor (non-saturated) hysteresis loops (Geshev, 2008b). Fita et al. (2013) have studied the magnetic properties of $La_{0.9}Ba_{0.1}CoO_3$ cobaltite, which at low temperatures exhibit the FM cluster-glass behavior and EB effect. They found that pressure increases the temperature of emergence of FM clusters ($T_C^{cl} \approx$ 190 K at ambient pressure) with the rate $dT_C^{cl}/dP \approx 0.5$ K/kbar and also enlarges the FM phase volume in the sample. The EB field H_{EB} increases sharply below the SG-like transition temperature, $T_f \approx 38$ K, in consequence of the exchange coupling at the FM/ SG interfaces. Applied pressure was found to strongly suppress the EB; in particular, H_{EB} is reduced by a factor of ~ 2 under pressure of 10 kbar at 30 K. It appears that H_{EB} varies oppositely to the changes in the FM phase volume, very similar to the behavior observed for phase-separated manganites with FM/AFM interfaces. Overall, the pressure-induced decrease in H_{EB} is explained by considering an increase of the FM cluster size and by related decrease of distance between clusters. The results showed that in low-doped $La_{0.9}Ba_{0.1}CoO_3$ cobaltite, the EB may be controlled by external pressure through variation in size of the FM clusters embedded in an SG matrix. In general, this behavior resembles the pressure-tuned EB observed in manganites exhibiting different morphology of phase separation, including the FM/AFM interfaces.

Magnetic properties of bulk polycrystalline Mn-doped $SrRu_{1-x}Mn_xO_3$ perovskite at doping range $0.2 \leq x \leq 0.3$ were studied by Fita et al. (2016a). $SrRuO_3$ is a unique FM metal among $4d$ transition-metal-based perovskite oxides, exhibiting an itinerant ferromagnetism with the Curie temperature, $T_C = 163$ K, and a considerably reduced moment in its ground state, ~ 1.6 μ_B per Ru atom. No signature of EB was detected in bulk $SrRuO_3$ (Klein, 2006). Conversely, a remarkable EB effect was observed in strained thin films (Sow et al., 2014) and in ultrathin films (Xia et al., 2009) below the critical

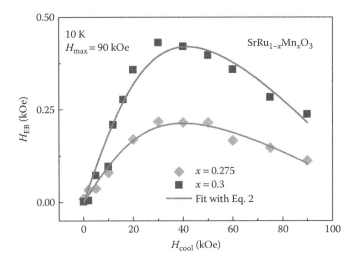

FIGURE 10.11 Dependence of H_{EB} at 10 K on cooling field, H_{cool}, for $SrRu_{1-x}Mn_xO_3$ with $x = 0.275$ and $x = 0.3$. The solid lines represent the best fit with Equation 10.1. (Reprinted from *Ceramics International*, 42, Fita I. et al., Exchange bias driven by the structural/magnetic transition in Mn-doped $SrRuO_3$, 8453–8459. Copyright (2016), with permission from Elsevier.)

thickness of 4 or 5 monolayers, suggesting the existence of both FM and AFM interacting regions. It was found that an apparent EB effect emerges with increasing Mn doping in $SrRu_{1-x}Mn_xO_3$ at $x = 0.25$ following the FM-to-AFM phase transition, which is accompanied by the change in structure symmetry, and then, the EB field, H_{EB}, increases significantly in doping range $0.25 < x < 0.3$. The EB was verified by both H_{cool} dependence of the H_{EB} and the training effect. As shown in Figure 10.11, a markedly nonmonotonic H_{EB} vs H_{cool} dependence, with maximum at around 40 kOe, was found, resembling the behavior of phase-separated EB system.

Moreover, a clear analogy with classic EB system of $Pr_{1/3}Ca_{2/3}MnO_3$ was found, strongly suggesting that the EB effect in $SrRu_{1-x}Mn_xO_3$ originates from exchange interactions at the interface of nanoscale FM clusters (size of ~ 1.5 nm at $x = 0.3$) coexisting together with the dominant AFM phase at the boundary of the first-order FM/AFM transition. The training effect observed is well understandable within the spin-configuration relaxation model (Mishra et al., 2009) that indicates important contribution from the AFM domains' rearrangement at the interface.

The EB reversal in magnetically compensated $ErFeO_3$ single crystal was recently reported by Fita et al. (2016b). The $ErFeO_3$ is a representative of the rare earth orthoferrites that have received renewed attention in the last years because of their attractiveness for applications, including ultrafast spin switching, spin reorientation transition, and multiferroicity (see, e.g., de Jong et al., 2011; Shen et al., 2013). With decreasing temperature, $ErFeO_3$ exhibits a sequence of magnetic transitions keeping the same orthorhombic *Pbnm* perovskite structure. Below $T_N \approx 636$ K, the Fe^{3+} spins demonstrate the *G*-type AFM order with slight spin canting, caused by the Dzyaloshinskii–Moriya interaction, resulting in a weak FM moment along the *c*-axis. With lowering temperature, the Fe^{3+}

spins spontaneously reorient following two successive second-order phase transitions starting at 97 K and ending at 88 K, leading to a reorientation of the FM moment from the c-axis toward the a-axis. At lower temperature, $ErFeO_3$ exhibits a magnetic compensation, which results from the strong AFM coupling between the Er^{3+} and Fe^{3+} magnetic moments. Owing to this coupling, the Er^{3+} spins, despite being in paramagnetic state, develop an alternative canted AFM order with an FM moment opposite to that of the Fe^{3+} spins. The induced Er-sublattice magnetic moment increases with lowering temperature and compensates the moment of Fe^{3+} spins at the compensation temperature $T_{comp} = 45$ K, and the net magnetization equals zero. At lowest temperatures, the long-range AFM order of Er^{3+} spins develops. Fita et al. (2016b) have shown that the EB appears in the vicinity of the compensation point, increases on approaching T_{comp}, and changes sign across T_{comp}. Both H_{EB} and H_C fields diverge below T_{comp}, and they are restricted above T_{comp}. The EB was found to depend critically on thermal history, as hysteresis loops were measured in field-cooled (FC) and field-cooling-warming (FCW) modes. The following features were observed: the EB sign is generally negative for T above T_{comp} in the case of FC and below T_{comp} in the case of FCW; on the other hand, both for FC and FCW, the EB sign is positive just after overpassing T_{comp}. Hence, as shown in Figure 10.12, changing the cooling/warming protocol switches the EB sign for the same $|H_{EB}|$ and H_C values.

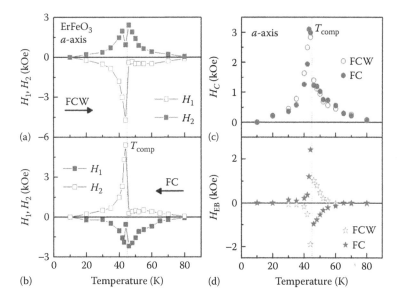

FIGURE 10.12 (a, b) Temperature variation of coercive fields H_1 and H_2 at the first and second magnetization reversals, respectively, obtained following different cooling procedure: FC in 10 kOe from 300 to 10 K and then warming to the given T—abbreviated in the figure as FCW, and FC in 10 kOe from 300 K to the given T—abbreviated in the figure as FC. (c, d) Average coercive field H_C (c) and exchange-bias field H_{EB} (d) around the compensation point T_{comp}, obtained for FCW and FC protocols. Remarkably, both coercive H_1 and H_2 fields and the EB field obtained with FCW exhibit the mirror behavior of those obtained with FC. (Reprinted with permission from Fita, I. et al., *Phys. Rev. B*, 93, 184432, 2016. Copyright 2016 by the American Physical Society.)

The negative EB is compatible with the equilibrium spin configuration, and the positive EB is compatible with the metastable state. This relevant feature apparently reminds one of the possibilities of electric-field-induced switching of the EB sign in a magnetoelectric Cr_2O_3/CoPt heterostructure (Borisov et al., 2005).

10.7 Concluding Remarks

The driving force for the EB effect in bulk manganites is an intrinsic phase separation. The simplified exchange interaction model proposed by Niebieskikwiat and Salamon (2005) is very useful in description of many aspects of the effect. In particular, predicted by the model, a simple linear relationship between the remanence asymmetry, M_{EB}, saturation magnetization, M_S, and the H_{EB}: $M_{EB}/M_S \propto -H_{EB}$ and a formula describing the dependence of the exchange field H_{EB} on the cooling field H_{cool} for systems with small-sized FM clusters embedded in an AFM matrix hold in majority of the cases. In general, the EB effect requires the interfaces or inhomogeneities. The proper estimation of magnetization curve parameters is fundamental in studying EB effect. In particular, if shifts of the hysteresis loop strongly depend on maximum applied magnetic field, H_{max}, and are completely eliminated if this filed is high enough, the observed shifts may be due to minor loop effect. However, the EB effect in bulk manganites may appear when the overwhelming magnetic volume of the studied samples is occupied by AFM matrix and hysteresis loops are not saturated at all. Then, one could conclude about the EB effect if the hysteresis loop is *effectively saturated*, for example, if the ascending and descending branches of hysteresis loop coincide for fields higher than the anisotropy field. Parameters of the EB effect may be very effectively tuned by even tiny change in the chemical composition and moderate external hydrostatic pressure, due to their impact on AFM/FM phases ratio. The EB effect in manganites may be induced by the Dzyaloshinskii–Moriya interaction across the interface between FM clusters and the surrounding AFM matrix. Both positive and negative EB effects are observed in perovskites. In some perovskite compounds, a spontaneous hysteresis loop shift after zero-field cooling (called spontaneous EB or zero-field-cooled EB effect) was found to occur and was very large (> 0.5 T). In several compounds, a connection between magnetization-reversal phenomenon (alignment of spins in the direction opposite to that of an applied magnetic field) and tunable EB effect was observed.

Acknowledgment

This work was partly supported by the Polish NCN grant 2014/15/B/ST3/03898.

References

Ahn CH, Rabe KM, Triscone JM (2004) Ferroelectricity at the nanoscale: Local polarization in oxide thin films and heterostructures. *Science* 303:488–491.

Allodi G, De Renzi R, Guidi G (1998) [139]La NMR in lanthanum manganites: Indication of the presence of magnetic polarons from spectra and nuclear relaxations. *Phys Rev B* 57:1024–1034.

Anderson PW, Hasegawa H (1955) Considerations on double exchange. *Phys Rev* 100:675–681.

Belik AA (2013) Origin of magnetization reversal and exchange bias phenomena in solid solutions of $BiFeO_3$–$BiMnO_3$: Intrinsic or extrinsic? *Inorg Chem* 52:2015–2021.

Belik AA (2014) Negative exchange bias in polycrystalline hexagonal $ScMnO_3$, $InMnO_3$, $YMnO_3$, $4H$-$SrMnO_3$, and $6H$-$SrMnO_3$ and perovskite $YMnO_3$: Effects of impurities. *J Phys Soc Jpn* 83:074703.

Bora T, Ravi S (2013) Sign reversal of magnetization and exchange bias field in $LaCr_{0.85}Mn_{0.15}O_3$. *J Appl Phys* 114:183902.

Borisov P, Hochstrat A, Chen X, Kleemann W, Binek C (2005) Magnetoelectric switching of exchange bias. *Phys Rev Lett* 94:117203.

Coey JMD, Viret M, von Molnár S (2009) Mixed-valence manganites. *Advances in Physics* 58:571–697.

Dagotto E, Burgy J, Moreo A (2003) Nanoscale phase separation in colossal magnetoresistance materials: Lessons for the cuprates? *Solid State Commun* 126:9–22.

Dasari N, Mandal P, Sundaresan A, Vidhyadhiraja NS (2012) Weak ferromagnetism and magnetization reversal in $YFe_{1-x}Cr_xO_3$. *Europhys Lett* 99:17008.

De Gennes PG (1960) Effects of double exchange in magnetic crystals. *Phys Rev* 118:141–153.

De Jong JA, Kimel AV, Pisarev RV, Kirilyuk A, Rasing Th (2011) Laser-induced ultrafast spin dynamics in $ErFeO_3$. *Phys Rev B* 84:104421.

Dorr K (2006) Ferromagnetic manganites: Spin-polarized conduction versus competing interactions. *J Phys D: Appl Phys* 39:R125–R150.

Fertman EL et al. (2015) Exchange bias phenomenon in $(Nd_{1-x}Y_x)_{2/3}Ca_{1/3}MnO_3$ ($x = 0, 0.1$) perovskites. *Low Temperature Physics* 41:1001–1005.

Fita I et al. (2012) Pressure-induced exchange bias effect in phase-separated $CaMn_{0.9}Ru_{0.1}O_3$. *J Appl Phys* 111:113908.

Fita I et al. (2013a) Pressure-tuned exchange bias and coercivity in Ru-doped $CaMnO_3$. *Phys Rev B* 88:064424.

Fita I, Puzniak R, Wisniewski A, Markovich V, Troyanchuk IO, Pashkevich YuG (2013b) Pressure enhanced ferromagnetism and suppressed exchange bias in $La_{0.9}Ba_{0.1}CoO_3$ cobaltite. *J Appl Phys* 114:153910.

Fita I et al. (2016a) Exchange bias driven by the structural/magnetic transition in Mn-doped $SrRuO_3$. *Ceramics Int* 42:8453–8459.

Fita I, Wisniewski A, Puzniak R, Markovich V, Gorodetsky G (2016b) Exchange-bias reversal in magnetically compensated $ErFeO_3$ single crystal. *Phys Rev B* 93:184432.

Geshev J (2008a) Comment on: Exchange bias and vertical shift in $CoFe_2O_4$ nanoparticles. *J Magn Magn Mater* 320:600–602.

Geshev J (2008b) Comment on: Cluster glass induced exchange biaslike effect in the perovskite cobaltites. *Appl Phys Lett* 93:176101.

Geshev J, Pereira LG, Skumryev V (2008) Comment on: Exchange bias dependence on interface spin alignment in a $Ni_{80}Fe_{20}/(Ni,Fe)O$ thin film. *Phys Rev Lett* 100:039701.

Geshev J (2009a) Comment on: Particle size dependent exchange bias and cluster-glass states in $LaMn_{0.7}Fe_{0.3}O_3$. *J Phys: Condens Matter* 21:078001.

Geshev J (2009b) Comment on: Exchange bias in the layered cobaltite $Sr_{1.5}Pr_{0.5}CoO_4$. *J Appl Phys* 105:066108.

Giri S, Patra M, Majumdar S (2011) Exchange bias effect in alloys and compounds. *J Phys: Condens Matter* 23:073201.

Giri SK, Das PT, Hazra SK, Nath TK (2014) Exchange bias effect concerned with tunneling magnetoresistance in $Sm_{0.35}Pr_{0.15}Sr_{0.5}MnO_3$ phase separated manganites. *IEEE Trans Magn* 50:2000504.

Giri SK, Sahoo RC, Dasgupta P, Poddar A, Nath TK (2016) Giant spontaneous exchange bias effect in $Sm_{1.5}Ca_{0.5}CoMnO_6$ perovskite. *J Phys D: Appl Phys* 49:165002.

Goodenough JB (2003) Rare earth - manganese perovskites in *Handbook on the Physics and Chemistry of Rare Earth*, K.A. Gschneidner Jr., J.-C.G. Bünzli, V. Pecharsky (Eds.), Vol. 33, Elsevier Science: Amsterdam, pp. 249–351.

Harres A, Geshev J, Skumryev V (2015) Comment on superspin glass mediated giant spontaneous exchange bias in a nanocomposite of $BiFeO_3$-$Bi_2Fe_4O_9$. *Phys Rev Lett* 114:099703.

Harres A, Mikhov M, Skumryev V, de Andrade AMH, Schmidt JE, Geshev J (2016) Criteria for saturated magnetization loop. *J Magn Magn Mater* 402:76–82.

Hennion M, Moussa F, Biotteau G, Rodrıguez-Carvajal J, Pinsard L, Revcolevschi A (1998) Liquidlike spatial distribution of magnetic droplets revealed by neutron scattering in $La_{1-x}Ca_xMnO_3$. *Phys Rev Lett* 81:1957–1960.

Hong F, Cheng Z, Wang J, Wang X, Dou S (2012) Positive and negative exchange bias effects in the simple perovskite manganite $NdMnO_3$. *Appl Phys Lett* 101:102411.

Huang WG et al. (2008) Intrinsic exchange bias effect in phase-separated $La_{0.82}Sr_{0.18}CoO_3$ single crystal. *J Phys: Condens Matter* 20:445209.

Huang XH, Jiang ZL, Sun XF, Li XG (2011) Exchange bias effect induced by the intrinsic inhomogeneity in $La_{1-x}Ca_xMnO_3$ ($0.55 \leq x \leq 0.95$) compounds. *J Am Ceram Soc* 94:1324–1326.

Karmakar A, Majumdar S, Giri S (2011a) Orthorhombic distortion and novel magnetic phase separation in $Pr_{0.5}Eu_{0.5}MnO_3$. *J Appl Phys* 110:063914.

Karmakar A, Majumdar S, Giri S (2011b) Structural and magnetic properties of spontaneously phase-separated $Eu_{0.5}Sm_{0.5}MnO_3$. *J Phys: Condens Matter* 23:136003.

Klein L (2006) Comment on: Exchange bias-like phenomenon in $SrRuO_3$. *Appl Phys Lett* 89:036101.

Kuhns PL, Hoch MJR, Moulton WG, Reyes AP, Wu J, Leighton C (2003) Magnetic phase separation in $La_{1-x}Sr_xCoO_3$ by [59]Co NMR. *Phys Rev Lett* 91:127202.

Kumar M, Pandey PK, Choudhary RJ, Phase DM (2014) Comment on: Zero-field cooled exchange bias in hexagonal $YMnO_3$ nanoparticles. *Appl Phys Lett* 104:156101.

Luo W, Wang F (2007) Cluster glass induced exchange biaslike effect in the perovskite cobaltites. *Appl Phys Lett* 90:162515.

Maezono R, Ishihara S, Nagaosa N (1998) Phase diagram of manganese oxides. *Phys Rev B* 58:11583–11596.

Maity T, Goswami S, Bhattacharya D, Roy S (2013) Superspin glass mediated giant spontaneous exchange bias in a nanocomposite of $BiFeO_3$–$Bi_2Fe_4O_9$. *Phys Rev Lett* 110:107201.

Mandal P et al. (2010) Temperature-induced magnetization reversal in $BiFe_{0.5}Mn_{0.5}O_3$ synthesized at high pressure. *Phys Rev B* 82:100416(R).

Manna PK, Yusuf SM, Shukla R, Tyagi AK (2010) Coexistence of sign reversal of both magnetization and exchange bias field in the core-shell type $La_{0.2}Ce_{0.8}CrO_3$ nanoparticles. *Appl Phys Lett* 96:242508.

Mao J et al. (2011) Temperature-and magnetic-field-induced magnetization reversal in perovskite $YFe_{0.5}Cr_{0.5}O_3$. *Appl Phys Lett* 98:192510.

Mao HJ, Song C, Cui B, Wang GY, Xiao LR, Pan F (2013) Room temperature spontaneous exchange bias in (La, Sr)MnO_3/$PbZr_{0.8}Ti_{0.2}O_3$/(La, Sr)MnO_3 sandwich structure. *J Appl Phys* 114:043904.

Markovich V et al. (2004) Effect of pressure on magnetic and transport properties of $CaMn_{1-x}Ru_xO_3$ ($x = 0$–0.15): Collapse of ferromagnetic phase in $CaMn_{0.9}Ru_{0.1}O_3$. *Phys Rev B* 70:024403.

Markovich V et al. (2015a) Unconventional exchange bias effect driven by phase separation in basically antiferromagnetic $Sm_{0.1}Ca_{0.6}Sr_{0.3}MnO_3$. *J Alloys Compd* 622:213–218.

Markovich V et al. (2015b) Exchange bias effect in $CaMn_{0.9}Nb_{0.1}O_3$. *Mat Chem Phys* 164:170–176.

Markovich V et al. (2015c) Evolution of magnetic properties of $CaMn_{1-x}Nb_xO_3$ with Nb-doping. *J Phys D: Appl Phys* 48:325003.

Meiklejohn WH, Bean CP (1956) New magnetic anisotropy. *Phys Rev* 102:1413–1414.

Mishra SK, Radu F, Dürr HA, Eberhardt W (2009) Training-induced positive exchange bias in NiFe/IrMn bilayers. *Phys Rev Lett* 102:177208.

Murthy JK, Venimadhav A (2013) Giant zero field cooled spontaneous exchange bias effect in phase separated $La_{1.5}Sr_{0.5}CoMnO_6$. *Appl Phys Lett* 103:252410.

Murthy JK, Chandrasekhar KD, Wu HC, Yang HD, Lin JY, Venimadhav A (2016) Antisite disorder driven spontaneous exchange bias effect in $La_{2-x}Sr_xCoMnO_6$ ($0 \leq x \leq 1$). *J Phys: Condens Matter* 28:086003.

Nayak AK et al. (2013) Large zero-field cooled exchange-bias in bulk Mn_2PtGa. *Phys Rev Lett* 110:127204.

Niebieskikwiat D, Salamon MB (2005) Intrinsic interface exchange coupling of ferromagnetic nanodomains in a charge ordered manganite. *Phys Rev B* 72:174422.

Nogués J et al. (2005) Exchange bias in nanostructures. *Phys Rep* 422:65–117.

Ren Y, Palstra TTM, Khomskii DI, Nugroho AA, Menovsky AA, Sawatzky GA (2000) Magnetic properties of YVO_3 single crystals. *Phys Rev B* 62:6577–6586.

Renner Ch, Aeppli G, Kim BG, Soh YA, Cheong SW (2002) Atomic-scale images of charge ordering in a mixed-valence manganite. *Nature* 416:518–521.

Shen H et al. (2013) Magnetic field induced discontinuous spin reorientation in $ErFeO_3$ single crystal. *Appl Phys Lett* 103:192404.

Sow Ch, Pramanik AK, Kumar PSA (2014) Exchange bias in strained $SrRuO_3$ thin films. *J Appl Phys* 116:194310.

Su Y et al. (2010) Magnetization reversal and Yb^{3+}/Cr^{3+} spin ordering at low temperature for perovskite $YbCrO_3$ chromites. *J Appl Phys* 108:013905.

Tang YK, Sun Y, Cheng ZH (2006) Exchange bias associated with phase separation in the perovskite cobaltite $La_{1-x}Sr_xCoO_3$. *Phys Rev B* 73:174419.

Tokura Y (2006) Critical features of colossal magnetoresistive manganites. *Rep Prog Phys* 69:797–851.

Tung LD, Lees MR, Balakrishnan G, McK Paul D (2007) Magnetization reversal in orthovanadate RVO_3 compounds (R = La, Nd, Sm, Gd, Er, and Y): Inhomogeneities caused by defects in the orbital sector of quasi-one-dimensional orbital systems. *Phys Rev B* 75:104404.

Wu W, Israel C, Hur N, Park S, Cheong S-W, de Lozanne A (2006) Magnetic imaging of a supercooling glass transition in a weakly disordered ferromagnet. *Nat Mater* 5:881–886.

Xia J, Siemons W, Koster G, Beasley MR, Kapitulnik A (2009) Critical thickness for itinerant ferromagnetism in ultrathin films of $SrRuO_3$. *Phys Rev B* 79:140407(R).

Xiao L et al. (2016) Doping induced zero-field cooled exchange bias effect in hexagonal $Y_{0.95}Eu_{0.05}MnO_3$ single crystal. *Ceramics International* 42:2550–2556.

Xu MH, Wang ZH, Zhang DW, Du YW (2013) Exchange bias effect in $Tb_{0.4}Dy_{0.6}MnO_3$. *J Magn Magn Mater* 340:1–4.

Yan LQ, Wang F, Zhao Y, Zou T, Shen J, Sun Y (2012) Exchange bias effect in multiferroic $Eu_{0.75}Y_{0.25}MnO_3$. *J Magn Magn Mater* 324:2579–2582.

Yoshii K (2011) Positive exchange bias from magnetization reversal in $La_{1-x}Pr_xCrO_3$ ($x \approx 0.7$–0.85). *Appl Phys Lett* 99:142501.

Zener C (1951) Interaction between the d-shells in the transition metals. II. Ferromagnetic compounds of manganese with perovskite structure. *Phys Rev* 82:403–405.

Ziese M (2002) Extrinsic magnetotransport phenomena in ferromagnetic oxides. *Rep Prog Phys* 65:143–249.

11

Exchange Bias in Bulk Heusler Systems

Jyoti Sharma and
K. G. Suresh

In this chapter, we present a consolidated picture on the investigations reported on exchange bias (EB) phenomenon in bulk and ribbon forms of various ferromagnetic (FM) shape memory alloys belonging to the full Heusler family, mainly from the Ni–Mn-based subclass. In the last decade, full Heusler alloys have become one of the most important classes of materials because of their interesting magnetic and other multifunctional properties such as shape memory effect [1], magnetocaloric effect [2], magnetic superelasticity [3], giant magnetoresistance (MR) [4], and EB [5]. Most of these properties are associated with their first-order structural (i.e., martensitic) transition [6]. Magnetic, magnetocaloric, and electrical transport properties of these alloys have been extensively studied by many researchers, which have resulted in a large number of papers. Among these alloys, Mn-based alloys have been found to be singularly important. EB effect is one of the most important properties exhibited by a large number of full Heusler alloys. Though it is quite difficult to take into account all the wealth of information, we have tried our best to present a comprehensive picture regarding the EB aspects of these alloys in the bulk and ribbon forms.

Initially, we present a brief introduction to Heusler alloys, their structural and magnetic properties. Subsequently, certain phenomenologies as well as selected theoretical models describing the EB phenomenon are discussed. Most of these models are developed for the understanding of EB in bilayers/multilayers [7,8] or core–shell nanoparticles [9,10] in which two magnetically competing phases are present in different regions of the system. However, such a scenario is not obvious in the case of bulk systems. Because of this, the understanding of the mechanism of EB in Heusler alloys is not as clear as in the case of bilayers or fine particles. Therefore, different models are used to get a basic picture of EB in these alloys and hence, a brief account of such models is presented here. Following this, the experimental tools used to measure the EB and the findings on EB in bulk Heusler systems of various stoichiometries are discussed. The main reason for EB in these systems is the presence of coexisting, competing magnetic phases such as FM, antiferromagnetic (AFM), spin-glass (SG), superparamagnetic (SPM), cluster glass (CG), and disordered magnetic states [5,11–14]. As the magnetic properties of these alloys are strongly dependent on the chemical composition [15], it is found that by modifying/changing the chemical composition and controlling the external parameters such as temperature, magnetic field, or hydrostatic pressure, a significant change in EB properties can be achieved.

11.1 A Brief Introduction to Heusler Alloys

In 1903, Heusler [16] discovered that addition of sp elements (such as Al, In, Sn, Sb, or Bi) in Cu–Mn binary alloys turns them into a FM material (e.g., Cu_2MnAl) even though the consisting elements were not FM in nature. These alloys were named as *Heusler alloys*. Later, it was found that all the atoms occupy specific positions in the lattice of these alloys, which can be described as a body-centered cubic with a face-centered superstructure [17]. The magnetic moment of these alloys resides solely on Mn atoms, which occupies the body-centered position in the lattice. Researchers have established a relationship between composition, chemical order, and magnetic properties, which showed that many Heusler alloys order ferromagnetically in their stoichiometric (2:1:1) composition [2,6,17,18].

Crystal structure, composition, and heat treatment are found to determine their magnetic and other physical properties.

11.1.1 Structural Properties

Heusler alloys are mainly divided into two categories according to their structure, one is half Heusler and the other is full Heusler alloys [19]. The former has the XYZ (1:1:1) type chemical stoichiometry and exhibits the Cl_b (prototype) structure (not discussed here). Full Heusler alloys form in X_2YZ (2:1:1) type chemical stoichiometry and exhibit the $L2_1$ (prototype) structure. In general, X and Y are the transition metals, and Z is either a semiconductor or nonmagnetic metal (from group III, IV, or V) and in some cases, Y can be replaced by a rare earth element or an alkaline earth metal. Unit cell of the $L2_1$ structure consists of four face-centered cubic sublattices, which can be characterized by positions X_1 (1/4,1/4,1/4), X_2 (3/4,3/4,3/4), Y (0,0,0), and Z (1/2,1/2,1/2). The same atomic positions hold for the half Heusler alloys as well, except that the X_1 positions remain empty. Some examples of full Heusler alloys are

1. Cu_2MnAl, Cu_2MnIn, and Cu_2MnSn
2. Ni_2MnGa, Ni_2MnIn, Ni_2MnSn, Ni_2MnSb, and Ni_2MnAl
3. Mn_2NiGa, Mn_2NiSn, Mn_2NiIn, Mn_2NiSb, and Mn_2PtGa
4. Pd_2MnAl, Pd_2MnIn, and Pd_2MnSn
5. Fe_2MnGa

11.1.2 Martensitic Phase Transition

In general, full Heusler alloys undergo a martensitic phase transition, that is, a first-order magneto-structural transition from a highly symmetric cubic austenitic phase to a low symmetry tetragonal/orthorhombic/monoclinic martensitic phase, as the temperature is lowered [6]. It should be noted that among these alloys, only Ni_2MnGa exhibits martensitic transition in the stoichiometric (2:1:1) form [20]. Therefore, all other Heusler alloys are usually studied in the off-stoichiometric form. Unlike the atomic order–disorder phase transitions, the martensitic transition is caused by a nondiffusional, cooperative movement of atoms. The strong coupling between the magnetic and structural properties leads to the interesting magneto-structural behavior, which results in various multifunctional properties such as shape memory effect [1], giant magnetocaloric effect [2,3], large MR [4], thermal conductivity [21], and EB effect [5].

11.1.3 Magnetism in Heusler Alloys

In general, it has been found that the magnetism in the austenite phase of these alloys is simple to understand, but the magnetic state in the martensite phase is quite complex. Because of this, these alloys show very interesting magnetic properties, and one can study a variety of interesting, diverse magnetic phenomenon such as itinerant and localized magnetism, ferromagnetism, antiferromagnetism, ferrimagnetism, or

Pauli paramagnetism in the same family of alloys [19]. Magnetism of these alloys was initially explained by the complex exchange interaction mechanism and later by the band structure calculations. These mechanisms are attributed to the variations in exchange interaction strengths. Among the Heusler alloys, Ni–Mn-based alloys are extensively studied in the literature. For example, in Ni–Mn-based Heusler alloys, the total magnetic moment arises from Mn atoms only as Ni atoms possess negligible moment and the nature of magnetism depends on the indirect exchange coupling (i.e., RKKY exchange) [22]. The sign of this exchange varies with distance, that is, positive for FM and negative for AFM and gives rise to a variety of magnetic ground states such as SG, reentrant SG (RSG), cluster SG (CSG), and long-range magnetic ordering [23]. Therefore, these alloys exhibiting the coexistence of different magnetic phases are potential candidates for investigating the EB effect.

11.2 Introduction to Exchange Bias Phenomenon

11.2.1 Discovery and Background

In 1956, Meiklejohn and Bean [24,25] discovered a new type of magnetic anisotropy, which was unidirectional in nature, as compared with the uniaxial anisotropy such as the magnetocrystalline anisotropy. This new anisotropy was referred to as exchange anisotropy, which has the most important manifestation as EB. It was discovered first in Co nanoparticles in which the core is FM, whereas the shell made of the oxide layer is AFM in nature. The exchange anisotropy observed was attributed to the interaction between the AFM and the FM regions of the material. Subsequently, EB was observed in bilayers and multilayers consisting of interfaces such as FM/SG, and AFM/ferrimagnetic (FI) phases etc [9,26]. Significantly, it is also observed in bulk alloys and compounds [22], which have competing magnetic interactions mentioned above. EB observed in Heusler alloys is the typical example of such a case.

EB has attracted a great deal of attention because it is an integral part of modern magnetism with its implications from the fundamental research point of view to various device applications. EB effect is manifested by the shift of magnetic hysteresis (M–H) loop along the magnetic field axis, when the materials having interfaces separating competing magnetic phases are field cooled through the Néel temperature (T_N) of the AFM phase (in which the Curie temperature (T_C) of FM phase is larger than T_N). As expected, EB is a temperature-dependent phenomenon and vanishes at T_N, but most of the times it vanishes at temperatures below T_N. The temperature at which EB effect vanishes is usually referred to as blocking temperature (T_B) (discussed in Sections 11.2.6, 11.3.1.1, 11.3.2.2, and 11.5.3.1).

While performing the experiment, Meiklejohn and Bean observed that the M–H loop of Co/CoO shifted along the field axis, when they were cooled in the presence of a magnetic field. Figure 11.1 [24] shows the Meiklejohn and Bean's original finding, in which the curve (a) shows the shifted M–H loop along the field axis [it was measured at 77 K after field-cooled (FC) in 10 kOe], and curve (b) shows the symmetric M–H loop along the field axis, obtained at the same temperature after cooling in zero field.

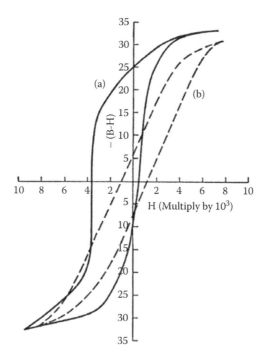

FIGURE 11.1 Magnetic hysteresis b-h??? loops measured at 77 K for Co/CoO system. Curve (a) shows the hysteresis loop measured after FC in 10 kOe, and curve (b) shows the loop when the system was cooled in zero field. (From Meiklejohn, W. H., Bean, C. P., *Phys. Rev.*, 105, 904, 1957.)

Meiklejohn and Bean [24,25] showed how the exchange interaction at FM Co/AFM CoO interface leads to the shift in the hysteresis loop and other manifestations of exchange anisotropy. They explained the loop shifting by considering a unidirectional anisotropy energy term in the free energy expression of a single-domain spherical particle at absolute zero. The single-domain particle was assumed to possess a uniaxial anisotropy with the easy axis oriented in the field direction. The field direction was assumed to be antiparallel to the particle's magnetization (I_s) direction. In such a case, the free energy of the particle can be written as [24].

$$F = HI_s \cos\theta - K_u \cos\theta + K_1 \sin^2\theta \qquad (11.1)$$

where:

θ is angle between the direction of easy axis and that of particle's magnetization
K_u and K_1 are the unidirectional and uniaxial anisotropy energy constants, respectively

Solutions of this equation can be expressed in terms of an effective field, which was given as [24]

$$H' = \frac{H - K_u}{I_s} \tag{11.2}$$

This suggests that the hysteresis loop gets displaced by an amount of K_u/I_s, along the negative field axis. After the discovery of EB in FM Co/AFM CoO core–shell particles system, it has been studied in a variety of systems having the AFM/FM interfaces such as nanostructures [10], layered structures [9,26], FM nanoparticles embedded in AFM matrix [27], thin films [28,29], and various magnetically inhomogeneous materials [24,30]. As mentioned earlier, in addition to the systems having AFM/FM interfaces, EB has also been studied in a variety of systems having different kinds of artificial interfaces such as FI/AFM FM/SG or diluted magnetic semiconductors/AFM [9,10,22,25–27,31,32]. However, investigations on EB in alloys and compounds having single-phase crystal structure have been found to be rare. One of the first reports on such systems was in the 1950s, when Kouvel et al. [33–36] reported the EB in binary alloys, involving different disordered magnetic states, such as SG or CG and coexisting FM/AFM phases. After a long gap, EB in structurally single-phase alloys and compounds having FM/AFM/FI core and disordered magnetic/SG/CG shell-like structures was reported in a few systems [22]. Subsequent to the detailed investigation in binary alloys by Kouvel et al. [33–36], EB attracted a great deal of attention in structurally single-phase alloys and compounds, Heusler alloys being one of them.

11.2.2 Applications

Till date, EB effect has been exploited in various technological applications such as magnetoresistive random access memories [37,38] and read heads of recording media [39]. It is being used in stabilizing the magnetization of SPM nanoparticles [40] or to improve the coercivity of permanent magnets [41]. For example, oxidized Fe–Co surface-modified nanoparticles exhibiting a coercivity enhancement have been used as hard magnets [42]. In recent years, the development of *spintronics* has attracted immense attention owing to its nonvolatility, reduced power consumption, and higher speeds in devices [43,44]. The spin-based electronics mainly consists of spin valves and magnetic tunnel junctions in which EB is used for pinning the magnetization of a FM layer by coupling it to an AFM layer [45].

11.2.3 Basic Phenomenology of Exchange Bias

As the microscopic mechanism of EB effect is not fully clear till date, a simple phenomenological interpretation of EB is generally used to explain the magnetic hysteresis loop shift along the field axis. This interpretation can be understood by assuming an exchange interaction at the FM/AFM interface. The schematic diagram with different stages (1)–(5) for explaining the EB mechanism in a prototype FM/AFM bilayer is shown in Figure 11.2. It is assumed here that the magnetic ordering temperatures of FM and AFM phases satisfy the condition $T_C > T_N$.

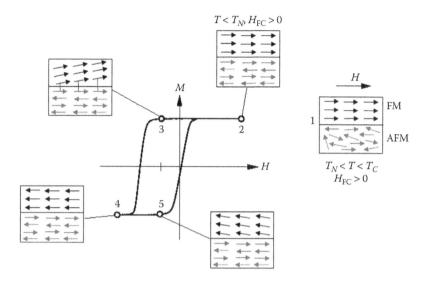

FIGURE 11.2 Schematic representation of EB mechanism in a FM/AFM bilayer under the strong FM–AFM coupling and an applied magnetic field, after field cooling.

1. When the system is field cooled to a temperature T such that $T_N < T < T_C$, FM moments are aligned along the field direction, whereas the AFM moments remain randomly oriented. The cooling field fixes the direction of FM layer magnetization during the cooling process.

2. When the system is field cooled to the temperature below T_N, AFM layer moments in contact with the FM layer align ferromagnetically and the remaining AFM moments follow the AFM order so as to produce zero net magnetization.

3. When the external field is reversed, the FM moments start to rotate, whereas the AFM moments do not switch because of the strong anisotropy. Thus, the interfacial interaction between the AFM and FM moments tries to keep FM moments aligned in their original direction (i.e., positive magnetization), or in other words, prevent the spontaneous reversal of FM moments along the (reversed) field direction. Therefore, the anisotropy of the AFM layer plays a crucial role in shifting the hysteresis loop.

4. A larger external field is needed to reverse the FM layer magnetization completely.

5. When the magnitude of the measuring field is reduced, again the anisotropy of the AFM layer becomes dominant, and the torque causes a reduced negative remnant magnetization as the field tends to zero. Thus, the material behaves as if there was an extra (internal) positive biasing field, which causes a shift in M–H loop along the negative field axis, i.e., EB. The value of EB field (H_{EB}) and coercivity (H_C) can be calculated by using the formulae $H_{EB} = -(H_L + H_R)/2$ and $H_C = |H_L - H_R|/2$, respectively, where H_L and H_R are the left and right coercive fields at which magnetization becomes zero.

11.2.4 Minor Loop Effect

Generally in nanocrystalline compounds, alloys and oxide materials (particularly having the disordered and/or glassy magnetic phases), magnetization does not often saturate even at high magnetic fields (e.g., $H_{\text{measuring}} = 50$ kOe). Therefore, a proper choice of maximum applied field for recording the hysteresis loop for EB measurements is required, as the small $H_{\text{measuring}}$ can give rise to the displacement of hysteresis loops even for FM and glassy magnetic systems. Thus, the H_{EB} estimated from these M–H loops will be somewhat overestimated because of presence of minor loop effect [22,46]. In order to avoid the minor loop effect, the $H_{\text{measuring}}$ should be chosen in such a way that $H_{\text{measuring}} > H_A$, where the H_A is anisotropy field. This has been investigated in several Heusler alloys to avoid the overestimation of EB (discussed in Section 11.5.8.1). In these reports, the plots for H_{EB} as a function of $H_{\text{measuring}}$ are shown, where H_{EB} was found to decrease with the increase of $H_{\text{measuring}}$ and stabilize at high $H_{\text{measuring}}$.

11.2.5 Training Effect

It is one of the characteristic features of EB and manifests as a decrease in the EB field caused because of cycling the system through several consecutive M–H cycles. It arises due to the fact that FM moments do not get reversed homogenously during the magnetization reversal, and it can be due to the rearrangement of AFM/frustrated moments at the interfaces. Therefore, the process results in a loss of net magnetization during the repeated field cycling process and results in the reduced H_{EB}. The evidence of training effect was first reported in a thin film, in which it was proposed that the decrease of H_{EB} satisfies the following empirical formula [47].

$$H_{\text{EB}} - H_{\text{EB}}^{\infty} \propto n^{-1/2} \tag{11.3}$$

where:

n is the number of field cycles
H_{EB}^{∞} is the value of EB field at $n = \infty$

There were two kinds of training effects suggested in these reports

1. When the EB effect decreases for $n < 2$ [48].
2. When the decrease of EB effect for $n \geq 2$ [9].

Recently, Mishra et al. [49] have proposed a modified explanation for the training effect, which is given in Equation 11.4. This equation considers the contributions of frozen as well as the unfrozen moments at the interface.

$$H_{\text{EB}}^{n} = H_{\text{EB}}^{\infty} + A_f \exp\left(\frac{-n}{P_f}\right) + A_i \exp\left(\frac{-n}{P_i}\right) \tag{11.4}$$

where:

A_f and P_f parameters are related to the change in frozen moments

A_i and P_i parameters are related to the interfacial magnetic frustration (i.e., unfrozen moments)

A_i and A_f have the dimensions of magnetic field

P_i and P_f are dimensionless

Ps' are representing the effect of relaxation process

Many reports are available on the training effect in bulk and thin films of Heusler alloys [9,22].

11.2.6 Double-Shifted Hysteresis Loop

In many of the EB systems, an interesting feature has also been observed, that is, the double-shifted hysteresis loops. Under certain conditions, the M–H loops get divided into two subloops, one gets shifted to the positive fields axis and the other one to the negative field axis. This double-shifted loop is usually found to occur in the FM–AFM systems, when it is either zero-field cooled in demagnetized state or cooled in presence of large magnetic fields [50]. If a FM system possesses a uniaxial anisotropy above T_B and demagnetized, it exhibits the magnetic domains structure with opposite magnetization direction along the direction of easy axis. Therefore, such domain structure of FM imprints onto the AFM during zero-field-cooled (ZFC) process and divides AFM into two types of regions, which are oriented in the opposite directions to each other. In this way, when the hysteresis loop is measured below T_B, these two AFM regions couple with FM in opposite ways and lead to the one subloop shifted toward the positive field axis and another one in negative field direction. It has been reported in the literature that for in-plane EB systems, the vertical amplitude of these dual loops depends upon the FM magnetization state above T_B [5,22,51]. In several Heusler alloys, this double-shifted loop has been observed, which also corroborates with the presence of EB in them. The EB induced by the FC, discussed in the earlier section, is these days termed as conventional EB. This is due to the observation of a rather new phenomenon of zero field cooled EB, also known as spontaneous EB. This new type of EB has been observed in many systems, including certain full Heusler alloys (discussed in Sections 11.5.3.3, 11.5.5.3, and 11.5.9.1).

11.3 Theoretical Models of Exchange Bias Phenomenon

Coehoorn [52] and Nogues [9] have classified the EB models into three categories: macroscopic, mesoscopic, and microscopic, according to the lateral length scales. Here, we discuss these models briefly. One or more of these models are needed to explain the EB phenomenon in bulk alloys even qualitatively.

11.3.1 Macroscopic Models

The macroscopic EB models are those in which the lateral magnetic structure of AFM and FM layers are not considered. The AFM and FM layers are assumed to be homogeneous in the x–y plane. Usually in FM/AFM layered systems, the AFM interfacial moments are considered to be uncompensated and aligned parallel to the interface plane. In some of the models that come under this category, the detailed moment structure in the z-direction is also considered.

11.3.1.1 The Meiklejohn-Bean Model

Meiklejohn and Bean [24,25] gave the first theoretical model, which accounts for the magnitude of the hysteresis loop shift in FM/AFM systems. They made the following assumptions:

1. Both AFM and FM layers are in a single domain state (DS).
2. The FM layer rotates rigidly as a whole.
3. The AFM layer is magnetically rigid, which means that the AFM moments remain unchanged during the rotation of FM spins.
4. The FM/AFM interface is assumed to be atomically smooth.
5. Moments of AFM interface layer are uncompensated, which means that the AFM layer as a whole possesses a net magnetic moment.
6. Exchange interaction at the FM/AFM interface is characterized by an interfacial exchange coupling energy per unit area, denoted as J_{EB}.

The Meiklejohn-Bean model gives the basic picture of EB, provided the given assumptions are taken care of. But in reality, for most of the thin film systems, the experimental value of hysteresis loop shift was found to be several orders of magnitude lower than predicted one by this model [7,25,26]. However, in some of the cases, experimental results are found to be in agreement with this model [53]. This model neglects many parameters, which have been found to be important in explaining the real picture of EB. There are some other consequences of this model, for example, H_{EB} is found to be zero for compensated AFM surfaces, which means that they possess zero net AFM surface magnetic moments [54] or H_{EB} is always negative [55], which are not true always. Similarly, as per this model, the coercivity of the magnetic layer would be the same, with and without the presence of EB, which is found to be not true in most of the experimental findings.

EB effect is mainly attributed to the magnetic order of AFM phase. Therefore with the increase of measuring temperature (near to T_N), EB effect disappears. However, in various systems especially the nanostructured systems and bulk alloys or compounds, it is often observed that EB disappears at temperatures far below T_N. These temperatures are termed blocking temperature, and it has been correlated with the finite size effects in the AFM phase [56] and recently, this effect has been reported to be more complex [57]. In high-quality thin films with thick AFM layers, T_B was found to be equal to T_N, whereas in very thin or polycrystalline AFM-layered systems, T_B was less than T_N.

11.3.1.2 Neel Domain Wall Model

To reduce the theoretical value of hysteresis loop shift obtained according to the Meiklejohn-Bean model [24,25], Neel [58] proposed a modification in that model. He introduced the concept of the formation of planar domain wall during the reversal of the magnetization of FM layer, rather than a coherent rotation of the domain. According to this approach, the AFM domain wall would store a fraction of exchange coupling energy, as a result of which H_{EB} would reduce. He assumed that magnetization of a layer (both in AFM and FM phases) is uniform within the layer and aligned in the parallel direction to the interface. According to this model, under appropriate conditions, domains develop both in the AFM and FM layers, provided the thickness of AFM/FM layers was at least about hundreds of nanometers. After Neel' model, similar arguments were made by Mauri et al. [59], Kiwi et al. [60], Geshev [61], or Kim et al. [62] in their EB models. The terms related to the formation of domains were considered in energy expression. Thus, interface domain wall can significantly decrease the energy of equilibrium magnetic configuration and consequently results in the decrease in values of effective coupling. It was found to be in good agreement for some of the thin film Heusler systems [63].

11.3.2 Mesoscopic Models

Mesoscopic models for explaining the EB effect are those which consider the possibility of different moment configurations in x–y plane. One of these models, which include some lateral moment distributions, was introduced by Kouvel et al. [31]. Another mesoscopic model was established by Fulcomer and Charap.

11.3.2.1 Kouvel's Ferromagnetic–Antiferromagnetic Model for Cu–Mn and Related Alloys

In the model, Kouvel [31] assumed that the magnetic unit consists of a small ensemble of the mutually interacting FM and AFM domains. The model was designed for inhomogeneous alloys such as Cu–Mn, Ag–Mn, and other related alloys. It was considered that each domain ensemble possesses zero net magnetization in the ground state. As the system is field cooled, some of the domain ensembles are forced to go into a different state (i.e., state of nonzero magnetization). This state is stabilized by the strong anisotropy of AFM domains. Thus, the magnetic hysteresis loop of the system shows the asymmetric shift along the field axis, after the FC process. Similarly, the torque curves show the single easy axis of magnetization even in very high fields. EB properties and their variations with field and temperature predicted by this model are found to be in agreement with the experimental observations for Cu–Mn, Ag–Mn, and other alloys [33–36]. As discussed later, this model is useful in explaining EB in some Heusler systems [13].

11.3.2.2 Fulcomer and Charap Model

In their model, Fulcomer and Charap [64] assumed that the AFM grains are distributed in size, and they considered the effects of grain size distribution on EB effect. They assumed that the AFM phase is made up of *single-domain* grains, in which all the moments switch collectively by coherent rotations. The simple approach used in this model was that the different AFM grains couple differently to FM ones, and those small AFM grains would tend to be SPM in nature. The AFM grains were supposed to possess a temperature-independent uncompensated interfacial moment due to the AFM sublattice's different contributions to the FM/AFM interface. Fulcomer and Charap model predicts the maximum coercivity near the blocking temperature, T_B.

Recently, models based on different properties of AFM grains, such as the direction of easy axes, distribution of the number of uncompensated spins, different degrees of coupling between AFM grains, AFM anisotropies, or FM/AFM interface coupling strengths have been developed [65,66]. Some other sophisticated models based on similar approaches in which the possibility of other effects such as partial domain walls in the AFM or perpendicular coupling inside the AFM grains have been established for polycrystalline AFM/FM systems. One of the examples is the Stiles and McMichael [67] model, which considers the AFM layer to be consisting of an ensemble of crystallites (of different sizes and AFM anisotropy directions). They have shown that, in the polycrystalline AFM/FM systems, the AFM grains with large AFM/FM direct coupling constants and the easy axes closer to cooling field direction would not easily reverse their magnetization, hence contributing to EB effect. The same AFM grains having the strong exchange coupling and the easy axes oriented at a certain angle from FM easy axis would switch together with FM one during the magnetization reversal of FM layer and contribute to the enhancement of H_C. The AFM grains, which have the weak direct coupling constant, would not strongly contribute to either H_{EB} or H_C. This model was found to be excellent in explaining the experimental results observed in AFM/FM fine particle nanocomposites, in which the different microstructures may contribute to the different observed effects [67]. Ideas of this model may be useful in explaining some EB features of Heusler alloys.

11.3.3 Microscopic Models

In microscopic models, the detailed moment configuration of each atom (or groups of atoms) in the volume of system, that is, in the x, y, and z directions, is taken into account for explaining the exchange anisotropy. There exist several microscopic models, which are based on Monte Carlo simulations, micromagnetic calculations, and different types of spin lattice models [68,69]. In spin lattice models, although the spin structures are taken into account, different kinds of approximations have been developed in order to simplify the search of minimum energy configurations. It is important to note that although the magnetic structure of the AFM phase has been taken into account in the microscopic models, most of these models assumed the simplest type of uniaxial AFM anisotropy.

11.3.3.1 Domain State (DS) Model

In DS model [70], it was proposed that the disorder is not only introduced via magnetic dilution at the interface but also in the bulk of AFM layer. The interesting feature of this model was that the AFM layer is assumed to be a diluted Ising antiferromagnet (DAFM). In the absence of a magnetic field, system undergoes a phase transition from a disordered PM state to a long-range ordered AFM state, which is dilution dependent. For small fields and at low temperatures, long-range-ordered phase is stable in three dimensions. As field is increased, DAFM develops a DS phase, which exhibits a SG-like behavior. AFM domains formation in DS phase is attributed to the statistical imbalance of the number of impurities of two AFM sublattices within any finite region of DAFM. This imbalance results in a net magnetization, which couples to the external magnetic field. Under this condition, a moment reversal of the region, that is, the formation of a domain, can lower the energy of the system. The domain wall energy can be minimized if the domain wall passes through the nonmagnetic defects at the cost of minimum exchange energy. DS model was used to explain the EB phenomenon in some of the Heusler systems [29].

11.4 Exchange Bias Measurement Techniques

EB and the related effects can be studied using various experimental techniques. The most commonly used techniques and the main information provided by them are discussed below.

11.4.1 Magnetization

As discussed above in the introduction part of the chapter, the M–H loops recorded after FC are the most commonly used technique to study the EB [24,25]. The M–H loops can be measured using vibrating sample magnetometer, superconducting quantum interference magnetometer, Kerr effect, torque magnetometry, or loop tracers. These techniques give the main information such as the loop shift (i.e., H_{EB}) and H_C. In addition, some information about the anisotropies may also be obtained from the shape of these hysteresis loops.

11.4.2 Magnetoresistance

In literature, MR measurements (i.e., the change in the resistivity with field) have been used to probe EB. MR isotherms measured up to the saturation of both FM layers of a spin valve devices (i.e., consisting of AFM–FM-nonmagnetic–FM layered structure) have been found to give both H_{EB} and H_C [26]. A typical example of the MR measurement to study the EB effect in a layered structure: Si/(150-Å NiFe)/(26-Å Cu)/(150-Å NiFe)/(100-Å FeMn)/(20-Å Ag) is shown in Figure 11.3a–b [45]. Figure 11.3a depicts the magnetization curve measured at room temperature, which shows two separate hysteresis loops. The hysteresis loop with smaller coercivity refers to the reversal of bottom NiFe layer, whereas the other loop shifted with H_{EB} ~90 Oe refers to the reversal of NiFe magnetization. Figure 11.3b shows the change in resistance relative to the parallel alignment at room temperature. Now, as the field H is swept such that 2 > H > 135 Oe, the magnetizations of two NiFe layers change from parallel to antiparallel alignment, which implies that the observed change in resistance is related to the change in relative magnetization orientation of two FM layers. Moreover, the MR measurements have also been used to determine the H_{EB} and H_C in AFM–FM bilayers systems [26]. As many Heusler alloys show the considerable MR, this technique is of great importance in revealing the EB properties exhibited by them.

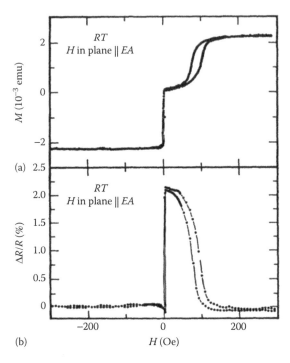

FIGURE 11.3 (a) The M–H loop and (b) change in resistance versus field curve for Si/(150-Å NiFe)/(26-Å Cu)/(150-Å NiFe)/(100-Å FeMn)/(20-Å Ag). Here, the magnetic field is applied parallel to the exchange anisotropy field. (From Dieny, B. et al., *Phys. Rev. B*, 43, 1297, 1991.)

11.5 Experimental Results on Exchange Bias in Full Heusler Systems

As discussed earlier in this chapter, the martensitic phase of Heusler alloys has a complex magnetic state, which gives rise to a variety of magnetic ground states such as SG, RSG, or CG and coexistence of AFM/FM phases, and so on [23]. Therefore, these alloys are considered as one of the potential candidates for investigating the EB effect. EB studies have been reported in several Heusler alloys. Here, we have reviewed the EB studies mainly in Ni-rich (Ni ~ 50 at%) Ni–Mn–X (X = Sn, In, Sb, Ga, and Al) and Mn-rich (Mn ~ 50 at%) Mn–Ni–X (X = Sn) full Heusler alloys in detail, in their bulk, nanoparticle, and ribbon forms.

11.5.1 Preparation Methods

As per the literature reports, bulk Heusler alloys are generally prepared by arc melting method, whereas fine particles are obtained after ball milling. For preparing the ribbons, melt spinning technique is commonly used.

11.5.2 Cu–Mn–Al Heusler Alloys

Koga et al. [71] first reported the EB anisotropy in Cu–Mn–Al alloy. $Cu_{44.7}Mn_{20.6}Al_{37.7}$ was found to show the displaced M–H after FC, which was ascribed to the exchange anisotropy associated with FM domains and the Mn-rich AFM domains. The characteristics of the magnetic behavior observed in the alloy were found to be similar to that reported for other binary alloys such as Cu–Mn alloys, which was explained by the Kouvel's [31,34–36] model.

11.5.3 Ni–Mn–Sn Heusler Alloys

11.5.3.1 In Bulk Form

Khan et al. [72] have reported a systematic study of EB effect in $Ni_{50}Mn_{50-x}Sn_x$ ($x = 11$–17) bulk Heusler alloys. The measured M–H loops at 5 K for these alloys, after zero FC and FC, are shown in Figure 11.4a–d [72]. The hysteresis loops after FC were found to shift along the field axis, whereas after ZFC, it remains symmetric along the field axis. They proposed that in these alloys, AFM interaction raised from the AFM coupling between

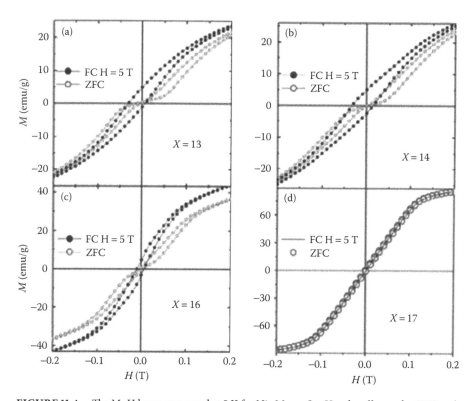

FIGURE 11.4 The M–H loops measured at 5 K for $Ni_{50}Mn_{50-x}Sn_x$ Heusler alloys, after ZFC and FC in 50 kOe for composition (a) $X = 13$, (b) $X = 14$, (c) $X = 16$ and (d) $X = 17$. (From Khan, M. et al., *J. Appl. Phys.*, 102, 113914, 2007.)

Mn atoms at the regular sites and Mn atoms at Sn sites, which plays an important role in the observed EB. The double-shifted hysteresis loops after ZFC were attributed to a striped-domain-type structure formed by FM regions. They explained the FC and ZFC hysteresis loops by considering that due to the presence of mixed AFM and FM regions, the magnetic domain structure after FC would be different from that formed after ZFC. It was explained by exchange spring effects, which imply that the regions of FM moments play the role of a soft magnetic phase, whereas frustrated one leads to the hard magnetic phase. Thus, the exchange coupling between these two phases occurs in the same way as in the case of exchange-spring magnets and results in the double-shifted loop. Therefore, double-shifted M–H loop after ZFC and shifted loops after FC clearly indicated the presence of EB in these alloys, which was attributed to the interfacial exchange coupling between two magnetically different phases.

Chatterjee et al. [11] have reported the EB effect in $Ni_2Mn_{1.36}Sn_{0.64}$ alloy, in which the low temperature RSG-like ground state was confirmed using the frequency-dependent ac susceptibility measurements. The ac susceptibility data indicated the onset of SG freezing as revealed by the frequency dependence. Therefore, the EB effect in the alloy was attributed to the pinning effect at the FM/SG-like interface, contrary to the other reports in Ni–Mn–Sn alloys, in which EB effect was suggested to arise at the FM/AFM interfaces [72].

In addition to the variation of Ni/Mn concentration, effect of doping (e.g., Co, Al, Ge, Pd, and Pt) on EB properties in the Ni–Mn–Sn alloys has also been studied by many researchers [73–75]. For example, Ghosh et al. [74] have investigated the EB effect in Co-doped $Ni_{48.5-x}Co_xMn_{37}Sn_{14.5}$ ($x = 0–2$) alloys, in which H_{EB} was found to decrease with the increase in x, which has been attributed to enhancement of FM interactions with increasing Co concentration. They observed that H_{EB} decreases with the increase of temperature, and it was found to vanish near T_B for all the compositions, which was attributed to the decrease in the size of AFM grains, which leads to the weakening of AFM–FM coupling at the interfaces. In contrast, another report on $Ni_{50}Mn_{36}Sn_{14-x}Ge_x$ ($x = 1$ and 2) [75] showed that the EB was found to increase with the addition of Ge, which has been ascribed to an increase of AFM interactions as a result of the decrease of cell volume due to the smaller ionic radius of Ge compared with that of Sn. Dong et al. [73] reported the EB effect in Pd/Pt-substituted Ni–Mn–Sn Heusler alloys, in which the EB was found to increase with the increase of Pd/Pt concentration. This behavior was attributed to the coexistence of percolating FM domains with the SG phase. They explained the observed behavior of EB by considering two aspects. One was the variation of FM/SG phase ratio on Pd/Pt substitution. They proposed that as in a FM/AFM bilayer system, H_{EB} is described by $H_{EB} = J_{int}/(M_{FM} \cdot t_{FM})$, where J_{int}, M_{FM}, and t_{FM} are the interface coupling constant, saturation magnetization, and thickness of FM layer, respectively. Similarly, in these alloys, $M_{FM} \cdot t_{FM}$ can be considered as the fraction of FM phase and J_{int} as the mean exchange energy at the FM/SG interface. FM phase fraction was seen to decrease on Pd/Pt substitution in place of Ni by observing the decrease of magnetization. Thus, it explained the enhancement of H_{EB} in these alloys. They proposed that the same mechanism can also explain the H_{EB} behavior with magnetization in Ni–Mn–Sb and Ni–Mn–Sn alloys. The other aspect considered was associated with stronger spin-orbital coupling of Pd/Pt atoms than that of Ni, which can lead to the stronger magnetic anisotropy and consequently can result in the increased H_{EB}. In Cu–Mn SG system, it was

TABLE 11.1 Maximum H_{EB} Reported for Ni–Mn–Sn Heusler Alloys

Sr. No.	Composition	H_{EB} (Oe)	Reference
1.	$Ni_{50}Mn_{50-x}Sn_x$ ($x = 11$)	225	[72]
2.	$Ni_{50}Mn_{36}Sn_{14}$	180	[76]
3.	$Ni_{50-x}Mn_{37+x}Sn_{13}$ ($x = 4$)	377	[77]
4.	$Ni_2Mn_{1.36}Sn_{0.64}$	120	[11]
5.	$Ni_{52}Mn_{34}Sn_{14}$	331	[78]
6.	$Ni_{45}Co_5Mn_{38}Sn_{12}$	306	[79]
7.	$Ni_{50}Mn_{34}Sn_6Al_{10}$	250	[80]
8.	$Ni_{50-x}Mn_{36}Sn_{14}Pt_x$ ($x = 3$)	609	[73]
9.	$Ni_{50-x}Mn_{36}Sn_{14}Pd_x$ ($x = 3$)	316	[73]

reported that the substitution of nonmagnetic Au/Pt impurities with strong spin–orbit coupling can significantly enhance the magnetic anisotropy, which was attributed to an additional term in RKKY interaction [73]. Similarly, in this Pd/Pt-substituted Ni–Mn–Sn alloys, Pd/Pt substitution can increase the unidirectional anisotropy of SG phase and subsequently results in the increase of EB. Table 11.1 lists the maximum values of H_{EB} for many Ni–Mn–Sn Heusler alloys, reported in literature.

11.5.3.2 In Ribbons

In addition to the investigations of EB effect in bulk Ni–Mn–Sn Heusler alloys, it has also been studied in their ribbon forms. In a report by Czaja et al. [81], the EB effect in $Ni_{48}Mn_{39.5}Sn_{12.5x}Al_x$ ($x = 0$–3) melt spun ribbons was investigated. At low temperatures (below 100 K), these alloys were found to exhibit EB, which increased with the increase of Al concentration. This was attributed to the slight difference in the interatomic distance between Mn and Al atoms, which was associated with the considerable difference between atomic radii of Sn and Al. They have also observed the double-shifted hysteresis loops after ZFC, which was attributed to the subdivision of AFM region magnetic structure into two types of subregions, which couple to FM in opposite directions (as explained in Section 11.2.6). Recently, Zhao et al. [82] studied the effect of Fe substitution on EB properties in $Ni_{46-x}Fe_xMn_{43}Sn_{11}$ ($x = 0$–3) ribbons, in which the H_{EB} was found to increase with the increase of Fe concentration as it varied from 469 to 534 Oe, as x changes from 0 to 3. This behavior was attributed to the change of exchange interactions at AFM/FM interfaces due to the slight change of distance between Mn atoms by the appreciable difference in the atomic radius of Fe and Ni.

11.5.3.3 Zero Field Cooled (Spontaneous) EB

Generally, EB effect is observed in conventional exchange coupled systems, which is observed only after FC process from a temperature above T_N. However, in some recent reports on Heusler alloys, EB has also been obtained after ZFC process, which implies that a large FM unidirectional anisotropy can be produced isothermally, and this unusual phenomenon was named as spontaneous EB. Liao et al. [83] reported the spontaneous EB in $Ni_{50}Mn_{36}Co_4Sn_{10}$ Heusler alloy, in which they observed a large H_{EB} of ~3 kOe after ZFC.

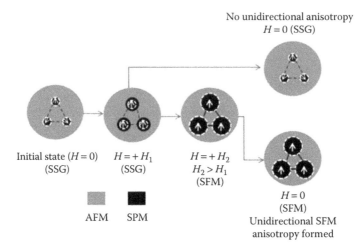

No unidirectional anisotropy
$H = 0$ (SSG)

Initial state ($H = 0$) $H = +H_1$ $H = +H_2$
(SSG)   (SSG) $H_2 > H_1$
                                         (SFM)

AFM   SPM

$H = 0$
(SFM)
Unidirectional SFM
anisotropy formed

FIGURE 11.5 A schematic diagram for the spontaneous EB mechanism. The SPM domains are shown, which are embedded in an AFM single domain in the presence of external magnetic field at temperature below blocking temperature. Here, the AFM anisotropy axis is parallel to the applied magnetic field direction. Initial state after ZFC is the SSG state (first from left). The arrows (inside the SPM domains) show the direction of super spin of SPM domains. Dashed circles represent the interfaces coupling between SPM and AFM. The solid lines refer to the coupling of SPM domains, that is, SFM exchange, whereas the dashed lines show the coupling between SPM domains, that is, glassy coupling. (From Wang, B.M., *Phys. Rev. Lett.*, 106, 077203, 2011.)

It was attributed to the enhancement of AFM coupling, which gives rise to a strong interaction with superferromagntic (SFM), super spin glass (SSG), and SPM states. The mechanism of spontaneous EB was first explained by Wang et al. [84] and the schematic diagram for the spontaneous EB mechanism explained by them is shown in Figure 11.5. EB was related to the newly formed interface among the different magnetic phases and explained in terms of the increase in SPM domain size in presence of external magnetic field, which gives rise to a transition from SSG to SFM state. According to these authors, at the same time, SFM unidirectional anisotropy can be induced isothermally, at the SFM–AFM interface at low temperatures. Thus, the interface transition takes places in the sequence SSG–AFM to SPM–AFM and finally to SFM–AFM. In this way, the observation of a spontaneous EB effect can be expected, and it has opened up a new direction in the field of EB.

11.5.4 Ni–Mn–Sb Heusler Alloys

Khan et al. [85] have first reported the EB effect in bulk $Ni_{50}Mn_{25+x}Sb_{25-x}$ Heusler alloys and attributed it to the coexistence of AFM and FM phases at low temperatures. In their another report on $Ni_{50}Mn_{25+x}Sb_{25-x}$ alloys [86], the increase of EB with the increase of Mn concentration was explained by the variation of AFM/FM exchange interaction and also supported by the computational results with the density-functional theory. They proposed that, according to theoretical calculations on Ni_2MnGa alloys, excess Mn atoms were found to align antiparallel to the neighboring Mn atoms, which results in the decrease of saturation magnetic moment of the alloy. Similarly, in these alloys, it can be considered that the Mn moments at Sb sites in $Ni_{50}Mn_{25+x}Sb_{25-x}$ alloys would align antiparallel to the Mn at regular sites, which caused the reduction of total magnetic moment, which leads to the increased EB with increase in Mn concentration.

After that, many researchers have studied the variation of Ni/Mn concentration in Ni–Mn–Sb alloys, and additionally, the effect of doping (such as Co and Fe) on EB has also been reported by some researchers. Nayak et al. [5] found that $Ni_{50-x}Co_xMn_{38}Sb_{12}$ ($x = 0$–5) alloys show large EB, and the H_{EB} was found to increase with increase in Co concentration. This was in contrast to the case of Co-doped Ni–Mn–Sn alloys [79], in which the Co doping reduced the EB. This enhancement in the case of $Ni_{50-x}Co_xMn_{38}Sb_{12}$ was attributed to the enhancement in the AFM coupling as a result of Co substitution. These alloys were also found to show large training effect, which was considerably larger than that observed in undoped Ni–Mn–Sb and Ni–Mn–In alloys, which was attributed to the anisotropy associated with Co. In their another work [87], a detailed ac susceptibility and dc magnetization measurements were performed in $Ni_{50-x}Co_xMn_{38}Sb_{12}$ ($x = 0$–7) alloys to understand the origin of EB, where they found the coexistence of RSG and FM phases, which in turn was found to play an important role in explaining the observed behavior of EB. The strength of RSG phase was found to decrease with increase in x and finally vanished for $x = 7$, which was in contrast to the observed variation of EB, which initially increased with x and showed a peak at $x = 5$. They suggested that the system actually enters into a frustrated FM state just below the magnetic order–disorder transition (T_C^M) of the martensite phase and the strength of RSG phase was found to be strongly dependent on the sharpness of T_C^M. Effect of Fe doping on EB in $Ni_{50}Mn_{38-x}Fe_xSb_{12}$ ($x = 0$–6) alloys was investigated by Sahoo et al. [88] where they observed that a small variation in Fe concentration led to a considerable change in EB. The values of H_{EB} obtained from the literature reports on Ni–Mn–Sb system are summarized in Table 11.2.

Nayak et al. [89] have reported the effect of change in the grain size on magnetic and EB behavior in ball-milled $Ni_{50-x}Co_xMn_{38}Sb_{12}$ ($x = 0$ and 5) alloys. According to these authors, the existence of magnetically soft and hard phases due to the wide range of grain sizes in the ball-milled samples resulted in dramatic changes in the structural and magnetic properties. For $x = 0$ alloy, the EB field was found to be 3.2 kOe at 3 K after ball milling, compared with the value of 245 Oe for the bulk sample of the same composition [85]. This increase was attributed to the enhanced exchange coupling between the soft and the hard magnetic particles.

TABLE 11.2 Maximum Values of H_{EB} Reported in Ni–Mn–Sb Heusler Alloys

Sr. No.	Composition	H_{EB} (Oe)	Reference
1.	$Ni_{50}Mn_{38}Sb_{12}$	245	[85]
2.	$Ni_{50}Mn_{25+x}Sb_{25-x}$	248	[86]
3.	$Ni_{45}Co_5Mn_{38}Sb_{12}$ ($x = 5$)	480	[5]
4.	$Ni_{50}Mn_{38-x}Fe_xSb_{12}$ ($x = 2$)	288	[88]

11.5.5 Ni–Mn–In Heusler Alloys

11.5.5.1 In Bulk Form

Pathak et al. [90] have reported the EB properties in $Ni_{50}Mn_{50-x}In_x$ ($14.5 \leq x \leq 15.2$) bulk Heusler alloys. A maximum H_{EB} of 120 Oe was obtained at 5 K. The observation of EB in these alloys was also attributed to the presence of AFM and FM exchange interactions at low temperatures in the martensite phase as proposed in other Heusler systems [76–77,85–86]. In this system also, in addition to the variation of Ni/Mn concentration, effects of various doping elements such as Si and Fe on EB were reported [91–92]. The maximum values of H_{EB} obtained in this system are summarized in Table 11.3. Generally, in undoped Ni–Mn–In Heusler alloys, EB was found to decrease with the increase in saturation magnetization, but in the Si-substituted system, EB was found to increase with increase in saturation magnetization. This increase in EB at low temperatures with Si concentration was attributed to the increase in the valence electron concentration (i.e., e/a ratio, where a is the number of atoms per formula unit, and e is the corresponding number of valence electrons) [91]. However, EB decreased at higher temperature with the increase in Si concentration, which has been related to the change in the blocking temperature with change in Si concentration.

11.5.5.2 In Ribbons

Zhao et al. [93] reported the effect of annealing on EB properties in $Ni_{48}Mn_{37}In_{13}$ ribbons, in which EB was found to increase after annealing. The observed EB in these ribbons was attributed to the exchange interactions between AFM and FM phases. It was explained by considering the fact that in heterogeneous systems, the fraction of high temperature phase is generally significantly reduced when the material is field cooled. However, in some cases, it does not reduce much even down to 5 K. Therefore, such first-order phase transition systems can show the EB behavior if the high-temperature phase is FM in nature and the low-temperature phase shows a transition to AFM. At the same time, the origin of EB in these ribbons can be understood by the fact that the rapid solidification process may result in heterogeneity between FM state of high temperature austenite phase and AFM state of low temperature martensite phase in the low-temperature region. On the other hand, rapid solidification method increases the site disorder and excess Mn at Ni or In sites couple antiferromagnetically. After annealing, these ribbons would undergo a structural relaxation and growth of grain size takes place, which would change the atomic site slightly and results in the change of degree of atom order, leading to further decrease in Mn–Mn distance. Thus, the AFM exchange coupling would strengthen after annealing and lead to increase in

TABLE 11.3   Maximum Values of H_{EB} for Ni–Mn–In Heusler Alloys

Sr. No.	Composition	H_{EB} (Oe)	Reference
1.	$Ni_{50}Mn_{50-x}In_x$	120	[90]
2.	$Ni_{50}Mn_{35}In_{15}$	160	[92]
3.	$Ni_{50}Mn_{35}In_{15-y}Si_y$ ($x = 4$)	170	[91]

H_{EB} in these ribbons. In another report by Sanchez et al. [94], $Ni_{50.0}Mn_{35.5}In_{14.5}$ ribbons were prepared by melt-spinning method. They have shown the effect of various short time annealing processes on the EB of these ribbons, in which the EB was also found to increase after annealing. They explained the observed EB behavior in terms of the decrease in the site disorder. As mentioned earlier, in Ni-Mn based Heusler alloys AFM interaction originates in the martensite phase from the RKKY type interaction between Mn at Z sites and those at Y sites. Recently, EXAFS measurements in these alloys have shown that the AFM interaction leads to the hybridization between Ni 3d and Mn 3d states [94]. In $Ni_{45.5}Mn_{43.0}In_{11.5}$ ribbons, H_{EB} of ~270 Oe was observed for the as-spun ribbon, which was found to increase up to 1.21 kOe after annealing. The EB in these ribbons has been attributed to SPM freezing of FM clusters, which are embedded in an AFM matrix.

11.5.5.3 Spontaneous Exchange Bias

Wang et al. [84] have first reported the spontaneous EB in bulk Ni–Mn–In Heusler alloys. They obtained a large value of H_{EB} of 1.3 kOe at 10 K after ZFC, which was attributed to the unidirectional anisotropy created isothermally below the EB blocking temperature (as explained in Section 11.5.3.3). They have observed that the spontaneously exchanged systems after ZFC show the same relationship of temperature dependence of H_{EB} and also the training effect, as observed in conventionally exchanged coupled systems after FC. In another report, a spontaneous EB was observed in Fe-doped Ni–Mn–In alloys [95] where a maximum H_{EB} of 1.5 kOe was obtained at 5 K after ZFC. They observed that with the decrease of temperature, these alloys show the SPM behavior (above 225 K) and SPM phase embedded in AFM matrix between 225 and 100 K, and below 100 K the coexistence of complicated magnetic states, that is, SPM, AFM, SSG, and SFM. These alloys were found to show the nonmonotonic trend of spontaneous EB with temperature, which was attributed to the competition between the volume fraction of SFM clusters and SFM/AFM interface coupling.

11.5.6 Ni–Mn–Ga Heusler Alloys

Han et al. [96] have reported the EB effect and magnetic phase separation in off-stoichiometric $Ni_2Mn_{1.4}Ga_{0.6}$ alloy. A maximum H_{EB} of 348 Oe was obtained at 10 K after FC in 10 kOe. They proposed that the coexistence of FM and SG phases at low temperature takes place due to the spatial composition fluctuation and competing FM and AFM interactions between Mn atoms, leading to the EB. In another report by Wang et al. [97], effect of Co doping on EB properties of Ni–Mn–Ga alloys was investigated. A maximum H_{EB} of ~70 Oe was obtained at 10 K, which was attributed to presence of AFM interaction inside the FM matrix.

11.5.7 Ni–Mn–Al Heusler Alloys

Recently Singh et al. [98] reported the EB in Si-doped Ni–Mn–Al Heusler ribbons. A large H_{EB} of 2.5 kOe was observed for $Ni_{55}Mn_{19}Al_{24}Si_2$. In another report by the same authors [99], in $Ni_{56}Mn_{21}Al_{22}Si_1$ ribbons, a maximum H_{EB} of 2.6 kOe was reported at 2 K.

These ribbons were found to show the SG state below 68 K and SPM state above 60 K. These authors suggested a new approach for the enhancement of EB effect in these ribbons by the manipulation of SPM (or SFM) cluster size using cooling field, in contrast to the defect mediated increase of EB mentioned in many reports (discussed in Section 11.3.3.1).

11.5.8 Mn–Ni–Sn Heusler Alloys (Mn-Rich Alloys)

Though studied much less compared with the Ni-rich alloys, results obtained in Mn-rich alloys of the type Mn–Ni–X are quite impressive, especially from the point of view of EB. In the following section, results of investigation on the Mn-rich alloys are discussed.

11.5.8.1 In Bulk Form

As discussed earlier, the magnetic properties of Mn-based Heusler alloys are very sensitive to the Mn content as the total magnetic moment arises from the Mn atoms, as Ni has negligible moment. The AFM/FM exchange interactions strongly depend on the Mn–Mn interatomic distance, which plays a key role in determining their various physical properties (mentioned earlier). Therefore, the AFM/FM exchange interactions are expected to be larger for the Mn-rich (Mn ~ 50 at %) Mn–Ni–X alloys, as compared with that in Ni-rich (Ni ~ 50 at %) alloys, which in turn would affect the magnetic and related properties.

Recently, Xuan et al. [100] reported the EB effect in bulk $Mn_{50}Ni_{40-x}Sn_{10+x}$ ($x = 0$ and 1) alloys, in which a maximum H_{EB} of 910 Oe was observed for $x = 0$, and it was found to decrease with the increase of x. According to these authors, as the Sn concentration increases, that is, the Ni concentration decreases, the excess Mn atoms (formerly occupying the Sn sites) would occupy the Ni sites. As the Mn moments occupying the Sn and Ni sites are different and the Mn–Mn distance being different for Mn at Sn and Ni sites, the AFM coupling would change accordingly in the martensite phase, which results in the reduced H_{EB} with x. In a similar composition $Mn_2Ni_{1.6}Sn_{0.4}$, Ma et al. [101] reported the H_{EB} of about 1.2 kOe at 5 K. They confirmed the coexistence of RSG and FM phases in the martensite phase from the dc magnetization and ac susceptibility measurements. The exchange anisotropy created at RSG/FM phase was found to be responsible for the observed large EB in the alloy. Sharma et al. [102] have observed a giant EB field of 7.1 kOe in $x = 1$ of $Mn_{50}Ni_{41.5+x}Sn_{8.5-x}$ alloys at low temperatures, which is significantly larger than that reported in Ni-rich Heusler alloys. The observed giant EB in these alloys can be explained by the Kouvel' model of EB developed for Cu–Mn and Ni–Mn binary alloys [31,34–36], in which the exchange anisotropy was found to arise from small regions of FM order (e.g., between Ni–Mn and Ni–Ni pairs) and the AFM order (between Mn–Mn nearest neighbors). In these alloys, excess Mn atoms at Sn and Ni sites were found to couple antiferromagnetically due to the decrease in Mn–Mn distance, and therefore, the AFM coupling would significantly enhance. Hence, the FM clusters (made up of small regions of FM order between regular Mn–Mn pairs, and up to some extent from regular Ni–Ni pairs) were thought to cooperatively freeze to the SSG phase. It was proposed that the SSG phase is embedded in a strong AFM matrix, which is giving rise to the strong unidirectional anisotropy at the SSG/AFM interface, resulting in giant EB.

In another report, Sharma et al. [103] have investigated the effect of Co doping on EB properties in $Mn_{50}Ni_{40-x}Co_xSn_{10}$ alloys. EB fields of 920 Oe and 833 Oe were observed for $x = 0$ and 1, which was attributed to the large exchange anisotropy created at the interface between frustrated and FM phases. With the increase in Co, EB was found to decrease, which was attributed to the decrease in the strength of the magnetically frustrated phase. Similarly in their work [104], $Mn_{50}Ni_{40-x}Fe_xSn_{10}$ alloys were found to show a decrease in EB with the increase of Fe concentration, which was also attributed to the change in exchange anisotropy at frustrated/FM interfaces. EB observed in these Co and Fe doped alloys are found to be significantly larger than that reported for other Co- and Fe-doped Ni–Mn–X alloys [5,79,88]. H_{EB} values observed in Mn–Ni–Sn alloys so far are listed in Table 11.4. As can be seen, in general, Mn-rich alloys show much higher EB fields as compared with Ni-rich alloys. Furthermore, the temperatures up to which reasonably large EB is retained are also high in the former one.

The observed giant EB in Mn–Ni–Sn alloys is attributed to the strong exchange coupling between the frustrated and FM phases, and this frustrated phase can have a different magnetic anisotropy. Thus, in order to confirm the true magnetic nature of the frustrated phase, Sharma et al. [13] carried out various ac magnetic susceptibility measurements to probe the origin of giant EB. A critical analysis of the temperature dependence of the ac susceptibility data using the critical slowing down relation has revealed a SSG phase at low temperatures. They further confirmed the SSG state with the help of memory effect and aging effect under various protocols. They have also investigated the minor loop effect (as discussed earlier in Section 11.2.4) in these Mn–Ni–Sn alloys to avoid the overestimation of EB. For example, Figure 11.6 [104] shows the M–H loops measured at 5 K after FC in 10 kOe, in field range of $H_{measuring} = \pm 10, \pm 20, \pm 30,$ and ± 40 kOe for a typical case of $x = 1$ of $Mn_{50}Ni_{40-x}Fe_xSn_{10}$ alloys. The dependence of H_{EB} on measuring field ($H_{measuring}$) is shown in the inset of Figure 11.6. It is clear from the inset that H_{EB} decreases up to 20 kOe field (as H_{EB} is ~782 Oe at $H_{measuring} = \pm 10$ kOe and decreases to ~743 Oe at $H_{measuring} = \pm 20$ kOe) and remains nearly unchanged for higher measuring fields; this suggested that the ± 20 kOe was the optimum field for EB measurements.

They have also studied the effect of cooling field strength on EB of these alloys, as a typical case for $x = 0.5$ is shown in Figure 11.7 [13]. In low cooling fields (up to 0.5 kOe), H_{EB} increases substantially with the increase in the cooling field and then decreases monotonically with the further increase in cooling field. They explained that in low cooling fields, Zeeman coupling between the field and the AFM matrix moments is not strong enough to

TABLE 11.4 Maximum H_{EB} Values Observed in Various Mn–Ni–Sn Heusler Alloys

Sr. No.	Composition	H_{EB} (Oe)	Reference
1.	$Mn_{50}Ni_{40}Sn_{10}$	920	[100]
2.	$Mn_2Ni_{1.6}Sn_{0.4}$	1170	[101]
3.	$Mn_{50}Ni_{41.5+x}Sn_{8.5-x}$ ($x = 1$)	7100	[102]
4.	$Mn_{50}Ni_{40-x}Co_xSn_{10}$ ($x = 1$)	833	[103]
5.	$Mn_{50}Ni_{40-x}Fe_xSn_{10}$ ($x = 0.5$)	890	[104]
6.	$Mn_{50}Ni_{41}Sn_9$	1165	[105]

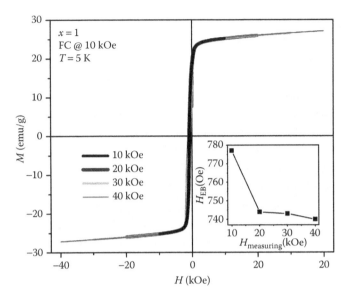

FIGURE 11.6 Minor loop effect: M–H loops at 5 K for $x = 1$ of $Mn_{50}Ni_{40-x}Fe_xSn_{10}$ alloys, measured in different fields after field cooling in 10 kOe. The inset shows the variation of H_{EB} as a function of measuring field.

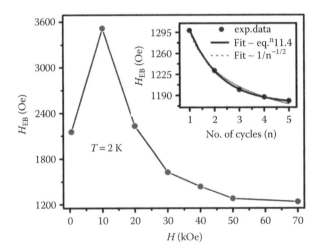

FIGURE 11.7 Cooling field dependence of H_{EB} at 2 K for $x = 0.5$. The inset shows the training effect of EB, the plot of H_{EB} versus the number (n) of field cycles. The solid circles show the experimental data, solid line shows fit to model (given by Equation 11.4), and dotted line shows the $1/n$ functional fit (given by Equation 11.3). (From Sharma, J., Suresh, K.G., *Appl. Phys. Lett.*, 106, 072405, 2015.)

compete with the exchange interaction between the SSG and AFM phases, which can lead to the alignment of more and more SSG moments along a preferential direction induced by cooling field, resulting in large H_{EB}. On the other hand, for higher cooling fields (>0.5 kOe) due to the increase in Zeeman coupling, the exchange coupling between the two phases decreases, which leads to a decrease in H_{EB}. The cooling field of 0.5 kOe was referred to as effective depinning threshold field, above which the magnetic interactions would be overcome by the Zeeman coupling. A similar behavior of H_{EB} with cooling field was observed in other Ni–Mn-based Heusler alloys [47].

A strong training effect was observed in these alloys and a typical case for $x = 0.5$ of $Mn_{50}Ni_{41.5+x}Sn_{8.5-x}$ alloys is discussed here [13]. The calculated H_{EB} from these field cooled M–H curves was found to decrease with number of field cycles (n), as shown in the inset of Figure 11.7 [13] It was found that the simple power law (Equation 11.3) that describes the purely AFM moments at the interface did not fit well with these data (shown as dotted line in the inset of Figure 11.7). However, they could fit it with a modified equation, given by Equation 11.4, which considers the contributions of frozen as well as the unfrozen moments at the interface (shown as solid line in inset of Figure 11.7). These observations gave another confirmation about the frustrated SSG moments at the interfaces and their role on the EB.

In addition to the effect of composition, temperature, and cooling field on EB properties in Mn-rich Mn–Ni–Sn alloys, Sharma et al. have studied the effect of pressure on EB in these alloys. H_{EB} was found to slightly decrease with the application of pressure, which was attributed to the decrease in AFM coupling with the pressure due to change in volume and bond lengths.

11.5.8.2 In Ribbons

In addition to the EB study in bulk Mn–Ni–Sn alloys, Sharma et al. also studied the EB in $Mn_{50}Ni_{50-x}Sn_x$ Heusler ribbons. They observed a maximum H_{EB} value of ~10.2 kOe at 2 K in these ribbons, which is one of the highest values observed so far for any Heusler ribbon system [81–82,93–94]. The observation of giant EB in these ribbons was also attributed to the strong exchange anisotropy at SSG/AFM interfaces (as discussed in their respective bulk compositions). Similar to the bulk alloys, they studied the nature of frustrated magnetic phase in these ribbons through the dc magnetization and ac susceptibility measurements. Frequency dependence of spin freezing temperature on the scaling law and the study of memory and aging effects under various aging protocols were performed to confirm SSG-like magnetic ground state, which confirmed the existence of cooperative freezing of the moments and/or moment clusters rather than uncorrelated dynamics of moments of SPM state. Recently, Zhao et al. [106] have studied the effect of Co doping on EB in $Mn_{50}Ni_{41-x}Sn_9Co_x$ ($x = 0$–6) ribbons, in which H_{EB} was found to be tuned from 345 Oe to 3154 Oe with Co substitution. They have explained the observed EB behavior in terms of a change in the frustrated phase from a canonical SG to a CSG and finally to a FM phase.

11.5.9 Other Heusler Alloys

In this section, we present some very important observations of EB phenomena reported in full Heusler alloys, which are not Ni–Mn based.

TABLE 11.5   Maximum H_{EB} for Fe–Mn–Ga Heusler Ribbons

Sr. No.	Composition	H_{EB} (kOe)	Reference
1.	γ-Fe$_2$MnGa	3.86	[109]
2.	Fe$_{52}$Mn$_{23}$Ga$_{25}$	2.11	[110]
3.	Fe$_{50}$Mn$_{24+x}$Ga$_{26-x}$ ($x = 3$)	1.74	[111]

11.5.9.1 Mn–Pt–Ga Alloys

Recently, Nayak et al. [107] have reported a giant EB of ~30 kOe in Mn–Pt–Ga alloys. They proposed that this large exchange anisotropy arises from the exchange interaction between the compensated FI host and FM clusters, which originates from the intrinsic antisite disorder in the alloy. In their another work [108], they have also reported a large spontaneous EB of 1.7 kOe in these alloys. They proposed that the FI ordering is essential to induce the exchange anisotropy isothermally, which causes spontaneous EB.

11.5.9.2 Exchange Bias in Fe–Mn–Ga Alloys in Ribbon Forms

Along with the most studied Ni-Mn-based Heusler systems in literature, another potential system of significant interest from the multifunctional point of view is Fe-Mn-based system. This system has been investigated only to a small extent, and only a few reports are available. For example, Fe–Mn–Ga Heusler alloy are found to show giant EB in the ribbon forms. The values of H_{EB} obtained from these reports are listed in Table 11.5. The mechanism of giant EB in these alloys is expected to be similar to that of Ni–Mn-based alloys.

11.6 Summary and Conclusions

FM full Heusler alloys have a special position in the field of multifunctional materials. Though only EB properties were discussed in this article, many systems mentioned here show promising magnetocaloric and MR properties as well. In all these alloys, EB arises due to the complex magnetic states arising from the first-order martensitic transition. It is true that the exact mechanism of EB in these alloys is not yet fully understood. Exchange coupling between competing magnetic phases appears to be the most basic mechanism in all these alloys. However, a better understanding of the exchange coupling is needed. From the application point of view, the main challenge is to develop the materials with large H_{EB} near room temperature. The promising aspect, however, is the tunability of EB with the help of various internal parameters (such as composition) and external parameters (such as cooling field, pressure, and temperature). Another positive aspect is the observation of EB in bulk, ribbons, thin film, and particle forms. In addition to the conventional EB, certain Heusler alloys are very promising from the spontaneous EB point of view. Based on the extensive studies being carried out on these types of alloys, it is expected that better EB materials suitable for room temperature applications would emerge in the near future.

References

1. K. Ullakko, J. K. Huang, C. Kantner, R. C. O'Handley, and V. V. Kokorin, *Appl. Phys. Lett.*, 69, 1966, 1996.
2. T. Krenke et al., *Nat. Mater.*, 4, 450, 2005.
3. T. Krenke et al., *Phys. Rev. B*, 75, 104414, 2007.
4. A. K. Pathak, I. Dubenko, C. Pueblo, S. Stadler, and N. Ali., *Appl. Phys. Lett.*, 96, 172503, 2010.
5. A. K. Nayak, K. G. Suresh, and A. K. Nigam, *J. Phys. D: Appl. Phys.*, 42, 115004, 2009.
6. Y. Sutou et al., *Appl. Phys. Lett.*, 85, 4358, 2004.
7. M. Kiwi, *J. Magn. Magn. Mater.*, 234, 584–595, 2001.
8. R. L. Stamps, *J. Phys. D: Appl. Phys.*, 33, R247–R268, 2000.
9. J. Nogues et al., *Phys. Rep.*, 422, 65–117, 2005.
10. O. Iglesias, A. Labarta, and X. Batlle, *J. Nanosci. Nanotechnol.*, 8, 2761–2780, 2008.
11. S. Chatterjee, S. Giri, S. K. De, and S. Majumdar, *Phys. Rev. B*, 79, 092410, 2009.
12. D. Y. Cong et al., *Appl. Phys. Lett.*, 96, 112504, 2010.
13. J. Sharma and K. G. Suresh, *Appl. Phys. Lett.*, 106, 072405, 2015.
14. M. K. Ray, K. Bagani, P. K. Mukhopadhyay, and S. Banerjee, *EPL*, 109, 47006, 2015.
15. J. C. McLennan, *Phys. Rev.*, (Series I) 24, 449, 1907.
16. F. Heusler, *Verh. Dtch. Phys. Ges.*, 5, 219, 1903.
17. P. J. Webster, *Contemp. Phys.*, 10, 559–577, 1969.
18. T. Krenke, M. Acet, E. F. Wassermann, X. Moya, L. Mañosa, and A. Planes, *Phys. Rev. B*, 72, 014412, 2005.
19. P. J. Webster and K. R. A. Ziebeck, *Landolt-Börnstein, Group III Condensed Matter*, vol. 19C, Springer, Berlin, 1988, pp. 75–184.
20. P. J. Webster, K. R. A. Ziebeck, S. L. Town, and M. S. Peak, *Philos. Mag. B*, 49, 295, 1984.
21. Y. K. Kuo, K. M. Sivakumar, H. C. Chen, J. H. Su, and C. S. Lue, *Phys. Rev. B*, 72, 054116, 2005.
22. S. Giri, M. Patra, and S. Majumdar, *J. Phys.: Condens. Matter*, 23, 7, 2011.
23. J. A. Mydosh, *Spin Glasses: An Experimental Approach*, Taylor & Francis, London, 1993.
24. W. H. Meiklejohn and C. P. Bean, *Phys. Rev.*, 105, 904, 1957.
25. W. H. Meiklejohn, *J. Appl. Phys.*, 33, 1328, 1962.
26. J. Nogués and I. K. Schuller, *J. Magn. Magn. Mater.*, 192, 203, 1999.
27. S. Das, M. Patra, S. Majumdar, and S. Giri, *J. Alloys Compd.*, 488, 27, 2010.
28. J. Dubowik, I. Goscianska, K. Załeski, H. Głowinski, and Y. Kudryavtsev, *J. Appl. Phys.*, 113, 193907, 2013.
29. I. Goscianska, K. Zaleski, H. Glowinski, Yu. V. Kudryavtsev, and J. Dubowik, *Acta Physica Polonica A*, 121, 5–6, 2012.
30. A. Yelon, in: M. H. Francombe, R. W. Hoffman (Eds.), *Physics of Thin Films*, vol. 6, p. 205, Academic Press, New York, 1971.
31. J. S. Kouvel, *J. Phys. Chem. Solids*, 24, 795, 1963.

32. K. F. Eid et al., *Appl. Phys. Lett.*, 85, 1556, 2004.
33. J. S. Kouvel and Jr. C. D. Graham, *J. Phys. Chem. Solids,* 11, 220, 1959.
34. J. S. Kouvel, *J. Phys. Chem. Solids,* 21, 57, 1961.
35. J. S. Kouvel and W. Abdul-Razzaq, *J. Magn. Magn. Mater.,* 53, 139, 1985.
36. J. S. Kouvel, *J. Phys. Chem. Solids,* 16, 107, 1960.
37. I. L. Prejbeanu et al., *J. Phys.: Condens. Matter,* 19, 165218, 2007.
38. Y. T. Chen, S. U. Jen, Y. D. Yao, J. M. Wu, J. H. Liao, and T. B. Wu, *J. Alloys Compd.,* 448, 59–63, 2008.
39. C. Tsang et al., *IEEE Trans. Magn.,* 35, 689, 1999.
40. D. Weller and A. Moser, *IEEE Trans. Magn.,* 35, 4423, 1999.
41. J. Sort, J. Nogues, X. Amils, S. Surinach, J. S. Munoz, and M. D. Baro, *Appl. Phys. Lett.,* 75, 3177, 1999.
42. M. P. Sharrock, *IEEE Trans. Magn.*, 36, 2420, 2000.
43. G. A. Prinz, *Science*, 282, 1660, 1998.
44. J. F. Gregg, I. Petej, E. Jouguelet, and C. Dennis, *J. Phys. D: Appl. Phys.*, 35, R121, 2002.
45. B. Dieny, V. S. Speriosu, S. S. P. Parkin, B. A. Gurney, D. R. Wilhoit, and D. Mauri, *Phys. Rev. B*, 43, 1297, 1991.
46. J. Geshev, *J. Magn. Magn. Mater.,* 320, 600, 2008.
47. D. Paccard, C. Schlenker, O. Massenet, R. Montmory, and A. Yelon, *Phys. Status Solidi* (B), 16, 301, 1966.
48. A. Hoffman, *Phys. Rev. Lett.,* 93, 097203, 2004.
49. S. K. Mishra, F. Radu, H. A. Durr, and W. Eberhardt, *Phys. Rev. Lett.*, 102, 177208, 2009.
50. S. Brück et al., *Adv. Mater.*, 17, 2978–2983, 2005.
51. H. W. Zhao, W. N. Wang, Y. J. Wang, W. S. Zhan, and J. Q. Xiao, *J. Appl. Phys.*, 91, 6893, 2002.
52. R. Coehoorn, in: K. H. J. Buschow (Ed.), *Handbook of Magnetic Materials*, vol. 15, North-Holland, Amsterdam, Chapter 1, 2003.
53. J. S. Jiang, G. P. Felcher, A. Inomata, R. Goyette, C. Nelson, and S. D. Bader, *Phys. Rev. B*, 61, 9653, 2000.
54. N. J. Gökemeijer, R. L. Penn, D. R. Veblen, and C. L. Chien, *Phys. Rev. B*, 63, 174422, 2001.
55. F. Canet, S. Mangin, C. Bellouard, and M. Piecuch, *Europhys. Lett.*, 52, 594, 2000.
56. T. Ambrose and C. L. Chien, *J. Appl. Phys.*, 83, 6822, 1998.
57. M. S. Lund, W. A. A. Macedo, K. Liu, J. Nogués, I. K. Schuller, and C. Leighton, *Phys. Rev. B*, 66, 054422, 2002.
58. L. Neel, *Ann. Phys. Paris*, 2, 61, 1967.
59. D. Mauri, H. C. Siegmann, P. S. Bagus, and E. Kay, *J. Appl. Phys.*, 62, 3047, 1987.
60. M. Kiwi, J. Mejía-López, R. D. Portugal, and R. Ramírez, *Europhys. Lett.*, 48, 573, 1999.
61. J. Geshev, *Phys. Rev. B*, 62, 5627, 2000.
62. J. V. Kim, R. L. Stamps, B. V. McGrath, and R. E. Camley, *Phys. Rev. B*, 61, 8888, 2000.
63. A. Hirohata, J. Sagar, L. Lari, L. R. Fleet, and V. K. Lazarov, *Appl. Phys. A*, 111, 423, 2013.
64. E. Fulcomer and S. H. Charap, *J. Appl. Phys.*, 43, 4184, 1972.
65. K. Takano, R. H. Kodama, A. E. Berkowitz, W. Cao, and G. Thomas, *Phys. Rev. Lett.*, 79, 1130, 1997.

66. C. Hou, H. Fujiwara, K. Zhang, A. Tanaka, and Y. Shimizu, *Phys. Rev. B*, 63, 024411, 2001.
67. M. D. Stiles and R. D. McMichael, *Phys. Rev. B*, 59, 3722, 1999.
68. V. Forename Buchelnikov, V. Sokolovskiy, I. Taranenko, S. Taskaev, and P. Entel, *J. of Phys.: Conference Series*, 303, 012084, 2011.
69. N. C. Koon, *Phys. Rev. Lett.*, 78, 4865, 1997.
70. U. Nowak, K. D. Usadel, J. Keller, P. Miltenyi, B. Beschoten, and G. Guntherodt, *Phys. Rev. B*, 66, 014430, 2002.
71. S. Koga, and K. Narita, *J. Appl. Phys.*, 53, 1655, 1982.
72. M. Khan, I. Dubenko, S. Stadler, and N. Ali, *J. Appl. Phys.*, 102, 113914, 2007.
73. S. Y. Dong et al., *Sci. Rep.*, 6, 25911, 2016.
74. A. Ghosh and K. Mandal, *J. Alloys Compd.*, 579, 295–299, 2013.
75. G. R. Raji, B. Uthaman, S. Thomas, K. G. Suresh, and M. R. Varma, *J. Appl. Phys.*, 117, 103908, 2015.
76. Z. Li, C. Jing, J. Chen, S. Yuan, S. Cao, and J. Zhang, *Appl. Phys. Lett.*, 91, 112505, 2007.
77. S. Esakki Muthu, N. V. Rama Rao, D. V. Sridhara Rao, M. Manivel Raja, U. Devarajan, and S. Arumugam, *J. Appl. Phys.*, 110, 023904, 2011.
78. D. Pal, A. Ghosh, and K. Mandal, *J. Magn. Magn. Mater.*, 360, 183–187, 2014.
79. C. Jing et al., *Advanced Materials Research*, 875, 272, 2014.
80. S. Agarwal, S. Banerjee, and P. K. Mukhopadhyay, *J. Appl. Phys.*, 114, 133904, 2013.
81. P. Czaja et al., *J. Magn. Magn. Mater.*, 358–359, 142–148, 2014.
82. X. G. Zhao et al., *J. Appl. Phys.*, 113, 17A913, 2013.
83. P. Liao et al., *Appl. Phys. Lett.*, 104, 092410, 2014.
84. B. M. Wang, *Phys. Rev. Lett.*, 106, 077203, 2011.
85. M. Khan, I. Dubenko, S. Stadler, and N. Ali, *Appl. Phys. Lett.*, 91, 072510, 2007.
86. M. Khan, I. Dubenko, S. Stadler, and N. Ali, *J. Phys.: Condens. Matter*, 20, 235204, 2008.
87. A. K. Nayak, K. G. Suresh, and A. K. Nigam, *J. Phys.: Condens. Matter*, 23, 416004, 2011.
88. R. Sahoo, A. K. Nayak, K. G. Suresh, and A. K. Nigam, *J. Appl. Phys.*, 109, 123904, 2011.
89. A. K. Nayak, R. Sahoo, K. G. Suresh, A. K. Nigam, X. Chen, and R. V. Ramanujan, *Appl. Phys. Lett.*, 98, 232502, 2011.
90. A. K. Pathak, M. Khan, B. R. Gautam, S. Stadler, I. Dubenko, and N. Ali, *J. Magn. Magn. Mater.*, 321, 963–965, 2009.
91. A. K. Pathak, I. Dubenko, S. Stadler, and N. Ali, *IEEE Trans. Magn.*, 45, 10, 2009.
92. C. J. J. Chen, Z. Li, Y. Qiao, B. Kang, S. Cao, and J. Zhan, *J. Alloys Compd.*, 475, 1–4, 2009.
93. X. G. Zhao et al., *Scr. Mater.*, 63, 250–253, 2010.
94. T. Sanchez et al., *J. Magn. Magn. Mater.*, 324, 3535–3537, 2012.
95. C. Jing et al., *Solid State Commun.*, 241, 32–37, 2016.
96. Z. D. Han et al., *Appl. Phys. Lett.*, 103, 172403, 2013.
97. B. Wang and Y. Liu, *Metals*, 3, 69–76, 2013.
98. R. Singh, B. Ingale, L. K. Varga, V. V. Khovaylo, and R. Chatterjee, *Phys. B*, 448, 143–146, 2014.
99. R. Singh, B. Ingale, L. K. Varga, V. V. Khovaylo, S. Taskaev, and R. Chatterjee, *J. Magn. Magn. Mater.*, 394, 143–147, 2015.

100. H. C. Xuan et al., *Appl. Phys. Lett.,* 96, 202502, 2010.
101. L. Ma et al., *Appl. Phys. Lett.,* 99, 182507, 2011.
102. J. Sharma and K. G. Suresh, (unpublished results).
103. J. Sharma and K. G. Suresh, *J. Alloys Compd.,* 620, 329–336, 2015.
104. J. Sharma and K. G. Suresh, *Solid State Commun.,* 248, 1–5, 2016.
105. J. Sharma and K. G Suresh, *IEEE Trans. Magn.,* 50, 4800404, 2014.
106. D. W. Zhao et al., *J. Appl. Phys.,* 116, 103910, 2014.
107. A. K. Nayak et al., *Nat. Mater.,* 14, 679–684, 2015.
108. A. K. Nayak et al., *Phys. Rev. Lett.,* 110, 127204, 2013.
109. X. D. Tang et al., *Appl. Phys. Lett.,* 97, 242513, 2010.
110. X. D. Tang et al., *Appl. Phys. Lett.,* 99, 222506, 2011.
111. C. W. Shih, X. G. Zhao, H. W. Chang, W. C. Chang, and Z. D. Zhang, *J. Alloys Compd.,* 570, 14–18, 2013.

Index